Water

Water

Our Sustainable and Unsustainable Use

Edward G. Bellinger
Central European University
Vienna, Austria

This edition first published 2025

Registered Office(s)
John Wiley & Sons, Inc., 111 River Street, Hoboken, NJ 07030, USA
John Wiley & Sons Ltd, The Atrium, Southern Gate, Chichester, West Sussex, PO19 8SQ, UK

For details of our global editorial offices, customer services and more information about Wiley products, visit us at www.wiley.com.

Wiley also publishes its books in a variety of electronic formats and by print-on-demand. Some content that appears in standard print versions of this book may not be available in other formats.

Library of Congress Cataloging-in-Publication Data
Names: Bellinger, Edward G., author.
Title: Water: our sustainable and unsustainable use / Edward G. Bellinger
Description: First edition | Hoboken, NJ : Wiley, 2025 | Includes index
Identifiers: LCCN 2023052802 (print) | LCCN 2023052803 (ebook) | ISBN 9781118797181 (paperback) | ISBN 9781118797174 (adobe pdf) | ISBN 9781118797167 (epub)
Subjects: LCSH: Water. | Water-supply. | Water quality.
Classification: LCC GB661.2 .B45 2025 (print) | LCC GB661.2 (ebook) | DDC 333.91–dc23/eng/20231122
LC record available at https://lccn.loc.gov/2023052802
LC ebook record available at https://lccn.loc.gov/2023052803

Cover Design: Wiley
Cover Image: © Lane V. Erickson/Shutterstock

Set in 9.5/12.5pt STIXTwoText by Straive, Pondicherry, India

Printed and bound by CPI Group (UK) Ltd, Croydon, CR0 4YY

C9781118797181_290824

Contents

About the Author

I am a graduate in Biological Sciences from the University of London. After graduating, I studied for a PhD at the Royal Holloway College (London University) on algal problems in water storage and treatment, investigating the causes of problems and their potential solutions at various waterworks of the Metropolitan Water Board (now Thames Water). On receiving my PhD, I was employed by the Metropolitan Water Board, which enabled me to gain much practical experience. In the early 1970s, I joined the Department of Engineering at the University of Manchester, lecturing on Public Health Engineering, and was also involved in setting up the Pollution Research Unit. This group started a master's degree course in Pollution and Environmental Control, probably the first in the United Kingdom, the format of which was followed by many other universities. In 1991, I accepted the post of Professor of Environmental Sciences and Policy at the Central European University (CEU) in Budapest (now in Vienna). This period involved travelling and presenting occasional lectures to students in many countries, including the former Soviet Union. I was also an occasional consultant for the United Nations, working with the World Health Organization and the International Atomic Energy Association, monitoring various water- and sanitation-related topics that involved travelling to different countries in Africa, Central and South America, the United States of America and Canada. I also worked with the UK Water Industries Training Board, running courses on different aspects of reservoir management and water and sanitation treatment for several UK Water Authorities. I was invited to become a member of the UK Department of the Environment Standing Committee of analysts, which was responsible for producing a series of 'Blue Books' on biological and chemical analytical methods for water. I have published several books on freshwater algae, water pollution control and the impact of climate and water on agriculture in the former Soviet Union as well as numerous scientific papers on these subjects. I am a Fellow of the Chartered Institution of Water and Environmental Management and a Fellow of the Royal Society of Arts.

The aim of this book is to provide an overview of water quality issues and what can be learned from the continuing history of water use in societies from ancient times to the present day. I also cover the main treatment methods currently being used. I have fully documented each chapter so that readers can follow up on any of the specific topics covered.

Our needs and the needs of ecosystems involve multiple pathways for water, which are still being added to. Some are changing and evolving, although others are constant. We can always learn from past experiences, and it is hoped that this will provide a basis for readers in their future work with water and in the water industry.

Preface

Our perception of water quantity and availability largely depends on where, in geographical terms, we live. If you live in a semi-arid or arid country or certain regions of some developing countries, you will certainly have a different approach to water use and treat it with more care when compared with a developed country with an established infrastructure for collecting, storing, treating and supplying water to its residents where water is taken more for granted. This often results in water being treated as an infinite resource at least from the point of view of humans. It is essential that all societies, both developed and developing, have a better understanding of the key role water plays. Water is essential to all life on this planet. There is no substitute for water in our lives and our requirements for living.

Humans need water either directly or indirectly in all sections of their lives, including our individual bodily requirements as well as the needs of society as a whole. We need it for the production of food, in food preparation, sanitation and waste disposal as well as economic and industrial activities.

Currently, this planet has an environment suitable for a large variety of living organisms found here. This has taken millions of years to evolve, and the essential components, including water, need to be used and managed properly. What we must realise is that water resources are finite and cannot be managed for humans to the exclusion of other organism's needs. Earth has evolved an environment suitable for life as we know it, and this variety of life in turn helps maintain a healthy environment for us, including a suitable atmosphere with the right proportion of oxygen and a suitable climate and temperature. Without these natural ecosystems, the environment would change for the worse. No one group of organisms should command right to an unlimited supply of water and other resources at the expense of others. All resources are finite, although some, e.g. certain minerals, are currently present in adequate amounts for human needs. What humans must realise is that available water is a key finite substance, and human needs cannot override all other needs without greatly modifying or even destroying the life systems on this planet.

Water resource management is being made both more difficult and more urgent because of the rapidly increasing human population and their increasing demands for more water. The aim of the leaders of countries is to improve the standard of living of the populations in those countries. Unfortunately, this also results in their expectations rising and inevitably them using and demanding access to more water for both their households, with things such as washing machines, baths and showers, and for gardens, swimming pools, parks and golf courses. Some municipalities also use water to wash streets. All of these factors not only increase the consumption of water but also increase the amount of wastewater to be treated and disposed of.

Groundwater reserves are being seriously depleted in some areas as is the storage capacity for water in reservoirs and the summer flows in rivers, all of which affect not only the reserves available to humans but also natural ecosystems on which we rely. Wetlands that help cleanse water draining through them are being destroyed worldwide in response to the need for more agricultural land. In addition to these impacts on water availability, our economic and industrial activities are gradually changing the global climate by causing global warming. The immediate effect of this is on the hydrological cycle and will probably result in more droughts and extreme storm events as well as some more subtle events. What is clear is that all communities and water users will need more informed management of water resources and their use by all aspects of society for communities worldwide to survive. Unfortunately, the required holistic view of water management frequently does not exist, with the responsibility being divided among many departments of government. As water is needed for all aspects of society, all users and user groups, including NGOs and local councils, must understand its management and have an input into the management strategies, including ways in which water use can be modified to minimise waste.

In this book, I have stressed the importance of good safe water in preventing disease and generally improving health as well as providing food and sustaining ecosystems. It is also important to have an understanding of the technologies associated with a clean water supply and the safe disposal of waste as well as the potential problems that will occur if these control measures are not followed. It is also important to understand that the technologies involved cost money to build and run, and although governments can help with capital costs, clean safe drinking water and suitable waste disposal will have a cost to users. All parts of society and industry must realise that their actions impact water quality. We must understand that you cannot just turn on a tap and expect there to be an inexhaustible supply of water. The hydrological cycle is global, and individuals do have a cumulative effect on this global cycle. This is why a sustainable future can only be achieved if everyone understands the role of this vital but finite resource, and that the technology involved in its management is quite complex. This book is based on my work with the water industry both here and abroad, and it is concluded that the basic principles are true for all countries.

Acknowledgements

Firstly, I would like to thank my wife PAM for her patience, encouragement and understanding during my writing of this book. I would also like to thank Neil Porter for his help with scanning figures and other electronic issues.

I would also like to thank the Thames Water Authority, with whom I first worked, although they were then called the Metropolitan Water Board and Severn Trent Water for permission to use some of their training figures. I have also worked with most of the other UK water authorities and would like to thank them for their help and experience with their technologies. I have always enjoyed working with them and many others working in the water industry both here and in many other countries, which enabled me to gain experience with a wide range of problems and solutions in water use and management. I would also like to thank Wiley-Blackwell staff for their help with occasional publication issues with the text.

Introduction

Water is an essential component of all living things including humans. The poet W.H. Auden said 'many have lived without love, none without water'. Water is the most widely occurring substance on the surface of this planet, and living cells depend upon it. We not only contain large amounts of water, about 70% of our body weight, but it also provides a medium for the passage of materials around our bodies. Ripl (2003) described water as 'the bloodstream of the biosphere' because of its importance as a major transport route for essential chemicals. It is a naturally occurring solvent capable of dissolving, to a greater or lesser extent, a very wide range of chemicals, which is perhaps its most important biological role (Sharp 2001). This includes both small molecules, such as nitrates, phosphates and sugars, and very large molecules, such as proteins and nucleic acids. Virtually, all biologically important chemical reactions need to be in solution to work. Water is also essential for the provision of food, as a drink, and is also needed for hygiene as well as providing a large amount of energy that we use. Water moves over the surface of this planet as well as through the atmosphere in a continuous cycle that is partly driven by gravity and partly by energy from the sun. This movement is called the *hydrological cycle* (see Chapter 4). This cycle involves water as a liquid, a solid and a gas. It is needed to sustain the myriad of ecosystems and provide the ecosystem services used by human societies (Jimenez-Cisneros 2015). Most of the water on this planet is saline and occurs in the oceans. Only a small amount is present as freshwater.

Water does not stay in any one state or location indefinitely. It does, however, remain in each location, or reservoir, for different lengths of time before moving to another reservoir either in the same state or a different one. For example, changing from liquid in a lake to water vapour in the atmosphere by evaporation. The amount of time that the water stays in any one reservoir before moving to another is called its residence time, typical examples of which are given in Chapter 4. From these examples, it is clear that water can exist in different forms and that availability for human and ecosystem use will vary greatly.

Water has many key properties that affect its behaviour and are also exploited by organisms. These properties both cause and allow water bodies, especially standing waters, to behave in certain ways, and these properties have been capitalised by many aquatic organisms to their advantage and moulded their behaviour (see Chapter 6). As land plants including trees rely on soil as a main medium in which they grow, although they require soil moisture for mineral transport, fish and aquatic plants rely on these water properties.

Water: Our Sustainable and Unsustainable Use, First Edition. Edward G. Bellinger.

Many aquatic organisms are equally dense or less dense than water that is then also used for physical support. Pure water is very transparent so the upper layers of a water body have reasonable levels of light intensity. The fact planktonic cells are continuously bathed in a medium with dissolved chemicals as well as being supplied with solar energy is only good for growth if they are all in the correct quantities. If some are present in excessive amounts, e.g. the major nutrients (see Chapter 6), it can lead to major imbalances in the biological response such as excessive growths of harmful algal blooms causing severe water quality problems and be dangerous for human health. Atmospheric moisture and its circulation play an important role in the movement of heat energy across the planet. Water movement is also important in shaping our landscapes by erosion, weathering of rocks and transporting minerals. Its interaction with the whole biosphere results, if the patterns of change are regular, in a self-supporting system (Ripl 2003). Human societies have traditionally developed close to readily available supplies of water such as rivers, lakes and springs. In the past, and even today, water has provided food in the form of fish and has been harnessed for that purpose. It is now commonly used for recreational activities as well as being significant in a number of religious beliefs and activities. Unfortunately, human interventions in the system can lead to destabilisation of these processes so that whole sectors of the system can be disrupted. It must always be remembered that this planet can support life as we know it because of the biosphere that is composed of many different ecosystems maintaining an environment suitable for human life. Green plants capture solar energy and absorb carbon dioxide from the atmosphere and with the aid of water produce more complex organic molecules, and as a by-product, pass out oxygen and water vapour into the atmosphere through the process of photosynthesis.

There are potential problems concerning the uneven distribution of freshwater over the surface of earth especially as human population growth rates are in regions of the world where water is, or will soon become, scarce and also suffer from low incomes making them less able to cope with future shortages. Some areas are naturally water abundant whilst others are water scarce or arid. With human populations rising globally, it is not just a question of sufficient water for basic human needs but also enough for food production. Many countries have the aspiration of being self-sufficient for food which means continuous increases in water demand. Water is also needed for economic activities. Although the problems of global climate change have come to prominence in the past few years, the problem of providing safe, adequate freshwater and proper sanitation to many people has been with us, and still is, for some time, and with some organisations it is still at or near the top of their list of problems of human quality of life. Water and climate change are inextricably linked together.

For many thousands of years, human settlements have used water and have developed technologies and strategies, but this has had only local impacts and did not have a wider effect on the hydrological cycle (Chapter 1). Since those times of small local communities when the view developed, at least in northern European countries and the northwest United States, water supplies were inexhaustible and, at least in the case of large rivers and the sea, they could be used for waste disposal because there would be infinite dilution that would render the waste harmless. In the past centuries, however, the global human population has grown exponentially and with it so has demand for more water per individual. With larger populations and greater industrial activities over the past hundred or more years, this has ceased to be true and anyway infinite dilution does not occur.

Warnings about water quantity and availability have been around for many years (Falkenmark 1997 and Chapter 4). Unfortunately, our attitude started to change with the onset of the industrial revolution. This new post-Holocene era has been termed the Anthropocene (Lewis and Maslin 2015). This was marked by both a rapid increase in population and an ever-increasing exploitation of natural water and other resources. Earth support systems are now being threatened. The IPPC (2007) has clearly pointed out the impacts of human activity on global climate systems. It is reasonable to assume that, in pre-Anthropocene times, as long as resource use was kept within sustainable limits, the stability experienced in the Holocene would continue for many thousands of years (Berger and Loutre 1991). If these limits are exceeded, then major changes could occur threatening human existence. It is therefore important to understand what these limits are.

Rockstrom et al. (2009) introduced the concept of 'planetary boundaries'. They describe these boundaries as 'human determined values of the control variables set at a safe distance from a dangerous level'. In other words, where possible we should determine the maximum limit for human exploitation of any particular resource and limit our use of it to be safely within that limit. The problem arises that our scientific knowledge of these boundaries and our ability to properly quantify them is often imperfect. Rockstrom et al. (2009) acknowledged the difficulty with some of these complex systems of assessing the effects of mechanisms, e.g. feedback mechanisms and self-regulation, in natural systems, and the timescales involved. Rockstrom et al. identified nine planetary boundaries.

Although because of the inter-relationships of all planetary systems consideration should be given to all of these when considering the hydrosphere, climate change, biogeochemical flows, including the nitrogen and phosphorus cycles, freshwater use, land system changes, chemical population in general and biodiversity loss will be examined in more detail in some of the following chapters. The main process of interest in this volume is global freshwater and factors that have a direct impact upon it. Shiklomanov and Rodda (2003) point out that human activities are the main driving force for change in global flows. They also have an important influence on the seasonal timing of cloud formation and precipitation. Molden et al. (2007) have estimated that 25% of the world's river basins run dry before they reach the sea because of over abstraction. Human alterations to the hydrological cycle affect many other areas of activity including food production, human health, climate, and ecosystem functioning. These adverse effects also impact human livelihoods. Postel (1999) estimated the upper limit for blue water resources to be between 12 500 and 15 000 km^3/yr. Actual physical blue water scarcity is reached when withdrawals exceed 5000–6000 km^3/yr (Raskin et al. 1997). Rockstrom et al. suggest a planetary limit of about 4000 km^3/yr. If this is exceeded, there could be major perturbations or even collapse of other systems.

Land use change has a major effect on the hydrological cycle as well as local ecosystems and biodiversity. Continued expansion of crop lands will have increased impacts on water quality, run-off, floods, and overall river flows. A planetary boundary of no more than 15% of ice free land could be used for agriculture. If this boundary were to be followed, its implementation must take into account local conditions so that the most productive lands were used for crop production and other less productive land if suitably used for forests and other ecosystems.

Water quality is a continuing issue and has been throughout the Anthropocene. Deteriorating water quality severely reduces the quantity of useable water. Unfortunately, one side effect of human industrial, agricultural and medical progress has been that, as well as more traditional pollutants such as sewage, heavy metals, pesticides, etc., many new compounds are finding their way into waters. These affect both human and ecosystem health. Initially, this was confined to local events, but now some are having global effects, for example microplastics. Many countries, especially in the developed world, have placed restrictions on the use and disposal of many toxic/undesirable chemicals in the aquatic environment. These restrictions are usually made without taking into account the cumulative global impact when viewed additively on a global scale. There is thus a need for setting a planetary boundary for these chemicals. Difficulties arise with this because of our imperfect scientific knowledge on individual chemicals but perhaps more importantly what the effects are with combinations of them. This is not surprising when one realises that there are an estimated 80 000–100 000 chemicals on the world's markets. The planetary boundary concept suggests two approaches. The first concentrates on persistent chemical pollutants with a global distribution, and the other is for chemicals having unwanted long-term and/or large-scale impacts on living organisms. Examples of each of these are discussed in Chapter 6. Currently, no definitive boundaries have been set. It should be remembered that, although Rockstrom et al. defined nine planetary boundaries, these do not act independently but impinge upon each other so that a significant change in one will impact several others. An example of this is the decline in tropical rain forests, which not only affects biodiversity but also carbon sequestration, the hydrological cycle and ultimately the climate. Whilst the concept of planetary boundaries has its uses for developing strategies, there are objections to applying all nine globally. Some, it can be argued, only apply at a local or regional level. Moreover, six of the nine do not have planetary biophysical thresholds in themselves. Global climate change (of which ocean acidification is a part), ozone depletion (which may have been partly, at least, addressed with the greatly reduced use of chlorofluorocarbon gases) and phosphorus concentrations in freshwaters whose threshold point is quite probably a long way off may still be regarded as having global effects. Other tipping points may not be global and are at the most regional. This being so, it is difficult to set global boundaries.

Answers to the above and many other questions have been discussed by many international organisations, and there have been a number of international conferences during the past 50 years, aimed at paving the way to sustainable water management. In the year 2000, the United Nations set out a series of goals covering all aspects of human well-being, including many of the points raised in the previous conferences. These were to be achieved within the following few decades. For safe drinking water and sanitation, targets were set to halve the number of people without proper access to both by 2015. Although some progress has been made, unfortunately much more is still needed. Many of the millennium development goals involve the supply of proper drinking water and sanitation before they can be achieved as water impinges on areas such as health, food supplies and schooling. In the latter case, many children are unable to attend school because of suffering from waterborne illnesses such as diarrhoea. Solutions require an integrated approach with understanding and cooperation from all sectors of government and also all stakeholders, but water has often not been the highest priority in many governments as focus being placed on economic development. Consideration of sustainable water use must go beyond the particular needs of one particular activity and must take into account the way that water, as a substance, behaves on this planet. In previous years, water was viewed as a largely local or regional issue and had a narrow sectoral approach often with the responsibility of a single government ministry whereas all ministries need to be involved. The approach must look beyond just human needs, but also allow for the needs of ecosystems as these are the life support systems on this planet. It is often difficult to break away from this old approach, but it must be done and coordinated approaches must be adopted. With the expansion of world trade, we are sometimes referred to as a global village, and water is now of global concern and is, to an increasing extent, a global commodity. Part of the realisation for this all embracing approach has come from the concept of the food–energy–health–environment nexus (nexus being the complex series of connections between these sectors). The global aspect is allowing communities with water scarcity to survive based upon both real and virtual water (see Chapter 4). Water thus transcends the concept of being a single responsibility, which in some respects it is, and becomes a vital consideration for all sectors as its proper management is of benefit to all. The hydrological cycle and its interaction with humans is extremely complex. Setting goals requires trade-offs between political and economic requirements and, importantly, ecosystem needs. General agreements on limits to water use are extremely difficult to obtain. It is certainly true that, until water is managed sustainably both locally and globally, global improvements in human well-being will be difficult to achieve. It would certainly help if all users had a better understanding of the complex technologies involved in providing clean safe water and disposing safely of waste without destroying ecosystems. All of these functions, together with their maintenance, come at a cost that must be shared, to an extent, with the users. The cost of drinking water and sanitation must not be prohibitive, but people must understand that there is one and both have to be contributed to by the user. It is also important that the public understand various ways in which water can be saved and waste can be reduced, which if followed will save both the consumer and the water provider money. This would then be a win/win outcome. Certainly, in developed countries, adequate water supplies and sanitation tend to be taken for granted by the public, but with an increasing global population and their demands together with the potential effects of climate change, which is likely to still be there for several decades, careful use and general conservation measures need to become automatic. Without this approach, water shortages in the future will be inevitable.

References

Berger, A. and Loutre, M.F. (1991). Insolation values for the climate of the last 10 million years. *Quaternary Science Reviews* 10: 297–317.

Falkenmark, M. (1997). Meeting water requirements of an expanding world population. *Philosophical Transactions of the Royal Society B* 332: 929–936.

IPCC (2007). Climate Change 2007. Impacts, Adaptation and Vulnerability. Contribution of Working Group II to the Fourth Assessment Report of the Intergovernmental Panel on Climate Change (ed. Parry, Canziani, Palutikof, et al.). Cambridge University Press.

Jimenez-Cisneros, B. (2015). Responding to the challenges of water security: The Eigth Phase of the International Hydrological Programme, 2014–2021. Hydrological Sciences and Water Security: Past, Present and Future. *Proceedings of the 11th Kovacs Colloquium*, Paris, France, June 2014.

Lewis, S.L. and Maslin, M.A. (2015). Defining the Anthropocene. *Nature* 519: 171–180.
Molden, D. (ed.) (2007). *Water for Food, Water for Life: A Comprehensive Assessment of Water Management in Agriculture*. London: Earthscan.
Postel, S. (1999). *Pillar of Sand. Can the Irrigation Miracle Last? Worldwatch Institute*. New York/London: W.W. Norton & Co.
Raskin, P., Gleick, P., Kirshen, P. et al. (1997). *Water Futures: Assessment of Long-Range Patterns and Prospects*. Stockholm, Sweden: Stockholm Environment Institute.
Ripl, W. (2003). Water: the bloodstream of the biosphere. *Philosophical Transactions of the Royal Society B* 358: 1921–1934.
Rockstrom, J., Steffen, W., Noone, K. et al. (2009). Planetary boundaries: exploring safe operating space for humanity. *Ecology and Society* 14 (2): art 32.
Sharp, K.A. (2001). *Water: Structure and Properties. Encyclopedia of Life Sciences*. Wiley.
Shiklomanov, I.A. and Rodda, J.C. (2003). *World Water Resources at the Beginning of the 21st Century*, International Hydrology Series. Cambridge University Press.

1

Our History with Water: What Can Be Learned from Past Water-based Communities

Water is an essential resource to humankind and all other living things. Because water is essential for life and no life can exist without it, human societies have tended to live and develop their communities within easy access to freshwater. When this is not available, nomadic people would migrate from one source to another, often seasonally, to find suitable supplies and food for themselves and their animals. Some animals were kept for food, meat, milk and bones for tools and others for their skins which were used for clothing. When settled agriculture developed, water was required on a more consistent basis. Water was not always continuously available so then, as now, they developed techniques for water harvesting, conservation and its transport. This became especially important as populations grew and larger communities developed, both of which needed larger supplies. Although larger structures like the Parthenon and Colosseum are widely admired, the driving forces behind the success of these and other ancient communities was the development of technologies both to provide water and sometimes to harness it for power. These advances have not been given enough attention by historians who tended to concentrate upon their politics and wars. Indeed, most historians and economists often assume that societal progress using technology only became important after the middle ages. The wealth and success of many ancient civilisations was, in fact, based upon the control of water using hydro-engineering and the invention of new technologies. Many ancient civilisations needed to use water sustainably to survive, so developed techniques to both support their needs and harmonise with nature. Sometimes this led to conflict between different communities, especially where water resources were limited.

It is worthwhile to consider some of these techniques and, in some cases, perhaps use them in modified forms to solve some present-day water issues particularly in semi-arid climates and developing countries. Although initially the management of water concentrated on controlling it for irrigation, as time passed water power started to be harnessed and used for a number of other uses, including grinding grain, driving bellows, grinding mineral ores, operating forge hammers and saws (Wilson 2002). Although precise details of water technology's impacts on the overall economy of ancient countries are not quantified, there is some information on the huge investment that was for olive oil production in Africa and Tripolitania (Mattingly 1988). There is evidence of large investments in aqueducts for transporting it and cisterns for water storage which were used for irrigated agriculture for fruit and vegetables, both for supplies to Rome and export. This sort of evidence

Water: Our Sustainable and Unsustainable Use, First Edition. Edward G. Bellinger.

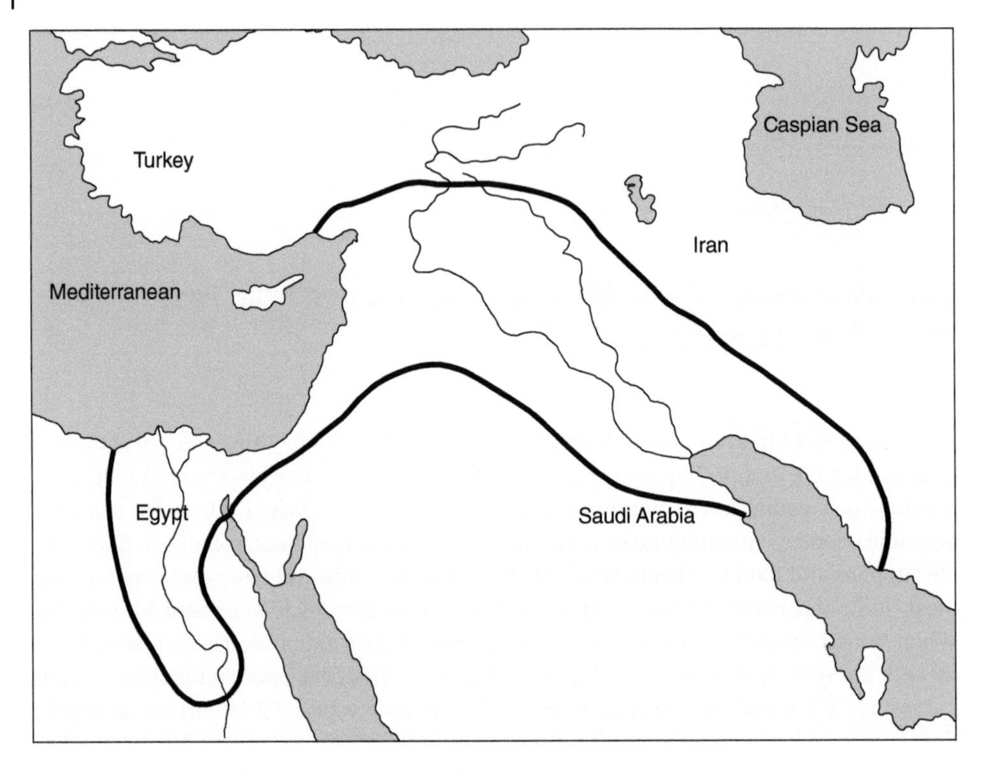

Figure 1.1 Map showing area of the Fertile Crescent.

illustrates the importance of water management and the amount of investment ancient societies were prepared to make to boost their economies. The region between the Rivers Tigris and Euphrates, part of which is in present-day Iraq, was where we believe settled agriculture developed and was known as the 'Fertile Crescent' (Figure 1.1). Agriculture also developed in other parts of the world, including Egypt, China and India where good water supplies were available. Water supplies were essential both for the populations and their economies, and if they failed starvation in parts of their populations was not uncommon. Early conurbations occurred in Jericho (8000 to 3000 BCE) and Egypt (3000 BCE). The first more intensive urbanisation occurred around inventions is covered in this chapter, and only selected example years to demonstrate the breadth of knowledge, engineering and invention of these societies are discussed below.

Mesopotamia

Settled agriculture was probably developed by people selecting seeds from plants that gave better yields, and some could be saved and used for planting and growing into new crops that gave higher yields. They also selected the best animals for breeding as well as for meat and milk production. This was basically the process of domestication. When settled agricultural communities developed a little over 8000 years BCE (Blockley and Pinhasi 2011), water was required on a more constant basis for raising their crops and livestock. As populations grew water was needed in ever larger quantities and its continued availability was essential for the economic and political development of those civilisations. One of the best known was that developed by the Sumerians, Assyrians and Babylonians in the 'Golden Triangle' between the rivers Tigris and Euphrates and extending up to the Upper Nile in which is now part of Turkey, Syria, Iraq, Israel, Palestine and down to Egypt. The river Euphrates, which is 2800 km long, originates to the north of Lake Van and winds its way down to the Syrian plateau. The Tigris (1950 km long) rises in Armenia near Mount Ararat and flows south-east roughly parallel to the Euphrates but does not merge with it until 100 km north of Basra to form the Shatt-el-Arab that eventually flows into the Persian Gulf. Mesopotamia is divided into upper and lower parts, the upper being mainly a piedmont zone below a mountain zone and supports dry agriculture and the lower is an alluvial plain. It is in this plain that settled irrigated agriculture developed. Across the plain, the rivers tend to flow between raised banks above the plain which, when the river banks are breached, form lakes and marshes. This crescent-shaped area is known as Mesopotamia and called 'the cradle of civilisation' (see Figure 1.1). People migrated from the highlands where the agriculture was rain fed to settle in the lowland plains where the soils were often quite rich. They became known as Sumerians. The area between the rivers was, however, prone to cycles of flood and drought which did not correspond to the cropping season. The flood periods tended to be April and June. In order to properly utilise the land, the Sumerians had to learn how to overcome the vagaries of climate, tame the floods and manage the droughts. Here, however, there were good soils, and the water began to be managed around 6000–7000 years ago for human and a little later for agricultural use. To overcome the uneven distribution of water supplies with time, they constructed a network of canals and reservoirs to supply water to the arable fields at the correct times. The provision of managed water extended the crop-growing season and increased yields. This also encouraged clearing of any suitable land of its natural vegetation for use in agriculture. There were strict laws governing the use of the water, the operation of the canals and reservoirs to regulate their use so as not to affect other water users and their crops. It is thought that writing was invented in Mesopotamia in the form of inscribed clay tablets. Agricultural yields began to increase faster than the population's needs leading to surpluses that could be used for trade. This resulted in the waterways being also used for transporting goods. As this use increased, there was a need to raise the river banks to allow increased water transport but this increased the risk of the banks failing, causing flooding as well as affecting local societies. Water use was so essential for human development that if the supply failed in any year people often starved and their economies suffered. In later years, poor crop yields arose not because of drought but through poor drainage of the soils leading to their salinisation. The black rich soils gradually turned white through salt accumulation. As populations grew, so did the development of the number of urban areas. This together with the irrigation needs for settled agriculture led to the development of growing expertise in aspects of engineering hydrology to manage any unpredictability of river flows and their flooding as well as the ability to produce food surpluses and an active food export business creating both wealth and power. The Mesopotamian society was at its peak between the third and seventh centuries CE. The ruling Sassanians were able to expand the hydrological schemes and irrigated most if not all of the cultivated land. It is estimated that this amounted to about 50 000 km^2 of land. More than double the area cultivated in modern Iraq with an estimated population of five million (Adams 1962). After the Arab conquest in 639 CE, the population declined as did work on the maintenance of the irrigation system. The estimated silt load carried by the Tigris and the Euphrates was as much as three million tons a day during floods. Much of this is deposited in the flatter parts of the plain where the river velocities decreased and in the irrigation channels. The silt also caused the river banks to be raised resulting in the river being much higher than the surrounding land. This silt in the irrigation channels had to be continuously removed to stop blocking, and this required much labour to keep them flowing. Salt build up gradually became a bigger problem reducing crop yields so that food surpluses were reduced which in turn reduced amounts for export so the wealth of the country steadily declined. There was even a danger of food shortages. By the tenth century CE, tax receipts had declined by 60%. Lack of income then forced the authorities to cut back on labour to do the maintenance. By the eighteenth century CE, revenues had declined by 87% (Christensen 1993). Within the next 200 years, salinisation continued making the area largely unsuitable for agriculture. The problem that arose in this irrigation system is that drainage was not properly allowed for causing the salt to build up. If the canals were not dug out continuously, blockage and reduced flows could require new canals to be built (Mays 2010). The behaviour of the sometimes erratic seasonal flows of both rivers is in contrast to the Nile whose flooding was fairly predictable in time if not entirely by volume. The Sumerians thrived during the period of plenty and developed a number of cities some of which had populations estimated at between 10 000 and 20 000.

There is evidence that the growth and development of the Mesopotamian civilisation was not continuous nor a steady one. Borrell et al. (2015) reported that a period of rapid cooling of the climate in the Northern Hemisphere resulted in a hiatus in development about 10 000 years ago. This interruption in development lasted for a short while before normal conditions returned. The Mesopotamians also developed other engineering technologies including the use of diversion dams, e.g. the Nimrud Dam (Butler 1960). In this, the river water was diverted through the Nahrawan Canal and was used to irrigate a large area of cropland. They also used lifting devices to take water from a river, canal or well for both human and agricultural use. The first were human powered and consisted of a bag and rope strung from a wooden beam with a counterbalance at the other end. These were called a shaduf (Figure 1.2). These were used as early as 2300 BCE. Early conurbations grew in Jericho (8000–3000 BCE), Egypt (3000 BCE) and Harappa in Pakistan (2300–1700 BCE). Other technological innovations included water tunnels and donkey powered chain lifts. More intensive urbanisation occurred around the Mediterranean at around 500 BCE to 500 CE. Historic written and archaeological material is more common for this region and period. In many areas, water was obtained by digging wells. Tapping into groundwater became more elaborate, and underground transport systems for moving this water allowed the irrigation of crops away from surface water supplies. The Sumerians did not need to understand the physics of their systems but by trial and error were able to work out the best ways to operate them. They did, for example, realise that by inflating a bag it could be used as a floatation device to help moving goods along a waterway. Mesopotamia also developed systems for wastewater disposal. Houses were provided with disposal systems (Stordeur 2000). The wastewater was either disposed directly to gutter or canal to pits or

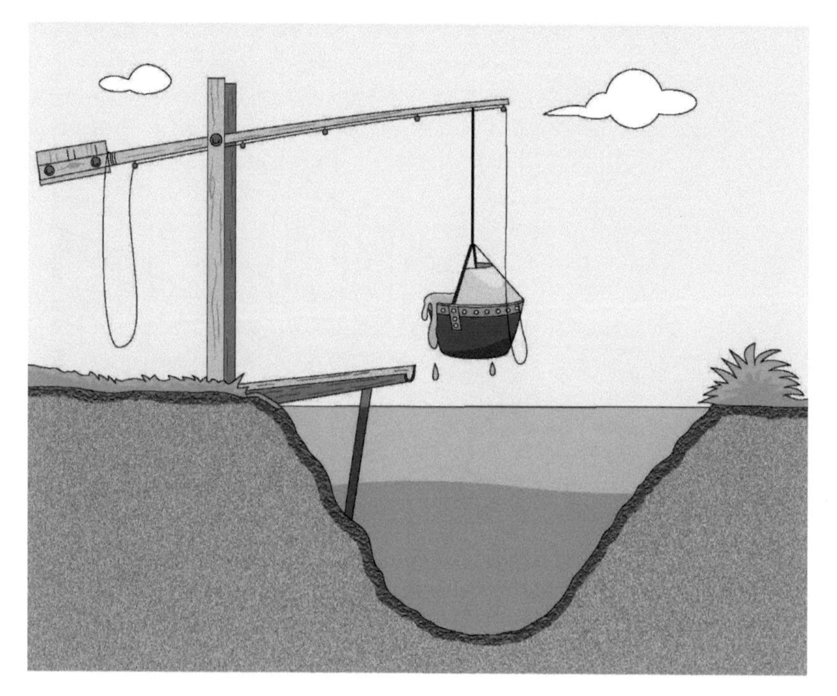

Figure 1.2 A typical shaduf for lifting water.

ponds in the city and then outside the city walls. Surface drainage systems were also provided in the cities to take away rain and floodwaters. Many houses were provided with toilets and pipe systems to take the waste away.

Qanats

Systems called qanats are still used in some countries, including parts of the Middle East, e.g. Iran. They were probably developed in ancient Persia over 3000 years ago (Remini et al. 2014). They were developed in semi-arid areas by Arab populations and from there have spread to many other semi-arid areas across the world. This method of harvesting groundwater is given different names in different countries, e.g. foggaras in Algeria, khettara in Morocco and falaj in Oman. Qanats are usually constructed where there is little or no obvious surface water. The technique involves constructing a gently sloping tunnel, sometimes many kilometres long, with a number of wells sunk into it acting as aeration sources (Figure 1.3). This allowed access to the groundwater that could then be used for irrigation. This system has been developed and expanded and used in many countries, mainly those within the zone encompassed by the tropics of Cancer and Capricorn. Water from this source allowed the development of palm oases particularly in North Africa, the Saudi Arabian Peninsula and Iran. Because groundwater has been a reliable source of good quality water, its use has been fundamental for human development in the area. There are several factors that need to be taken into account in constructing and managing a qanat. These include (i) To make sure the water table is adequate for the life of the qanat.

Figure 1.3 Diagram of a qanat system. (1) Mother well, (2) vertical shafts and air shafts, (3) outlet and main sloping horizontal shaft, (4) irrigated agricultural area, (5) higher land or foothills forming infiltration zone, (6) normal water table and sub-surface water, and (7) impervious bed rock.

Local knowledge of qanat diggers is essential in choice of location. (ii) Knowledge of groundwater recharge is essential in order to manage abstraction rates otherwise artificial recharge may be required. (iii) Recharge can be helped by constructing wide wells, up to 7 m wide and 3–5 m deep which were filled with large stones. These could capture any flood waters that occurred in that area and would help recharge the aquifer. This ancient system is still in use. (iv) Modern techniques such as Geographic Information Systems and remote sensing can help in maintaining and restoring qanats. The use of qanats extended over the centuries to more than 34 countries. There are about 32 000 of these systems in Iran discharging about nine billion cubic metres of water annually. Unfortunately, the use of qanats is generally declining through lack of maintenance and partly because modern wells and pumping are now much cheater and convenient and thus taking over groundwater abstraction. Care is still needed however not to over abstract!

Egypt

The ancient Egyptian civilisation can be traced back for as much as 11 000 years, and by 5000 years ago, agricultural production concentrated around the valley of the River Nile. This river, which is over 6670 km in length, is one of the longest in the world. It drains about 3 350 000 km^2 amounting to about one-tenth of the African continent (Mays 2010). The annual flood lasts, on average, about 110 days starting in June and peaking in late September. The Nile has two branches, the white and blue branches. The white originates in Burundi whilst the blue originates in the Ethiopian Highlands (Shiklomanov and Rodda 2003). The two branches merge at Khartoum (see Figure 1.4). The fertility of the lands adjoining the river depended upon the annual flooding which not only brought water but also deposited considerable quantities of silt which replenished the soils. If the floods

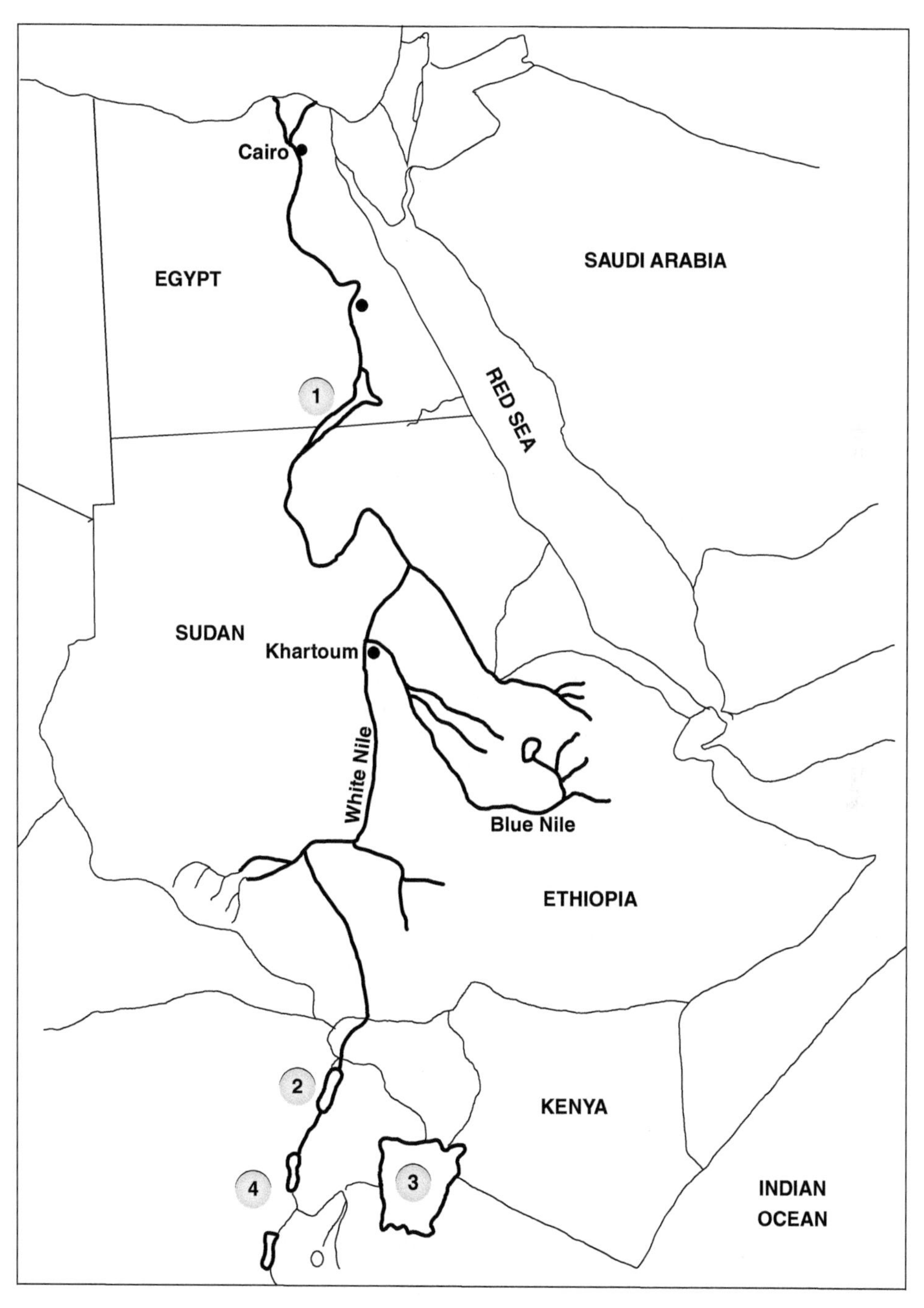

Figure 1.4 Map of river Nile. (1) Aswan High Dam, (2) Lake Albert, (3) Lake Victoria, and (4) Lake Edward.

failed, agricultural production was severely depleted resulting in starvation in the population and even civil unrest. It was thus important to know how strong or weak the flood was, and to this end a measuring device was installed in the river called the Nilometer. Records began with carvings on a large stone called the Palermo Stone recording the flood level as long ago as 2480 BCE. Other points of measurement were set up at Kernak where these were marked on the walls of the great temple. This dates from 800 BCE. At Roda, to the south of Cairo, the Nilometer recorded over 800 floods of which 73% were normal, 22% were low and 5% were dangerously high (Said 1993). Lake levels in the region also provide evidence of flood levels in both prehistoric and Pharonic times. When floods were low, famine was common. As the river enters the Mediterranean, the waters fan out to form a large triangle, the Nile delta, sometimes called 'Little Egypt'. Along its front with the sea, the shore is nearly 200 km wide with its apex at Cairo, 160 km inland. To the south of this apex is the Fayum Depression. The first recorded evidence of water management in Egypt is the mace head of King Scorpion, the final pre-dynastic king which has been interpreted as the first breaching of the river dyke to allow water to run into fields to irrigate crops. This was about 5000 years ago. Canals were constructed to allow water to flow where the floods would not normally reach. This could also provide some water at times when the floods were low. It was also realised that, for strategic, political or agricultural reasons, it could be necessary to transport water from a location of plenty to one of scarcity. The system of canals provided this option and, as engineering skills were developed, aqueducts carrying the water over large distances were built. A system of centrally managed canal construction and organised maintenance was developed although this management was often the responsibility of local groups (Strouhal and Forman 1992). Canals were also important for transport and, to an extent, helping with draining the marshes. One canal was used for transporting materials for building the pyramids in the old kingdom (Strouhal and Forman 1992). Higher land was watered using a shaduf. The Nile floods were quite regular until the end of the Neolithic wet phase (about 2350 BCE). There was occasional rain but this, together with the reliability of the flood volumes, led to a series of dry years leading to famine. The amount of arable land was increased in the Delta region through drainage. The combination of these water management strategies and some others were estimated to have increased the arable area to 8000 km^2 with an overall increase in crop production. They cultivated onions, garlic, peas, beans, lettuce, cabbage, asparagus and cucumbers among many others. As with many nations with successful irrigation, they were able to obtain revenue from exporting food.

Ancient Greece

Many of the communities in ancient Greece were affluent enough to be concerned about sanitation and valued access to clean water. Hence, many of their freshwater sources were from underground supplies rather than from surface waters. Ancient Greece, from the Minoans who flourished during the Bronze Age in Crete, was responsible for the development of water management technologies not seen previously. They did use some small rivers, unlike the Romans and Egyptians who made greater use of larger ones. They were concerned with the quality of the water, as were the Romans, and so made provision for proper sanitation including sanitary installations. They also made several technological advances and improved water management. There are many remains of ancient lavatories, both in private dwellings and certain public places such as Gymnasia. Antoniou (2010) suggested that concern of hygiene was related to the standard of living and their economic prosperity for both public and private installations. Public toilets were used by many people at the same time, up to 60 people at once. Lavatories were used throughout the Greek and Roman empires, the earliest Minoan ones dating from 3000 BCE. The Greeks also recognised that water quality had an influence on people's health. This impact was recorded as early as 470 BCE by a doctor, Alcmaeon of Croton. They recognised quality based upon their senses, taste, appearance, smell and temperature. The Greeks had different methods to the Romans for improving water quality. They used settling tanks, filters and sieves as well as boiling, the most widely recommended. In their climate, with a semi-arid ecosystem, the requirement for wood to burn and boil the water did, in some locations, become a problem causing scarcity. This is still a problem in many African countries at the present time. Waterborne infections were a problem in the ancient world, in particular malaria, which was common in Mediterranean countries.

In addition to sanitation, the provision for control of storm waters and floods was made. These structures date from the Minoan and Mycenaean periods. Clay containers for defecation were placed over cesspits or drainage channels constructed for the removal of the waste. These seats were remarkably similar in general shape to modern toilets (Figure 1.5). There is some evidence that some may even have had wooden seats. Seats have been found in Epidaurus that were made of stone rather than clay. Public toilets were often provided with a flow of water through a channel beneath the seats to take the waste away. Where no flow through could be organised, water carried in buckets was poured down to wash away waste. In many dwellings, the water from both the bathroom and the kitchen was used to augment the flow to remove the waste. In public toilets, the seats were arranged on three sides of the room. Often the waste flow from the toilets was routed to a small sedimentation tank.

Greek toilet design was influenced by Prolemean Egypt and later by Roman architecture. Public toilets were also a source of income to the municipality as there was an entrance fee for their use. The reuse of water for flushing was quite important as parts of classical Greece were semi-arid especially on some islands, and this situation still exists today. Around 1500 BCE, Knossos was at the height of its development and had several tens of thousands of inhabitants. Water was obtained through wells and also from rainfall collectors, roof tops, courtyards, or stone chambers. In some parts, terracotta flanged and cemented pipes were used to bring the water to dwellings. These methods were used for over 1000 years before the classical Greek period. Another important city on Crete was

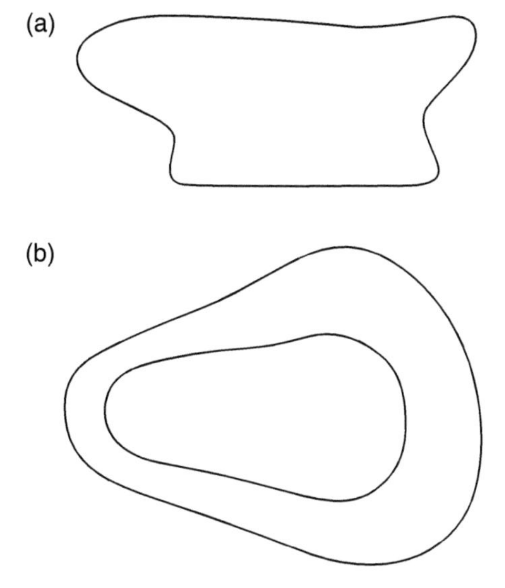

Figure 1.5 A Minoan toilet seat: (a) side view; (b) top view. Note the similarity to a modern toilet seat.

Tylissos. Here, the water simply came along an aqueduct from a spring with a tank to allow for sedimentation (Skivaniotis and Angelakis 2006). They also used a charcoal filtration at the spring to purify the water. Water was transported around the town using closed pipes and stone channels. Between 1600 BCE and 1100 BCE, the Mycenaean civilisation flourished in southern Greece. They built several long (up to 2500 m) low dams 24 m high to trap water. In some places, diversion canals were constructed to take flood water to other areas where it was needed. They also constructed surface and underground cisterns for storage.

Classical Greece and Roman

This period runs from 750 BCE to 323 BCE and includes the more familiar section of the Greek civilisation. During this period, the Greeks built wells, cisterns and aqueducts not unlike those built buy the Minoans. Among the new things they did contribute however was the development and refinement of those structures. This was helped by their advances in science and engineering. This included the construction of underground deep tunnel aqueducts, the best known being that of Eupalinos (530 BCE). This provided water for Samos and was made possible because of advances in mathematics and, in particular, geometry, in ancient Greece. The city of Athens had its water supplied by an aqueduct, the Peisistean aqueduct. This was made of terracotta pipes recessed into a channel with parts in a tunnel sometimes as deep as 14 m.

Perhaps the best known inventor in ancient Greece is Archimedes who was born in Sicily. Although some authorities attribute the invention of the screw pump to others, it is generally agreed that it was invented by Archimedes in the period 100–200 BCE (Koetsier and Biauwendraat 2004). The pump consists of a continuous helical blade contained within a cylinder. This cylinder must be tilted at an angle with one end dipping into the water (Figure 1.6). The screw pump was recorded in several texts from ancient times as being in use in many applications. It was, for example, used as a bilge pump in ships for mine drainage and as a lift pump for irrigation in the Nile delta. Interest in the design grew again in the late Middle Ages and was recorded by Leonardo Da Vinci in his Codice Atlantico where in place of helical screw blades it was constructed as a tube wound around a central core within the tube (Beck 1900; Beitrage zur Geschichte des Maschinenbauses). Cardano in 1550 described a multistage version of the screw pump to lift water to much higher elevations. Although complicated, the machine actually existed (Koetsier and Biauwendraat 2004). Galileo in the late sixteenth century called the screw pump 'not only marvellous but also miraculous'. Because of their efficiency and ease of use, screw pumps are still in use in many places at the present time although they are constructed of more modern materials. A well-known example of the screw pump is still being used at the Daveyhume sewage works in Greater Manchester, UK.

Roman Empire

The countries surrounding the Mediterranean do not generally have abundant water resources, so agriculture often relied upon irrigation. Some of the areas were in hilly regions, so they needed the water to be lifted. Even if the water was supplied from wells,

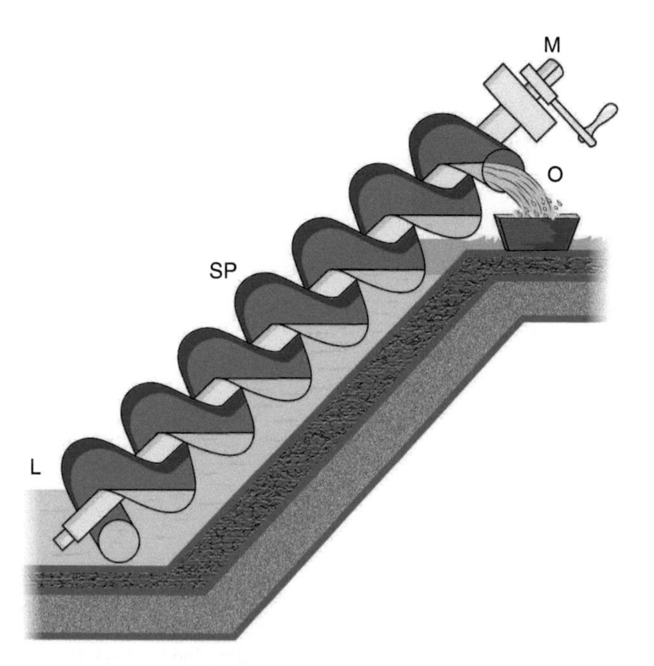

Figure 1.6 An Archimedes screw pump. L, liquid/slurry to be lifted; O, outflow; SP, rotating screw; M, motor to drive screw.

it still needed lifting (Wilson 2002). The earliest method was with the shaduf, but these were both labour intensive and of relatively low capacity. To overcome this water lifting, wheels were invented. These wheels had buckets or troughs attached to their rim (Figure 1.7). The wheel could be operated like a treadmill or by animals. As the wheel turned, the lower part would dip into a stream or other water source and collect the water in a bucket that was then rotated to the required upper level and discharged. Many of these seem to have been invented around the third-century BCE, so may have been of Greek origin and were embraced by the slightly later Roman Empire. They were used, for example, in Roman Egypt. Rome's control extended beyond Italy to much of the Mediterranean from Spain in the west to Asia and the Aegean in the east and to Britain in the north. Bucket chain lifts have also been discovered in London. Their success can be judged by the fact that they have been used up until the present day and in many countries. The Romans constructed many buildings and cities whose remains can still be seen. Because these conurbations all needed water to thrive, many were located along rivers or had access to springs. Where these were not available or could not meet the increasing demand, water had to be transported from outside. The problem increased as populations grew, and water supplies needed a proper public infrastructure. There was not only a need to supply the domestic needs of the population but also for fountains and public baths. The populations were also provided with toilets and disposal of their wastes. Initially, communities supplemented and surface water supplies with rainwater collected in cisterns. These cisterns could be for individual houses, groups of houses or for communities. They could be on the surface or underground. Some were fairly small whilst others were very large, e.g. the Piscina Mirabilis in the Bay of Naples. The most favoured source of water was from springs, but they also built dams and created reservoirs. Dams were often

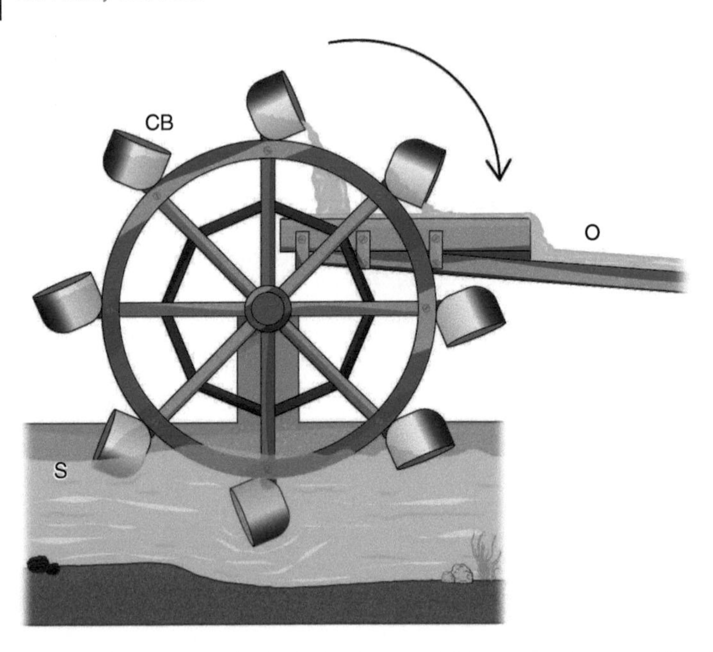

Figure 1.7 Rotating wheel with buckets for lifting water. S, stream; CB, collecting buckets; O, outflow.

earthen and could be quite large. One of a pair near Merida in Spain was about 427 m long and 12 m high (Mays 2010). The Romans also constructed a number of dams in southern Portugal. They were either used for water supply to the population or for irrigation. These were needed as precipitation in that area was often low. Many of these dams were quite low (>5 m) and often had a masonry wall with or without buttresses (Quintela et al. 1987). Occasionally, some showed evidence of a double masonry wall with mortar being used. The dams seemed to have bottom or middle level outlets. Unlike the ancient Greeks, the Romans predominantly used surface waters for their supplies. Adequate supplies of these often needed to be transported over large distances so the Romans developed a strong civil engineering culture to build dams, water supply networks for towns and cities and quite striking aqueducts. These transported water for both municipal and agricultural uses. Aqueducts were usually constructed as open channels or pipes. They could involve the construction of tunnels, bridges and inverted siphons (Figure 1.8). Aqueducts were often more than just a simple channel along which water flowed. Fahlbusch (2004) identified the following three areas of engineering skills that were used in aqueduct construction. (i) The size of the aqueduct canal was chosen according to the estimated discharge of water which it would carry. The size could vary along the course of the aqueduct. (ii) The cross section was large enough for people to walk along or through the canal for purposes of repair and maintenance, especially to remove calcareous deposits and silt. (iii) The cross section was kept as constant as possible, thus allowing manifold uses for encasings, especially the soffit scaffoldings for the vaults in a kind of industrialised construction. One of the most spectacular aqueducts is the Pont du Gard aqueduct near Nimes in France. This was part of the aqueduct of Nemausus, which was about 50 km long and took water from Uzes to Nimes. The average slope is 0.00085 m/m. The Pont du Gard is 275 m long and

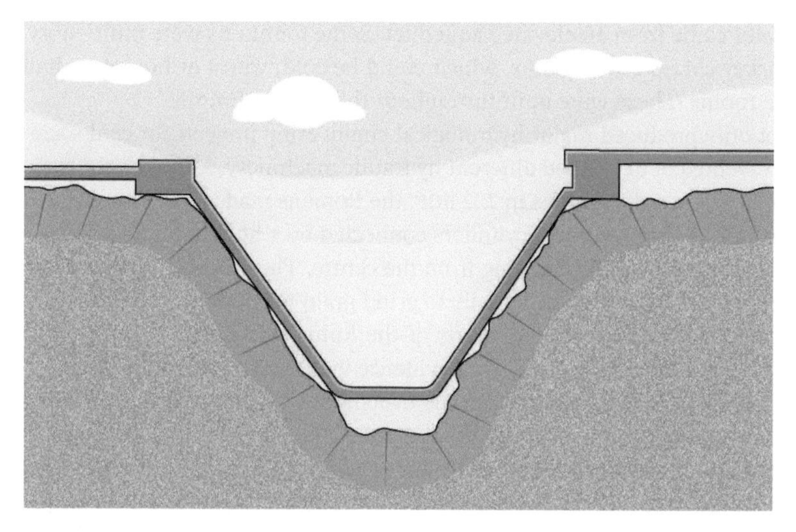

Figure 1.8 An inverted syphon.

48.4 m high. In an aqueduct, where slopes were steep, energy had to be dissipated with devices such as cascades (Chanson 2004). Other methods of flow regulation were also used such as constructing a large basin that could hold the flow and store it until it was acceptable to release it. The flow into and out of the basin was controlled by sluice gates. Siphons were frequently used, for example with the aqueduct supplying the Roman city of Lyon in France. This was 2600 m long and 123 m deep with an estimated discharge of 25 000 m^3/24 h (Hodge 1992). Siphon technology was important in providing Roman cities with much needed water. One of the largest siphons was at Madradag, which was 3000 m long and 190 m deep.

The ancients, including the Greeks and Romans, were aware of the importance of sanitation and providing clean water to households. This resulted in the development of quite sophisticated urban water distribution systems. The source was usually an aqueduct bringing water to the conurbation. Conduits, usually sub-surface but sometimes surface, conducted the water by gravity flow to a distribution tank. Depending upon the water source, there may have been settling tanks along the aqueduct or conduit to remove unwanted solids. The distribution tanks were connected through pipes to houses. The pipes were underground and, in the case of Pompeii, were made of lead and can still be seen just below the pavements along the streets. The pipework sometimes consisted of larger trunk pipes dividing into smaller service pipes to the dwellings. Not all dwellings had individual pipes but had to use taps placed in strategic positions. Less important dwellings often had their water tapped from a water tower with a lead lined storage tank supplying a water fountain. Rainwater was frequently collected off roofs through gutters leading to cisterns beneath the buildings via terracotta pipes. The Romans made an extensive use of cisterns for both collecting and storing water. Their use, however, dates back over 5500 years and has been used by many other civilisations. Roman society also used water for aesthetic purposes. They often constructed ornamental fountains and pools and public baths. Fountains, some of which were quite large, relied upon water pressure as the water came from an elevated aqueduct. Some fountains were multi-story, e.g. Miletus in Turkey (Mays 2010). Baths, which could be cold, warm or hot, often had public and private rooms. These were built throughout the Roman Empire.

The Romans not only produced major hydrological engineering projects for civil societies, but they also invented or developed different hydraulic machinery. Although the force pump was invented in Greece by Tesibus in 222 BCE, the Romans made more complicated versions. These consisted of two vertical cylinders connected by a horizontal pipe at their base. There was a single delivery pipe arising from the centre. They also developed water mills to use the power of running water initially to grind grain for making flour. Remains of these water mills have been found in all parts of the Roman Empire. Although these seem mainly to use vertical wheels, there is some evidence that horizontal wheels were also used (Lewis 1997). Vertical wheels could either be overshoot with the feedwater flowing over the top of the wheel or undershoot with the bottom of the wheel partly immersed in the stream. The remains of numerous water mills have been uncovered in the British Isles with the majority being along Hadrian's Wall (Spain 1984). Horizontal wheels have mostly been found in hilly or mountainous areas. Water power was also used in water lifting devices. The earliest types consisted of earthenware pots arranged at the periphery of a wheel picking up the water and discharging it at a higher level. These could also be powered by animals or humans. It is thought the water lifting wheel made its first appearance in Egypt at the time of Ptolemy II when the Fayum depression was irrigated to provide more food for the growing Egyptian population. In some locations, complexes where multiple wheels were installed have been found (Wilson 2002). There are also records of the wider use of water power, for example for water powered saws (Wilson 2002) and fulling Mills (Lucas 2011). Fulling is part of a particular part in woollen cloth-making, and the crushing and processing of mineral ores including gold. The use of water was widespread in gold and silver mining, and extraction was important in ancient economies. Mays (2010) argues that the Romans did not add as much as the Greeks to science, but what they did was to make major contributions to the development of ideas and engineering enabling inventions to be put into use.

Ancient American Civilisations

At similar times to the development of civilisations in the Middle East and Asia, civilisations were also developing in South-Western United States of America and Central and South America (see Figure 1.9). These also developed technologies and approaches to water management to suit their environmental conditions. There were three main cultures in the south-western states of the United States of America. These were the Chaco Anasazi, the Mogollon and the Hohokam. The area where these existed was either arid or semi-arid. Here, we consider just two of these societies and how they managed their freshwater. The Hohokam society lived in the south of Arizona and northern Mexico and flourished in the first millennium CE. (Andrews and Bostwick 2000). They were centred in the area of the Salt River Valley and practised floodwater farming. This was on the river floodplain and used the overspill of the river flood water and capturing rainwater runoff to irrigate the land. They created a productive agricultural region in the Arizona desert. Dry farming was also

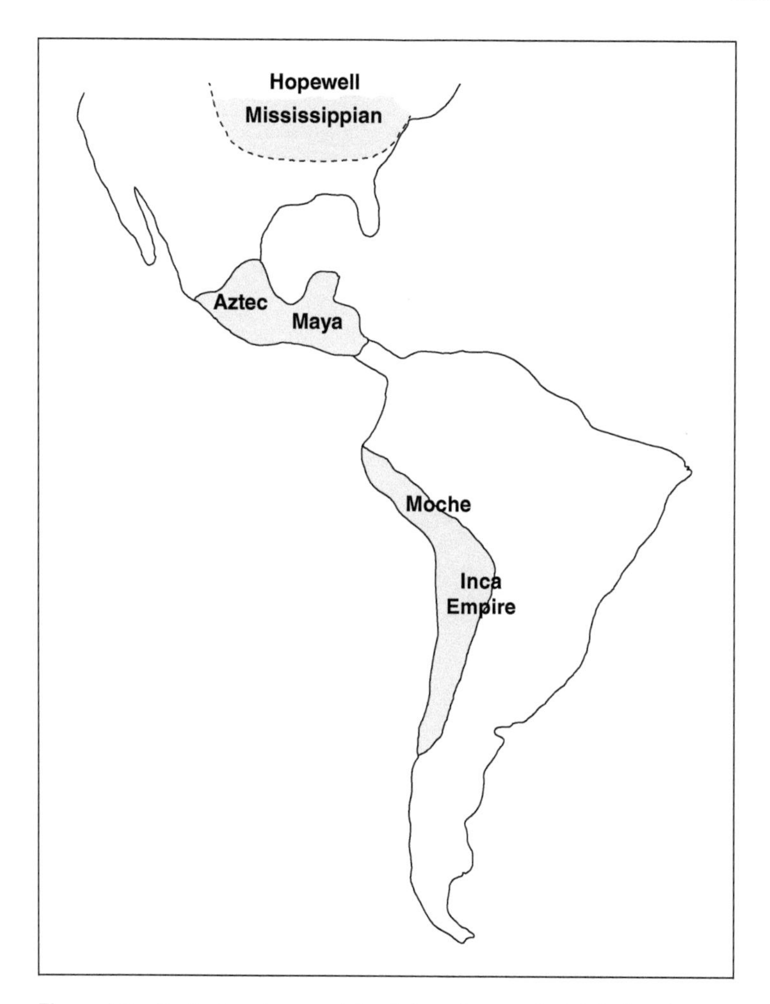

Figure 1.9 North American and South American ancient tribal regions.

used capturing direct rainfall. They developed a large complex system for irrigation consisting of hundreds of kilometres canals (483 km major canals and 1126 km of distribution canals) in the Salt River Valley (Mays and Gorokhovich 2010). To construct so many canals would have required a considerable labour force. There is no evidence of specialised tools being available, so digging was performed by hand using flat-wedge-shaped pieces of stone (Howard, www.waterhistory.org). The system was complex consisting of main canals, distribution canals and smaller laterals (Figure 1.10). They also had a system of gates made of logs woven with brushwood and reeds to control flows. The canals did not appear to have permanent linings or anything to prevent silting. Linings, if they had been present, would have slowed leakage and prevented damage from floods, waterlogging and salinisation of the soil. In 899 CE, a large flood caused the Hohokam population to move away from the Salt-Gila River Valley to land where they could practise dry farming. Further major floods in the following two centuries also contributed to the collapse of the Hohokam society that culminated in the major flood of 1358 and then declined during the next century.

Figure 1.10 An example of a typical irrigation canal arrangement as used by the Hohokams in America. MR, main river; A, diversion dam or weir; B, main canal; C, main flow control gates; D, control gates for individual canal branches; E, irrigated crops. *Source:* Adapted from Masse (1981).

Further south in Mexico from 1150 to 1519, the Aztec empire developed (Figure 1.9). Their capital was Tenochtitlan situated on a reclaimed island in the saline lake Texcoco. The Aztecs expanded their empire to conquer several other areas in Mesoamerica. They overcame the lack of good agricultural land by building what is called floating fields, chinampas, in Lake Texcoco. These were built in shallow areas of the lake and made of earth and vegetation (Woolf 2005). They were constructed by driving wooden poles into the lake bottom and filling the spaces between earth. In deeper areas, floating structures were anchored in place. In the thirteenth, fourteenth and early fifteenth centuries, many important advances in water hydrology occurred. One major advance was the development of aqueducts transporting water for both irrigation and domestic supplies. Tenochititlan became connected to an elaborate system of aqueducts bringing water from higher lands in the southern Mexico Basin to the city. Doolittle (1990) stated that 'The aqueducts built in the Basin of Mexico during the pre-Hispanic times were so numerous and outstanding that they were one of the first indigenous engineering accomplishments that the Spanish noted'. There were several aqueducts across the lake connecting Tenochititlan with settlements around its margin. One, from Chapultepec to Tenochititlan, was given sacred significance as its source was a spring gushing water from under a large stone (Brundage 1979). This aqueduct and probably many others were built on a series of connected islands. These were built by floating reed rafts into place and then anchoring them with wooden stakes to the lake bottom and then piling rocks and soil upon them until they sank. More rocks and soil were then added creating an island. A number of islands were built 3–4 m apart forming a chain (Bribiesca-Castrejon 1958; Pena Santana and Levi 1989). On each island they then constructed an aqueduct of clay, and each island aquifer was connected to the next by means of a hollowed log. Floods in the fifteenth century destroyed this aquifer, but it was quickly rebuilt with larger islands and a stone aquifer on each island. It is thought that the water was used for domestic purposes, although some have suggested some agricultural use as well.

The Inca empire spread along the west coast of South America from Columbia, Ecuador, Peru, Bolivia, Chile and Argentina. Of all of the Inca sites, the most famous is Machu Picchu that was built in the middle of the fifteenth century. It was constructed on a mountain ridge 2438 m above sea level (D'Altroy 2003) between two prominent peaks. The site overlooks the Urubamba valley and river that runs along three sides of the ridge. The site is located in Peru about 80 km north-west of Cuzco, the original Inca capital. Machu Picchu was the royal estate of the Inca ruler Pachacuti (Rowe 1990). Water was needed both for domestic supplies and to irrigate numerous agricultural terraces. This was supplied by means of a canal and a series of springs feeding fountains. The average annual rainfall was, at that time, calculated to be about 1940 mm/yr. This average was calculated from ice core data obtained from snow-capped adjacent mountain (Thompson and Mosley-Thompson 1989). Only a small amount, if any, was needed for agricultural irrigation, and none was found (Wright et al. 1997). This spring provided a perennial source of water for the inhabitants and gave an estimated annual yield of about 40 000 cubic metres, so water shortage was not considered a reason for the site being abandoned after a century of occupation even though up to 1000 people lived there and the permanent population was about 300. Any precipitation onto hard surfaces was collected as well as unused domestic water. There is still a stone spring collection system on the north slope of the peak to the south of the city. This is a properly planned and well-constructed piece of engineering as it is built of a permeable stone wall set into the hillside. It was about 14.6 m long and 1.4 m high with a collection channel at its base. Water quality of the main spring was very good (see Wright et al. 1997). If the dry season was prolonged and water supplies became more difficult, it could be supplemented by collecting water from the Urubamba river in large jugs (aryballo) carried up to the city. The water from the springs flowed across the surface in an open channel. Fountain 1 appears to be for the ruler, and other fountains were for other residents. Provision of adequate water supplies for human consumption and agriculture was essential for the success of the communities. To achieve this human capital, resources had to be invested in hydrological technology. It was also necessary to have proper water management for all uses, and this was organised at both state and local levels.

Some of the Changes from the Thirteenth to Nineteenth Centuries

Many changes in water technology and management occurred in Europe and the Middle East after the Roman Empire collapsed. Several of the techniques and structures introduced by the Greeks and Romans were still used although some fell into disrepair. As the population grew so did the size of towns, but the provision of sanitation did not keep pace and disease was common. Agriculture expanded as did the amount of contamination in many rivers. In London from the thirteenth century to the time of the great fire of London (1666), the city's population had grown from about 40000 to 0.5 million. Towards the end of the period, the development of industrialisation had a profound effect. As the city grew, it was still initially dependent upon water availability, so its expansion was determined by the existing small rivers all running into the River Thames and marshy areas just to the north. Sanitation was poor in the crowded houses, and much waste was thrown into the streets or dumped into these rivers that then became clogged and foul. This, together with the horse and other animal droppings, made the streets equally foul. Houses near to the Thames used this as a source of water. Others used shallow wells. There was also a number of springs that were also used. The continuous population increase however continued to require even more water. This was obtained by installing piped conduits into the city from further away. Water keepers were appointed to restrict industrial users and eventually to levy a charge on their use. Illegal users were prosecuted. Most individual houses did not have piped water or taps, so the water had to be carried to them. Later pipes were connected to the springs taking the water to cisterns or tanks that had taps for dispensing the water to water carriers. The sixteenth century witnessed a further large expansion of industry and the population. New management was organised for the water, and this was awarded to Hugh Middleton and Peter Morice. To meet the increased demand, they built a waterwheel beneath one of the arches of London Bridge. This was powerful enough to pump over 132000 gallons of water an hour around the city. Waterwheels were also used in several European cities such as Hanover and Bremen. Middleton realised that more water was needed and it would have to be brought in from further afield. He constructed a canal, completed in 1613. This was called 'The New River' and was 60 km long and is still used today. Initially, people were hesitant to connect to this supply because of cost, and it was not until some 20 years later that it was able to break even financially. There were also problems of wasteful use of the water and also letting taps run continuously. This as well as other problems, such as the ever-increasing growth of the population and greater demands from industry, meant that there would have to be constraints on the amounts of water being used. Movement of water around the city both to supply existing buildings and meet the demands of an expanding city still posed problems. This transport of water was achieved in London by using hollowed out wooden pipes. Although these were used for many years in many other parts of the world, the scale of their use in London was exceptional. This use required a considerable increase in pipe production. Originally, they were made by boring out tree trunks by hand, but the invention of a boring machine made mass production possible. Larger diameter pipes were used for mains and smaller ones for laterals. A further new introduction, if not innovation, was the introduction of the ballcock valve. This simple device helped overcome the wastage caused through taps not being turned off (Tomory 2012). In the past century, both the population and industry have continued to grow. To cope with the increasing demand for water, more storage reservoirs had to be built. As London was situated in a river valley, there were no suitable locations nearby to build dams across suitable valleys. This resulted in the development of pump storage reservoirs being built on suitable flat areas. These were constructed by building a watertight containing wall and filling them from the nearest river when flows were high enough. In the case of the Thames Valley reservoirs, the nearest river was the non-tidal part of the Thames.

In more recent years, Thames Water, the London water company, constructed a large drinking water ring main and, in order to cope with the increasing volumes of wastewater, a new ring main for collecting and managing this waste was constructed. This type of construction will probably be the trend for many major cities in the future.

What Can We Learn from These Ancient Approaches to Water Management?

Currently, in many parts of the world, there is a water crisis with over 1 billion people not having access to safe drinking water and over 2.5 billion with no access to proper sanitation. These problems have, in some past civilisations although the populations were smaller, had to be overcome for them to survive. Many of the solutions used then could be adapted and used today. These past approaches were often more sustainable. Many of their structures were adapted to suit local environmental conditions as well as meeting the human needs at that time but in a sustainable way. The United Nations Convention to Combat Desertification (UNCCD) developed this definition of traditional knowledge:

> *Traditional knowledge consists of practical (instrumental) and normative knowledge concerning the ecological, socio-economic and cultural environment. Traditional knowledge originates from people and is transmitted to people by recognizable and experienced actors. It is systematic (inter-sector and holistic), experimental (empirical and practical), handed down from generation to generation and culturally enhanced. Such a kind of knowledge supports diversity and enhances and reproduces local resources.*

Seven groups of practices were identified for which traditional knowledge had developed some answers for combatting these problems. These are water management for conservation, improvement to soil fertility, protection of vegetation, to fight against wind or water erosion, silviculture, social organisation, architecture and energy.

Although the convention focused upon desertification, several, if not all, of the seven practice groups can apply to water sustainability. Because of the speed of technological development during the past century, our approach to environmental problem-solving turns immediately to the latest technological solutions. If we need more water, sink deeper wells and pump more water. This is often carried out without considering the long-term sustainability of the aquifer, and probably it only solved the problem in the short term. Perhaps other more traditional approaches might have helped the original problem and saved the aquifer, for example better harvesting of meteoric water (rainwater) or only abstracting water at a sustainable rate. This approach does not necessarily directly apply technologies from the past, but instead considers the rationale behind their approach and see if this could be adapted to the present problem. Laueano et al. (2001) stated that 'Modern technological methods operate by separating and specializing, whereas traditional knowledge operates by connecting and integrating'. Good examples of this are the use of qanats and harvesting the power of water to do work. Management of water should not only consider quantity but also the security and sustainability of supplies. Major cities such as Rome did not rely upon a single source but had 11 aqueducts bringing water into the city. Ancient communities also had good collection methods for rainwater and storage as well as flood control and storage when water was in excess in order to be prepared for leaner times. They were also concerned with sanitation. One thing they did not have to deal with is the amount and range of pollutants, especially chemicals, that we have today. Few of these ancient civilisations collapsed suddenly. Mostly they declined slowly over a period of time. Extended droughts could cause a collapse, e.g. in the Indus valley (Pakistan) where the course of the Indus river was thought to have changed causing supply problems (Mays 2010). Prolonged droughts, earthquakes and salt accumulation in the irrigated soils all caused lower agricultural productivity, resulting in a decline in the wealth of a country often causing their decline. Land use changes where forests were cleared in favour of agriculture could have led to soil erosion and changes to the regional climates. When considering the collapse of societies, five factors possibly leading to the collapse can be identified. These are as follows:

1) Damage that people inadvertently inflict upon the environment
2) Changing climate
3) Society's responses to its problems
4) Hostile neighbours
5) Decreased support by friendly neighbours

Of these, the first two directly affect water resources and the third will either directly or indirectly affect resources and availability. Changes in the climate have occurred in the past including, for example, prolonged droughts such as occurred with the Mayan civilisation in Central America which lasted from CE 125 until CE 250. Environmental degradation also occurred with the clearance of the rain forest in order to increase agricultural land (Mays 2010). The Chacoans in north-western New Mexico collected and diverted runoff into canals, but this needed vegetation to be cleared leading to lasting deforestation and changes to the climate, a short-term remedy causing long-term problems.

Although the ancient Egyptians could rely on the predictability of the annual flood (if not its magnitude), the requirements of an increasing population meant that more agricultural land was needed. The water for irrigating the land required more artificial irrigation which led to the development of lifting devices such as the shaduf and saqiya (waterwheel) to bring water to the fields. All of this changed in modern times with the construction of the Aswan High Dam that stopped the annual flood meaning that silt deposits and soil replenishment also stopped. It compromised water quality through the need for more artificial fertilizers and pesticides. The aquatic ecology also changed and resulted in large growths of harmful aquatic algae. The situation has been made worse by the large growth in population along the lower stretch of the river and the increased discharges of domestic and industrial wastes into the water.

Ancient water management, for example in Greece, can provide pointers to sustainable water use for present societies. Water security and sustainability was required then as it is now (Koutsoyiannis et al. 2008). We are now hopefully waking up to the ideas of social and environmental responsibility promoted by the UN Conference in Rio as well as the Bruntland Commission 'Our Common Future' in 1988 (Keeble 1988). It is now recognised that access to adequate supplies of freshwater and proper sanitation is a cornerstone to future sustainable development. The provision of these must be achieved with due recognition of the 'knock on' effects of our actions. This is especially true of many of our large infrastructures that we embark upon in the field of water management. Some hydraulic engineering is being carried out on an unprecedented scale with long-term effects that are not yet known. Changes to human adaptability could keep pace with development until the middle of the nineteenth century, but since then the speed of some changes may have been too fast with unconsidered long-term implications. In the ancient world, humans adapted to water use in the context of the then prevailing environmental conditions often working with nature. Over time our attitude has changed to trying to fight and to conquer nature instead of trying to work in harmony with it as was often the case with ancient societies. Our consideration of traditional and ancient technologies that developed with this in mind need not be fossilised but with the aid of modern materials and knowledge adapted and built upon to provide efficient, environmentally friendly and sustainable ways ahead, making it a win-win situation.

References

Adams, R.M. (1962). Agriculture and urban life in early South-Western Iran. *Science* 136 (3511): 109–122.

Andrews, J.P. and Bostwick, T.W. (2000). *Desert Farmers at the Rivers Edge: The Hohokam and Pueblo Grande*. Phoenix, AZ: Recreation and Library Department, Pueblo Grande Museum and Archaeological Park, City of Phoenix.

Antoniou, G.P. (2010). Ancient Greek lavatories: operation with reused water. In: *Ancient Water Technologies* (ed. L.W. Mays), 67–86. Springer Science + Business.

Beck, T. (1900). Heinrich Zeising (gest 1613) *Beitrage zur Geschichte Des Maschinenbauses*. Berlin: Springer.

Blockley, S.P.E. and Pinhasi, R. (2011). A revised chronology for the adoption of agriculture in the Southern Levant and the role of late glacial climatic change. *Quaternary Science Reviews* 30: 98–108.

Borrell, F., Junno, A., and Barcelo, J.A. (2015). Synchronous environmental and cultural change in the emergence of agricultural economies 10,000 years ago in the levant. *PLoS One* 10 (8): e0134810. https://doi.org/10.1371/journal.pone.0134810.

Bribiesca-Castrejon, J.L. (1958). El agua potable en la Republica Mexicana: Los abastecimientos en la epoca prehispanica. *Ingeniera Hidraulica en Mexico* 12 (2): 69–82.

Brundage, B.C. (1979). *The Fith Sun. Aztech Gods. Aztec World*. Austin, TX; London: The University of Texas Press.

Butler, S.S. (1960). Irrigation systems of the Tigris and Euphrates valleys. *Journal of Irrigation Drainage Division* 86 (4): 59–70.

Cardano, G. (1550). *Desubtillitate Libri XXI*. Basel.

Chanson, H. (2004). Hydraulics of rectangular dropshafts. *Journal of Irrigation Drainage Engineering* 130 (6): 5230–5290.

Christensen, P. (1993). *The Decline of Iran-Shar*. Copenhagen: Museum Tusculanum Press, University of Copenhagen.

D'Altroy, T. (2003). *The Incas*, 389pp. Blackwell Publishing.

Doolittle, W.E. (1990). *Canal Irrigation in Prehistoric Mexico. The Sequence of Technological Change*. Austin, TX: University of Texas.
Fahlbusch, H. (2004). Men of dikes and canals, in Men of dykes and canals, The archaeology of water in the Middle East (ed. B. Hans-Dieter and H. Jutta). International Symposium held in Petra, WadiMusa (HK Jordan) (15–20 June 1999). Rahden/Wesf: VerlagMarie Leidorf GbH.
Hodge, A.T. (1992). *Roman Aqueducts & Water Supply*, 2e. London: Duckworth.
Keeble, B.R. (1988). The Bruntland report: 'Our common future'. *Medicine and War* 4 (1): 17–25. https://doi.org/10.1080/07488008808408783.
Koetsier, T. and Biauwendraat, H. (2004). The Archemedean Screw Pump. A note on its invention and the development of the theory. *International Symposium on History of Machines and Mechanisms: Proceedings HMM2004*, 181–194. Netherlands: Springer.
Koutsoyiannis, D., Zarakoulas, N., Angelakis, A.N., and Tchobanoglous, G. (2008). Urban water management in ancient Greece. Legacies and lessons. *Journal of Water Resources Planning and Management* 134 (1): 45–54.
Laueano, P., Cirella, A., and Whitehouse, A. (2001). *The Water Atlas: Traditional Knowledge to Combat Desertification*. Copyright by Peatro Laureano (trans. C. Anna and W. Angela).
Lewis, M.J.T. (1997). *Millstone and Hammer: The Origins of Water Power*, 20–28, 37–57. University of Hull.
Lucas, A. (2011). *Fulling Mills in Medieval Europe; Comparing the Manuscript and Archaeological Evidence; Archaelogie Des Moulins Hydrologiques a Traction Animale et a Vent, Des Origins a l'epoque Medieval*. Besancon Presses Universitaire de Franche-Cme.
Masse, B. (1981). Prehistoric irrigation systems in the Salt River Valley, Arizona. *Science* 214 (4519): 408–415.
Mattingly, D.J. (1988). 'Oil for Export?' A comparison of Libyan, Spanish and Tunisian olive oil production in the Roman Empire. *Journal of Roman Archaeology* 1: 33–56.
Mays, L.W. (ed.) (2010). *Ancient Water Technologies*, 1–180. Dordrecht, Heidelberg: Springer Science & Business Media.
Mays, L.W. and Gorokhovich, Y. (2010). Water technology in the Ancient American Societies. In: *Ancient Water Technologies* (ed. L.W. Mays), 171–220. Dordrecht, Heidelberg: Springer.
Pena Santana, P. and Levi, E. (1989). *Historia de la hidraulica en Mexico. Abastecimiento de agua desde la epoca prehispanica hasts el Porfiriato*, 1–168. IMTA Series 25.
Quintela, A.C., Cardoso, J.L., and Mascarenhas, J.M. (1987). Roman dams in southern Portugal. *Water Power and Dam Construction* 39 (5): 38–40.
Remini, B., Kechad, R., and Achour, B. (2014). The collecting of groundwater by the Qanats: a millennium technique decaying. *Larhyss Journal* 20: 259–277.
Rowe, J.H. (1990). Machu Picchu a la Luz de documentos de siglo XVI. *Historica* 14 (1): 139–154.
Said, R. (1993). *The River Nile, Hydrology and Utilization*. New York: Pergamon Press.
Shiklomanov, I.A. and Rodda, J.C. (2003). *World Water Resources at the Beginning of the 21st Century*. International Hydrology Series. Cambridge University Press.
Skivaniotis, M. and Angelakis, A.N. (2006). Historical developments of the treatment of potable water. *Proceedings of the 1st IWA International Symposium on Water and Wastewater Technologies in Ancient Civilizations*, Iraklion, Greece (28–30 October 2006).
Spain, R.J. (1984). Romano-British watermills. *Archaeologica Cantiana* 100: 101–128.
Stordeur, D. (ed.) (2000). *El Kowm 2: une ile dans le desert – La fin du neolithique preceramique dans la steppe syrienne*. Paris: CNRS Editions.

Strouhal, E. and Forman, W. (1992). *Life in Ancient Egypt*, 173. UK: Cambridge University Press.
Thompson, L.E. and Mosley-Thompson, E. (1989). One half millennia of tropical climate variability as recorded in the strategy of the Quelccaya Ice Cap, Peru, American Geophysical Union, Aspects of climate variability in the Pacific and the western Americas. *Geophysical Monograph* 55: 15–31.
Tomory, L. (2012). *Progressive Enlightenment: The Origins of Gaslight Industry. 1780–1820*, 1–347. MIT Press.
Wilson, A. (2002). Machines, Power and the Ancient Economy. Society for the Promotion of Roman Studies. *The Journal of Roman Studies* 92: 1–32.
Woolf, G. (ed.) (2005). *Ancient Civilizations*. London: Duncan Baird Publishers.
Wright, K.R., Witt, G.D., and Zegarra, A.V. (1997). Hydrogeology and paleohydrology of ancient Machu Picchu. *Groundwater* 38 (4): 660–666.

2

Water Movement in Time and Space

This chapter considers the cycling of water in general and, in particular, freshwater across the planet including its movement in the biosphere, its geographical distribution on the continents and variations with time (seasons) and implications for water availability.

Within human or geological timescales, the water on this planet is in continuous movement (see Figure 2.1, the hydrological cycle). Various forms of water together form the hydrosphere that is thought to have arisen on this planet between 3.5 and 4.0 billion years ago Klige et al. (1998). Geological evidence indicates that the hydrosphere has existed through all geological periods (Strakhoiv 1963). The current volume of the hydrosphere is estimated at 1370 million km^3 with the largest volume of freshwater being in glaciers and ice sheets (60 million km^3). The Antarctic has the largest amount of ice with Greenland a distant second. All of the ice in glaciers and ice fields will be affected in the future by the warming climate. Ice and snow reflect more solar energy, so when they melt bare soil is exposed, which absorbs more heat energy that in turn can accelerate the warming process. If all of the ice and snow were to disappear the earth's surface would receive 3–4 W more heating per square metre than it currently does (UNEP 2007). As a result of this and multiple other sources of heating, including greenhouse gas emissions from land-based sources, melting ice could cause sea levels to rise by as much as 4.4 m by the end of the century. Warming of the planet will also result in the oceans' warming that will cause thermal expansion of the sea water, which will raise the level by 1.6 mm/yr. The Antarctic ice sheet melting contributes 0.2 mm/year, the Greenland ice sheet 0.2 mm/year and glaciers and ice caps 0.8 mm/yr. These all add up to a rise of just over 2.8 mm/year (these numbers were increased between 1993 and 2003 based on IPCC 2007). The observed sea level rise as measured by precision satellite altimeters indicates that the global average sea level rise between 1993 and 2006 was 3.1 mm/year (UNEP 2007). The rise will continue throughout the twenty-first century. Figure 2.2 gives examples of two more frequent types of aquifers.

The oceans are saline so cannot be used by humans without suitable treatment. The question has been posed: 'why are the oceans salty and where does the salt come from?' It was originally thought that rivers erode rocks and soils and transport minerals to the sea, but more recently other sources have been highlighted (Church 1996). Hydrothermal recirculation along ocean ridges can dissolve basaltic rocks and pass the minerals into the water. A further source is by atmospheric deposition of dust and trace metals, some of

Water: Our Sustainable and Unsustainable Use, First Edition. Edward G. Bellinger.

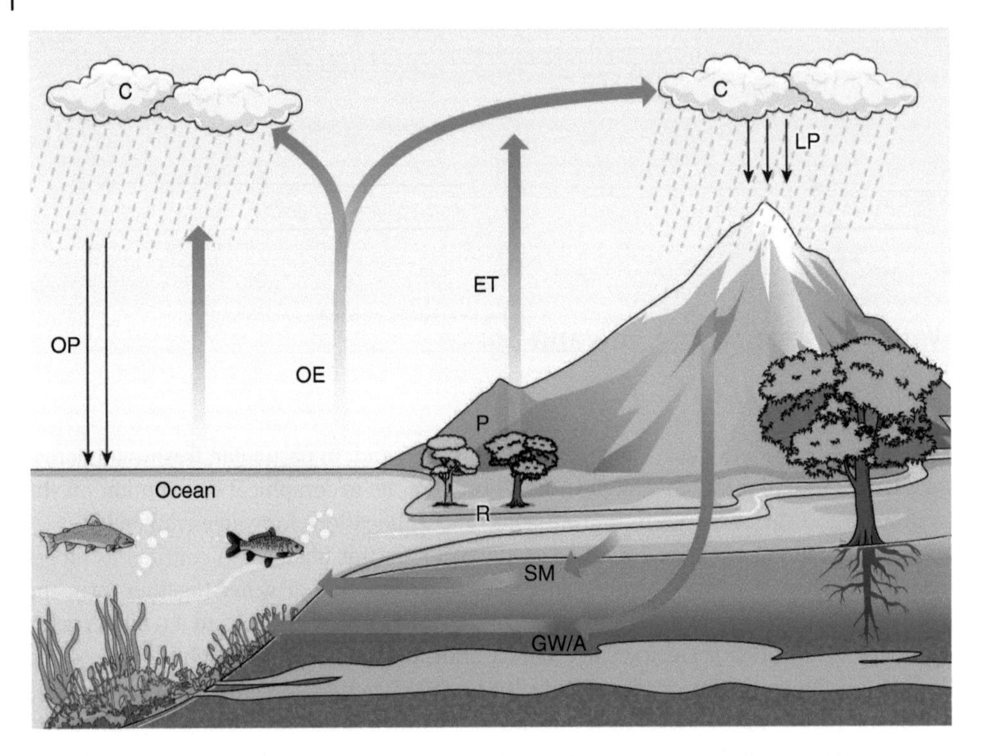

Figure 2.1 An outline diagram of the hydrological cycle. C, Rain cloud; OP, precipitation onto the oceans; OE, evaporation from the oceans; R, river; SM, soil moisture; P, plants; ET, evapotranspiration; SM, soil moisture; GW/A, ground water; LP, high level precipitation.

which can be essential nutrients for planktonic organisms. This source is likely to have increased since the industrial revolution. A third source is thought to be submarine groundwater discharge (SGWD) that has been suspected for some time but only in recent years supporting evidence has been found. SGWDs occur when fresh or nearly freshwater percolates through porous rocks and sediments from sub-surface freshwater aquifers into the overlying seawater (Moore 1996). He reported that there are, for example, several hundreds of kilometres of coastline in the south-eastern United States from which SGWD discharge is comparable to that observed from rivers. The base-water flow in many rivers is mainly from groundwater seeping from aquifers, so it contains dissolved weathered rock minerals. SGWD is also thought to be one of the sources of minerals in coastal lagoons. Some sub-surface aquifers are known to directly discharge to the ocean, especially when they originate in porous limestone rocks, e.g. in Florida, parts of the Mediterranean, The Red Sea and the Persian Gulf. In some of these regions, the limestone rocks contain submarine rivers and produce offshore freshwater upwellings where ancient mariners could replenish their freshwater supplies. Estimates of the amount of SGWD range from 0.01% to 10% of the surface water run-off although in inland seas and confined coastal regions they amount to 10% of stream flow inputs. Tracers have been used to more accurately estimate the amount of the SGWD input. The main one is radium selected for its long half-life (1620 years). Measurements of radium in near-shore waters by Moore (1996) concluded that there must be as much as 40% submarine groundwater discharging as there is river flow. He suggested that the pumping action of the rise and fall of the tide helped the flow

(a)

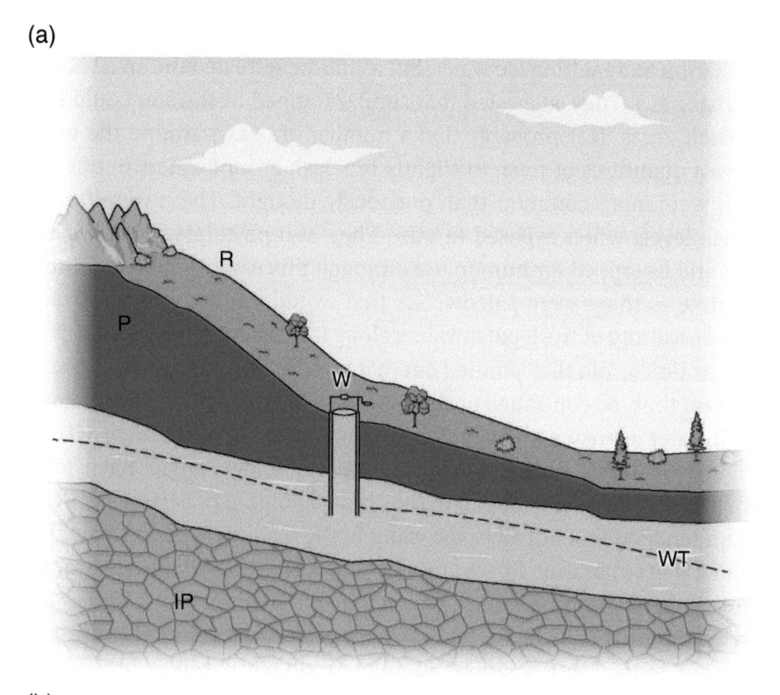

(b)

Figure 2.2 Main types of aquifer. (a) Unconfined aquifer. (b) Confined aquifer. IP, impermeable base rocks; R, recharge zone; P, permeable rocks; WT, water table; W, well.

of water from the SGWD. Kohout et al. (1977) working with a deep-water drilling project on Nantucket Island, just over 60 km from the New England coast (USA), found an aquifer of freshwater at a depth of 158 m, and three other freshwater zones at around 240, 280 and 450 m, the latter of which contained slightly salty water. They suggested that large areas of continental shelf were exposed due to low sea levels during the Pleistocene when the area was exposed to precipitation and melting ice water, but as the ice retreated the area became again covered by the rising sea. They suggested that similar trapped freshwater could exist in other continental shelf areas. It is probable that a number of coasts around the world would possibly have vast quantities of fresh to slightly brackish groundwaters in offshore regions and that these were more common than previously thought. These were formed during the past low sea levels when exposed to rain. They also postulated that some of these stores of water could be tapped for human use although this would equate to mining a non-renewable resource as these were paleowaters that would not be replaced. Cohen et al. (2010) modelled the salinity of fresh paleowaters along the Atlantic Continental Shelf of the United States of America, and they pointed out that these freshwaters were not being replenished at the present time, so continued abstraction, as happens with the Long Island aquifer, will lead to salt water intrusion and depletion of the aquifer. Some of these SGWDs extend many kilometres offshore. Although oil drilling technology might have to be employed to exploit them, e.g. near New England, they have a volume of between 500 and 2000 km^3. It must be understood that not all of the water in these aquifers can be extracted. Church (1996) estimated that submarine fluxes amounted to at least 5 l/m^2 per day over the inner shelf in the South Atlantic Bight. He also suggested that in modern times dredging of shipping lanes to keep them clear may help breach the sub-surface aquifers. He also pointed out that at the end of the last glaciations aquifers would have been recharged producing a large hydrostatic head that would have increased groundwater inputs into the sea. Increased dredging causing breaches could allow seawater contamination to occur. It is thus important to manage this activity to prevent such breaches and preserve the aquifer. It has also been pointed out that in coastal regions where freshwater is scarce careful tapping into an SGWD could provide a useful source, especially for megacities located on the coast such as Karachi.

As might be expected, the oceans, being the largest volume of water on this planet, have the greatest impact on the hydrological cycle. Most evaporation takes place from the ocean's surface and most precipitation falls upon them. The oceans play a major role in determining the climate that is experienced by the land masses. Within the oceans, currents play a major role in transferring heat from tropical regions to colder ones, e.g. the Gulf Stream from the Caribbean to the North Atlantic and to the west coast of the United Kingdom. Although there is considerable movement of water through evaporation and precipitation, liquid water movement is a surface and sub-surface phenomenon. Water in the ocean depths and in deep aquifers tends to be less mobile and can have a residence time amounting to thousands of years (see Table 2.1). Because of the properties of water, in particular its high specific heat and poor thermal conductivity, water temperatures change relatively slowly when compared with changes in atmospheric temperatures. This, particularly when considering large bodies of water, tends to ameliorate the climate of adjacent land masses.

Depending upon where it is held, its residence time may vary between minutes, days or tens of thousands of years (Table 2.1), and its continuous movement is termed the hydrological cycle in which water moves from oceans to the atmosphere and then to land and back to the atmosphere and/or oceans, which enables water to be transported around the globe. The movement is driven either directly or indirectly by both solar energy and gravity.

Table 2.1 Residence times and volumes of water in different sections of the hydrosphere.

Location	Volume ($km^3 \times 10^3$)	% Freshwater	Residence time
Oceans	1 338 500	—	Approx. 4000 yrs
Ice caps, glaciers and snowfields	24 064	68.7	10–100 000 yrs
Fresh groundwater	10 530	30.1	Weeks, days and thousands of years
Soil moisture	16.5	0.05	Weeks to years
Permafrost	300	—	10 or more years
Atmosphere	12.9	0.04	Approx. 9–10 d
Swamps and wetlands	11.47	0.03	Between 1 and 10 yrs
Rivers	2.12	0.006	About 14 d
Biota Logical water	1.12	0.003	About 1 week

Source: Data from Holden (2014), Gupta (2011) and Shiklomanov and Rodda (2003).

The mean residence times for water molecules, that is how long they remain in a given part of the hydrological cycle, can be estimated by dividing the volume of water in the reservoir by the mean flux in and out of it (Oki and Kanae 2006).

Water can take many pathways during its cycling or movement, but the overall volume is near enough constant. There is a finite amount of water present on this planet, but only a small portion of it is both suitable and available for human use, so our increasing and changing demands must be managed sustainably so as not to exceed this volume. Most of the water on this planet is in oceans, but this is saline. Some saline water can also occur in lakes. The water that occurs in the atmosphere, most lakes, rivers and groundwaters is freshwater, also called meteoric water. There is a general pattern of water movement called the hydrological cycle, but it must be understood that there is considerable variation within each component of the cycle. On a global scale, the oceans are relatively constant on a human timescale, although global warming is now continuously adding snow and ice meltwater to them changing their overall volume. There are many large- and small-scale ocean currents moving large volumes of water around the planet. There is a continuous loss of water from the ocean's surface through evaporation, but this is balanced by the inflow of water from rivers, rain and meltwaters. 90% of the evaporation falls back onto the sea as rainfall, and only a small amount (10%) eventually falls onto the ground as rain recharging the freshwater part of the system and then flowing back into the sea or seeping into the ground.

Various components of meteoric water vary greatly both geographically and seasonally in both intensity and volume. These variations have had a large influence on human development and ecosystem functioning over the past one or two million years. During the ice ages, vast quantities of water were trapped in the polar ice sheets that spread over many land masses. This, together with the uneven distribution of the land masses between the southern hemispheres and, in particular, those in the northern hemisphere (North America and Eurasia), affects the circulation of the ocean currents and the movement of warm water to the polar regions. Climate change is now affecting this ocean current transfer by melting considerable amounts of the polar ice caps and potentially affecting the balance of the hydrological cycle.

As may be expected, the oceans, being the largest volume of water on this planet, have the greatest effect on the hydrological cycle. Most evaporation (about 86%) takes place from the ocean's surface, and most precipitation falls upon them. The oceans determine the climate that is experienced by the land masses. Within the oceans themselves, currents play a major role in transferring heat from tropical regions to colder ones, e.g. the Gulf Stream from the Caribbean to the Arctic, which warms the north-western coast of the United Kingdom. In recent decades, the amounts of snow and ice have decreased. This is more pronounced in the northern hemisphere than the southern. There are potentially other important knock on effects of this reduction in snow and ice fields. If the uppermost layers of the frozen tundra were to thaw, more greenhouse gas would be released. Sea ice is, and is increasing to, thaw, affecting the heat exchange with water and the strength of the major ocean currents. One of the factors driving ocean circulation is the deep and dense water in the Greenland Sea and the Weddell Sea in Antarctica (IPCC 2007). Water density increases as it gets colder and more saline, so it sinks to the ocean floor where it flows forming cold currents. Warm water, being less dense, tends to flow near to the surface to the colder regions where it releases its energy. These flows form the major circulation of ocean currents. These patterns of circulation are likely to be affected by ice and snow melt coupled with the resulting inputs of large amounts of freshwater. Some currents will be weakened as the sea level rises, which would have a devastating effect on many small islands in the Indian and Pacific oceans and the Caribbean as well as many low-lying countries and coastal areas.

Although there is considerable movement of water through evaporation and precipitation, this is a surface and shallow water phenomenon. Water in the ocean depths is less mobile and can have a residence time of more than 10 000 years. Because of the properties of water, in particular its high specific heat, thermal capacity and poor thermal conductivity, water temperatures change relatively slowly when compared with changes in atmospheric temperatures. This, particularly when considering large bodies of water, tends to ameliorate the climate of the adjacent land masses.

Atmospheric Water, Clouds and Precipitation

Water vapour in the atmosphere, often resulting from evaporation and, where vegetation is present, evapotranspiration, is an important greenhouse gas. By reducing the earth's infrared heat loss, it helps maintain a surface temperature on earth and the immediate atmosphere, the biosphere, some 33 °C warmer than it would otherwise be. Without this effect, the average temperature at the earth's surface would be around −16 °C, not very suitable for life as we know it. Indeed, most of the liquid water that we see would cease to exist and become ice.

Atmospheric water (water vapour) is the most mobile component of the water cycle. This mobility means that it is also able to transfer considerable quantities of heat from one location to another, e.g. from the tropics to temperate regions. Heat energy is needed to drive evaporation, and this heat energy is released as an air mass cools, forming clouds and rain. This latent heat, if suddenly released, can be the instigator of storms, hurricanes and typhoons. The transfer of water from the oceans to land is not only the driving force for rivers but also, either directly or indirectly, the main source of water for humans and ecosystems. Ice, rain, standing and flowing waters are also the major shapers of land features. The amount of water being transferred to land by evaporation from the oceans is about 40 000 km^3/yr (Jones 2010; Shiklomanov and Rodda 2003). This is driven by the amount of incoming solar radiation. Hence, the greatest zone of precipitation is normally in the equatorial tropics. There are also smaller precipitation peaks in the mid-latitudes where, because of the availability of moisture and more favourable, less extreme, temperatures, the regions of greatest human activity occurs.

There are huge variations in the distribution of freshwater globally, seasonally and temporally. These variations not only lead to differences in human and agricultural activity but can also result in episodic disasters such as droughts and floods, all causing problems for water management.

Precipitation is the ultimate freshwater resource as it is responsible, either directly or indirectly, for recharging aquifers and providing surface run-off. The topography of a land area influences the amount of precipitation. Mountain ranges usually have a dominating effect on both precipitation and run-off. Two main zones of response can be identified based upon the influence of mountain ranges. Where the run-off flows to the sea, the zone is called exorheic. Where it flows to inland areas, it is called endorheic. The largest endorheic zones are generally in mid-continent areas and are usually drier, whereas exorheic zones, being closer to the sea, are moister. While it is important to quantify the amounts of water in different zones, it is also important to know how much of it is accessible to humans. The exorheic zones generate about half of the renewable freshwater resources. About 75% of the total run-off is accessible to humans. We must realise, however, that just because we can potentially access the resource, it does not mean that we can use as much as we want for the resource is also needed to support natural ecosystems and has a finite volume. Initially, in the geological past, the waters were devoid of life, but during the Proterozoic Era, photosynthetic organisms appeared in the hydrosphere. These photosynthetic organisms raised the atmospheric oxygen concentration from about 2% over 2 billion years ago to its current concentration of 20% today, making life as we know it possible. The oceans were larger until the Cretaceous and since then have retreated. In later years, cooling led to increases in the polar ice caps binding up large amounts of water.

Spatial Variations in Precipitation and Run-off

There are large spatial variations globally in the amount of precipitation. Latin America has the most precipitation with just over 30% of the global total. Asia has about 20% as does Sub-Saharan Africa and OECD (Organization for Economic Cooperation and Development) countries. Eastern Europe has about 6% and the Middle East and North Africa, the driest general area, under 2%. Some 80% of the global population obtain their water from accessible supplies, but this still leaves approximately 20% who do not. These people obtain their water from aquifers, many of which are ancient (thousands of years old) and should, in human timescales, be regarded as non-renewable. Other sources available to this 20% are from inter-basin transfers and desalination. It is surprising, considering the absolute requirement for water, that many people live in semi-arid and arid areas having limited access to renewable water.

Apart from developed countries and rivers that are relatively large or of economic/societal importance, monitoring river flows is quite fragmentary. Although thousands of small rivers exist globally, most discharge to the oceans comes from the world's largest rivers, such as the Amazon, La Plata, Congo, Niger, Ganges, Irrawaddy, Yangtze, Mekong and the Mississippi (Table 2.4). There is also a considerable volume of discharge from medium to large rivers. There are also a small number of significant rivers that do not discharge to the oceans or large seas, such as the Mediterranean, Caspian and Baltic but to inland enclosed basins. The three main examples of this are the Great Salt Lake in Utah (USA), Lake Eyre (Australia) and the Aral Sea (Central Asia). Modern human usage has meant that the Great Salt Lake area has been successfully occupied by people, whereas the Aral Sea basin has become a major disaster area through reduced inflow and salt accumulation.

Many industries in the past have grown up where there was a plentiful supply of water, e.g. the cotton and wool textile in northern England. Economic constraints already exist in at least some parts of developed countries because of water scarcity. Adaptation has had to take place both in the sustainable management of water use in these industries and/or their geographical location. Many future economic developments have been predicted to occur in the BRIC countries (Brazil, Russia, India and China). Each of these countries has major rivers flowing through them, but often part of their river system is in an adjoining country so they may not have complete control of it. Not all of these countries manage their rivers sustainably, especially where they are likely to be affected by future global climate change. Management is sometimes made worse by the fact that political boundaries often have no relation to hydrological boundaries. Transboundary waters, including groundwaters as well as surface flows, and their management can lead to disagreement and possible conflict.

The Sahel region of Africa has a fairly low rainfall but is also subject to large variations. Compare this to relatively wet regions in England and Wales where the average rainfall is much greater but less variable, although in some areas periods of drought may in the future become more common. An example of possible changing conditions is the reduced winter rainfall in parts of eastern United Kingdom in 2023 where the growth of certain crops is threatened by lack of rain during some of months of the year. The distribution of the precipitation, month by month, throughout the year varies regionally in most countries. For example, Niger has its rainfall typically concentrated in the period May to September, but in the United Kingdom it is spread over the whole year. Different areas tend to have characteristic rainfall patterns over each year, although these are, at least in some locations, changing possibly due to climate change. When precipitation falls to the ground, a proportion, depending upon the climate of that area, will evaporate back into the atmosphere. Additionally, depending upon the vegetation of that area, water will pass back into the atmosphere by means of transpiration. The two effects added together give evapotranspiration. This tends to be greatest during the plant-growing season (the summer months in the northern hemisphere) and can often exceed the amount of precipitation. In the non-growing season, precipitation can exceed evapotranspiration and in that case the excess water can either recharge sub-surface aquifers or form surface run-off.

Some regions of the world are subject to occasional very heavy downpours of rain that can sometimes cause catastrophic flooding. These events have recently been associated not with scattered spread-out events or low-pressure storms but with relatively narrow regions in the atmosphere carrying large quantities of water vapour from the tropics to temperate regions. These have been called 'atmospheric rivers (ARs)'. When these rivers, driven by strong winds, reach land areas with watersheds prone to flooding, they may release their vapour as torrential rain and lead to floods. These ARs can carry as much water as the River Amazon. The ARs can account for as much as 30–50% of the annual precipitation for the west coast of the United States of America. Although their existence has only recently been confirmed (Zhu and Newell 1998, MIT), the existence of such a phenomenon has been known on the west coast of the United States of America and was called the 'Pineapple Express' as it brought water vapour from near Hawaii to the Californian coast. In the United Kingdom, ARs have been linked to the 10 largest winter floods since the 1970s. The 2009 flood in Cumbria (NW England) coincided with an extra-tropical cyclone that tunnelled an AR towards the west coast of the United Kingdom. The Cumbrian mountains caused the AR to rise, cool and deposit its moisture as rain for several days resulting in floods (Lavers et al. 2011). There are wide variations in river flows geographically and temporally. These variations tend to be greatest in the drier regions, for example the ephemeral rivers of Namibia. Water entering a river does not only come from surface run-off, precipitation and springs, but also arises from underground flows. Throughflow arrives in a river by infiltration through the aeration zone without penetrating the main water table. Any groundwater that discharges into the river is called the base flow, and this can vary greatly with season, the geology and the topography of an area. In an area where the rocks are impermeable, base flow will be small and throughflow and surface flow more important. If the rocks are very permeable, there may be little or no surface flow as all of the precipitation quickly soaks through the rocks to the water table. About 90% of groundwater discharge globally feeds into streams and in total this represents almost 30% of global run-off (Margat 2008).

When considering biodiversity on this planet, most people think in terms of land- and aquatic-based systems. However, for nearly 200 years, the presence of organisms in the air has been recognised. Charles Darwin on his voyage on the Beagle in 1846 collected dust from the atmosphere and noted that it contained microorganisms (Darwin 2003). Over the last 50 years, atmospheric scientists have realised that there is a great variety of organisms living in the atmosphere, which they called the Aerobiological community and this biodiversity links to the earth's surface biome. Hundreds of thousands of microbial cells can be found in a cubic metre of air (Burrows et al. 2009a,b). These come from a range of taxa including bacteria and fungi (Fierer et al. 2008). Examination of the dust collected by Darwin was sent to Christian Gottfried Ehrenberg in the mid-1800s, and the collection is now housed in The Museum of Natural History in Berlin. Great care was taken to avoid contamination from modern dusts, and it was found that they could, quite remarkably, germinate in culture, a number of microbiological spores found in the samples even though they were over 150 years old. They were able to identify several species of bacteria and fungi present. They concluded that the origin of the organisms was North Africa coming from Bodele Depression in Northern Chad and the Sahara-Sahel region. They concluded that they travelled across the Atlantic mainly in the troposphere (2–10 km high) in dust storms, a journey taking 10–11 days. Biological particles represent a substantial proportion of particulates in the atmosphere, especially in the size range 1–3 μm. The behaviour and importance of these organisms, both for their interactions with climate and directly to humans, is now becoming better understood (Womack et al. 2010). The activities of these organisms play an important role in atmospheric processes including bioprecipitation. They play an important role in cloud formation and precipitation in the lower troposphere and act as ice nuclei (ICs) at temperatures as high as −20 °C. This is much higher than non-biological ICs. Some can act as cloud condensation nuclei (CCNs), (DeLeon-Rodriguez et al. 2013). We depend upon air around us but were, for a long time, not properly aware of the beneficial and pathogenic organisms that might be present. Airborne bacteria might have important impacts on human health not only because some might be pathogenic but also by triggering allergic reactions. These workers found that bacterial pathogens could be transported through the atmosphere from agricultural crops and livestock. The particle fractions of 2.5 μm diameter or less are the ones of most concern in air pollution. It has been known that particles can travel large distances in the air moving them from one location to another. Kellogg and Griffin (2006) pointed out that desert winds can blow several billions tons of dust every year from Africa. This dust also carries with it many microorganisms and pollen some of which could be pathogenic as well as interacting with the climate. These various dusts have been travelling in the atmosphere across different continents depending upon the prevailing winds for many hundreds of years expanding the geographical ranges of many different organisms. Bacteria in the atmosphere have been identified at a number of locations as diverse as Boulder, Colorado, United States of America (Fierer et al. 2008), Antarctica (Hughes et al. 2004) and France. A number of species, including *Enterobacteriaceae, Sphingomonas, Comamonadaceae, Pseudomonas* and *Sphingomonas*, were found. Bacteria are universal in their presence in the atmosphere occurring in concentrations of between 10^4 and 10^6 cells per cubic metre of air. There is little information of how land use changes affect the bacteria and other organisms in the atmosphere. Not only airborne dust but also leaf surfaces can affect the atmosphere near to the land surface. Different plant species harbour different species of bacteria and other microorganisms. Precipitation is the ultimate freshwater resource as it is responsible either directly or indirectly for recharging aquifers and providing surface run-off. Rain occurs when water molecules condense around CCNs in the atmosphere. Not all particles in the air can act as these nuclei. The most efficient ones are hygroscopic particles such as common salt or ice crystals. The raindrops that are formed are called cloud droplets. These collide with each other and coalesce to form bigger rain droplets. The highest concentrations of these bacteria seem to occur over vegetated areas. Bacterial particles seem to be more evenly distributed in the atmosphere compared with larger biological particles such as pollen, fungal spores and plant debris that show changes in abundance with changes in atmospheric conditions. The number of high temperature ICs that freeze at temperatures >10°C are affected by land type. There were significantly more nuclei, two to eight times more, in suburban and agricultural areas (Bowers et al. 2011). Many of these nuclei are found on our more common crops whereas conifers have very few. All of these observations show the close links between the biology and ecosystems of the land and the micro-ecology of the atmosphere as well as their influences and interactions. Burrows et al. (2009b) used a global chemistry-climate model to investigate the transport of bacteria sized 1 or 3 μm in the atmosphere and different contributions of a range of ecosystems. They found that the bacteria could remain in the atmosphere for three to eight days enough time for trans-continental travel. They also found that those from deserts, shrubby areas, crops and grasslands had the longest residence times in the atmosphere. With the exception of crops, these are dry or relatively dry zones. Residence times were also highest during the summer months. They pointed out that not all bacterial particles found in snow or rain could be nucleating types. It is still not clear whether certain bacteria are able to reproduce in the atmosphere and if the atmosphere can provide a suitable environment for their growth. As many bacteria in the atmosphere are viable, the question arises as to whether they actively break down organic compounds that may also be present and so influence the chemistry of liquid particles they come into contact with. Interestingly, Conrad (2009) suggested that they may be responsible for as much as 70% of the greenhouse gas methane in the atmosphere in some areas. They are also responsible for the production of nitrogen oxides and the precursor of dimethylsulphide (Bates et al. 1992). They and several others emphasise the importance of microorganisms in climate processes. Many land types still have not been fully investigated from the aerobacterial point of view including rain forests, wetlands and sandy deserts. It is generally accepted that airborne bacteria could have an influence on cloud formation and precipitation. Microorganisms have long been known to be active in the geochemistry of all habitats. The possibility of human-induced changes in the landscape that change plant communities and their associated microbial populations will also change some feedback cycles with the atmosphere and climate. More investigations are needed to assess these impacts if action is to be taken to avoid unexpected consequences. Both plants and the associated microbial growths can feed back into the climate cycle. Wet conditions favour the growth of microorganisms such as bacteria and fungi. The land masses are the main source of microbiological particles in the atmosphere. The main source is vegetation more than bare soil and wind-blown dust.

Not all precipitation seeps into the soil or runs off into streams, rivers, lakes and the oceans. Some evaporates back into the atmosphere. What does not evaporate amounts to about 40 000 km^3/year. Not all of this is accessible to humans, and they only use about 15% of it. A large proportion of that used by humans is returned to the hydrological cycle, unfortunately often contaminated to some extent. Generally, most precipitation occurs in the tropical regions with a secondary peak occurring in the mid-latitudes. Between these two peak zones is a region of low precipitation. Land topography influences the amount and location of precipitation. Mountain ranges usually have a dominating effect on both the amounts and the run-off. Two main zones of response can be identified based upon the impact of the mountains. Where run-off flows to the sea, the zone is called exorheic. Where it flows to inland areas but not to the sea, it is called endorheic. The largest endorheic zones are generally in the mid-continents and are frequently drier, whereas the exorheic zones are usually moister. While it is more important to quantify the amounts of water in different zones, it is also important to know how much of it is accessible to humans. The exorheic zones generate about half of the freshwater resources. About 75% of the annual run-off is accessible to humans. It must be understood however that just because we can potentially gain access to these resources, it does not mean that we should use them unsustainably as that resource also has to support natural ecosystems.

Globally, and even locally, there are large spatial variations in the amount of precipitation. Latin America has the most precipitation with just over 30% of the global total. Asia has about 20% as does Sub-Saharan Africa and OECD countries. Eastern Europe has about

6% and the Middle East and North Africa, the driest general area, has precipitation below 2%. Some 80% of the global population obtain their water from accessible supplies, but this still leaves approximately 20% who do not. These people obtain their water from aquifers, many of which contain ancient water often thousands of years old and should, in human terms, be regarded as non-renewable. Additional water sources for this 20% are from inter-basin transfers and desalination. Surprisingly, considering the absolute requirement for water, many people live in zones where renewable water supplies are not abundant. About 1 billion, for example, live in semi-arid and arid areas having little access to renewable water.

Many industries in the past have grown up where a water supply was plentiful. Examples include the cotton and woollen textile industries in northern England. Countries without large amounts of water could have constraints on industrial economic development for certain industries. Adaptation has to take place both for sustainable management of water in these and other industries or convert to technologies that use less water if they are able to survive. Major future economic developments are predicted to occur, all other factors permitting, in the BRIC countries (Brazil, Russia, India and China). Each of these countries has major rivers in their territories although some parts of some rivers are in adjoining countries so they do not have complete control of their use. Not all of these countries manage their rivers sustainably, especially where they are likely to be affected by global climate change. Management is sometimes made difficult because political boundaries do not correspond to hydrological boundaries. Transboundary waters, including rivers, lakes and aquifers, can lead to disagreements and even conflict over management. The Sahel region of Africa has fairly low rainfall and is subject to large variations when compared with England where the average rainfall is usually much greater and less variable although climate change may alter this. In Africa the Niger has its rainfall typically concentrated in the period from May to September, whereas in England and Wales it is spread more evenly over the whole year. Different regions tend to have characteristic rainfall patterns over each year although these may, at least in some locations, change with the changing climate. When precipitation falls upon the ground, a portion, depending upon the location, will evaporate back into the atmosphere. A proportion of that which is taken up by plants will be returned to the atmosphere by evapotranspiration. This tends to be greatest during the growing season (the summer months May to September) in the northern hemisphere and the opposite (November to April) in the southern hemisphere. In the non-growing season, precipitation can exceed evapotranspiration in which case excess water will either recharge groundwater or form surface run-off.

Groundwater

Groundwaters provide a more stable and reliable source of freshwater as long as they are used sustainably and not overexploited. Deep, ancient or fossil groundwaters tend to be more stable than more shallow ones as the latter are usually more dependent upon regular recharge, whereas the former may have limited or little recharge so their exploitation tends to be unsustainable and if overused they are easily depleted. It has been recently estimated that there is a mean renewable freshwater groundwater resource of 2091 m^3 per person per year potentially making it an important contributor to water resources (Döll and Fiedler 2007).

Many groundwaters have a relatively large storage capacity and a slow throughput, which gives rise to long, sometimes tens of thousands of years, residence times. Groundwaters with these long storage times are much less affected by short-term fluctuations in climate when compared with river flows. They can provide a stabilising influence for both humans and ecosystems, that is, as long as humans do not overexploit them. Groundwaters also need to be protected from contamination. Groundwaters provide a considerable and increasing amount of water for human resource needs. They contribute between 20% and 50% of municipal water supplies (Morris et al. 2003).

Any of these hydrological variations can be large enough to have a dramatic impact on the economics of those regions, causing reductions of their Gross domestic product (GDP) in those areas by as much as 38% for drought and 43% for floods. Management of surface flows can often involve costly engineering, so frequently a more economic remedy to maintain manageable supplies is to use groundwaters.

Of the freshwaters present on this planet, the largest proportion is in glaciers, ice caps, snow fields and permanently frozen ground or permafrost (the Cryosphere). While these are not generally available or convenient for human use, they do have a huge influence on the rest of the hydrological cycle and, if global climate change progresses at the rate predicted, will be subject to transformation from solid state to liquid water that will affect the amount of freshwater available to humans and ecosystems. Because of its colour, snow and ice reflect the sun's radiant energy back into space resulting in a lowering of the earth's temperature, but if there is a rise in the global temperature for other reasons, then the snow and ice will gradually melt and the albedo effect will be reduced, not as much heat will be reflected back into space but will be absorbed into the land and sea adding to the warming effect, an important feedback mechanism in global climate change. The ice and snow perform an important cooling function for this planet, preventing too high a general temperature rise. In recent decades, the reductions of ice and snow have been greater in the northern hemisphere than southern (IPCC, Climate Change 2007). There are potentially other knock on effects of snow and ice field reductions. If the uppermost layers of the frozen tundra thaws, the stored organic matter would break down more quickly releasing large quantities of carbon dioxide and methane, important greenhouse gases, accelerating global warming. A further effect is that large volumes of melting ice, whether from ice fields, glaciers or sea ice reduction, could affect the heat exchange and strength of the major ocean currents. One factor driving ocean circulation is the deep and dense water in the Greenland Sea and the Weddell Sea in Antarctica. Water density increases as it becomes saltier and colder, so it sinks to the ocean floor where it flows forming cold currents. Warm water, being less dense, tends to flow near to the surface to the colder regions where it releases its heat energy, thus becoming colder and at the same time more saline. This circulation forms the major circulation patterns of the ocean currents. The North Atlantic current is part of this system and helps maintain a more equable climate in north-west Scotland. These patterns of circulation are likely to be affected by ice and snow melt and the resulting inputs of large amounts of freshwater. Any weakening of the formation of deep cold water currents will weaken the North Atlantic Current by as much as 25%.

The water arising from the melting ice caps, glaciers and sea ice significantly contributes to the rise in sea level. The global sea level is currently rising by about 2.8 mm/year. This is partly due to melting snow and ice but also due to the expansion of water as sea temperatures rise (UNEP 2007; Barnett et al. 2005; Barlow et al. 2005). The IPCC calculates that the sea level could rise by as much as 50 cm this century, if even 20% or less of the Greenland and Antarctic ice melts, possibly a worse-case scenario, then the rise could be as much as 4–5 m (Barnett et al. 2005). This could have a devastating impact on many small islands and cities as mentioned previously. Many coastal cities would also be affected as well as considerable infrastructures built around coastal areas.

Many areas depend upon seasonal snow and its subsequent melting to provide much of their water. Examples include parts of the American west and Central Asia where about 75% or more of the water supply for agriculture and municipalities depend upon snow melt water. Hydroelectric power in some parts of the world such as Canada and Europe can depend upon snow melt water to drive them. Where seasonal snow is common, it acts as an important store of water during the cold/winter season which is then released as the temperature rises during the spring. This both feeds rivers and lakes and also plays an important role in recharging sub-surface aquifers. The length of snow cover has a marked effect on crop growth either decreasing or increasing the growing season in that particular area. Snow cover can also have a beneficial effect of insulation, that is, the ground and plants underneath insulating them from freezing winds. Early autumn snow or prolonged cover in the spring can also have a dramatic effect on reducing the growth period of crops which in turn reduces yields. Nowadays winter sports are an important leisure activity that is dependent upon a good covering of snow. If this does not occur, then there is a marked drop in the local economy.

Ice, or water in its solid-state covers, either seasonally or permanently, covers large areas of land. There are not only the two polar ice fields but also glaciers and ice fields in some mountain areas. The behaviour of these has an effect on the general hydrological cycle. As mentioned previously, the melting or freezing of the polar ice caps affects the sea level and hence potential flooding of land. Increased meltwater on land can affect river flows. The continuingly observed shrinking of glaciers can strongly affect the seasonal availability of freshwater. Glacial retreat allows the exposed land to be colonised by vegetation as does the thawing of permafrost. Although this could be regarded as an advantage in some ways, such as some new areas of agricultural land can be created, there are disadvantages in that melting permafrost can release more greenhouse gases that were bound in the frozen rotting vegetation adding to a warming climate. In addition, melting glaciers and ice caps add to sea level rise inundating existing lands.

Water Movement Through Evaporation and Precipitation

The most mobile aspect of the hydrological cycle is arguably evaporation and precipitation. The global water cycle is driven by the energy from the sun. It is this energy that is the main driving force that transfers water from the oceans, lakes, etc. by evaporation into the atmosphere where it can travel over large distances before being deposited both on land and other waters. Most evaporation occurs over the oceans and is estimated to amount to over 430×10^3 km^3/year. The amount evaporated over land is somewhat less and is estimated at about 65×10^3 km^3/year and includes transpiration losses from vegetation to the atmosphere. This is termed evapotranspiration. Not all water associated with vegetation has to be taken up by plants. Some precipitation is intercepted by leaves and lays on their surface

Table 2.2 Average precipitation in different continental regions.

Region	Precipitation (mm/yr)	Precipitation volume (km^3/yr)
Africa – Northern	96	550
Sub-Saharan	3304	16 048
Total	3400	16 598
Americas – Northern	1383	13 881
Central and Caribbean	5279	1515
Southern	6022	29 012
Total	12 684	44 048
Asia – Middle East	1409	1422
Central Asia	273	1271
Southern and Eastern Asia	6316	24 162
Total	7998	26 855
Europe – Western and Central	3266	4099
Eastern	1045	8462
Total	4311	12 561
Oceania – Australia and New Zealand	574	4598
Pacific Islands	2055	4733
Total	2629	9331

Source: Adapted from FAO (2014). Figures for regions were calculated by adding together data from sub-regions.

from where it directly evaporates. Transpiration is the passage of water through the soil into the roots and through the plant tissues to aerial parts, and from there it is evaporated into the atmosphere. Most of the water evaporated from the oceans falls back as rain onto their surface, but once in the atmosphere, water vapour represents the most mobile phase of the hydrological cycle. Of the water that evaporates from the oceans, approximately 10% falls as precipitation on land. The amounts of precipitation vary across the globe. In Mediterranean climates most precipitation occurs in the winter, whereas in countries with a monsoon climate most occurs in the summer. Table 2.2 indicates the average precipitation in the main regions of the world.

Glaciers, Ice Fields and Snow Melt

Glaciers and snow are the major stored reserves of freshwater. High mountain ranges can be regarded as 'water towers' because of their contribution to river flows from ice and snow melt. The growth and shrinkage of ice sheets and glaciers over many thousands of years has profoundly affected the global water cycle. If glaciers continue to shrink or even disappear, meltwater streams, both large and small, could potentially dry up completely during prolonged hot dry summers. This would impact both humans and ecosystems. Currently, glaciers are estimated to contain more than 60 million km^3, and as a result sea levels were lowered by about 100 m (Shiklomanov and Rodda 2003). There have been several ice ages in the past. The last one began about 120 000 years ago covering most of Canada and the United States of America down to Manhattan as well as more than half of the United Kingdom, most of Siberia and large parts of northern Europe. The water incorporated in this and other smaller ice sheets lowered the sea level by as much as 120 m. Britain was joined to Ireland and mainland Europe. Florida was much larger than at present, and in the southern hemisphere, Australia, Tasmania and New Guinea were all part of one land mass (Ananthaswarmy 2012). At the time of the last ice age, about one-third of the land was covered by ice (Benn and Evans 1998). About 20 000 years ago, the ice and snow started to melt resulting in large quantities of freshwater pouring into the oceans affecting the overturning and heat transfer of the circulation which virtually shut down. No heat was then transferred from the tropics to the arctic resulting in cooling in the northern hemisphere and a warming in the southern hemisphere. During the next 10 000 years, the global temperature gradually rose by about 3.5° centigrade, which restored the overturn circulation. Warm periods occurred as a result of carbon dioxide being released into the atmosphere from the southern ocean. This resulted in much of the ice melting although there were times, for example, when some glaciers increased. The changes in length of many glaciers have been made since the nineteenth century. These show a general global shrinking and that there were significant periods of retreat in the 1920s, 1940s, and 1980s (UNEP World Glacier Monitoring Service 2008). These retreats are generally still occurring and are more pronounced in the northern hemisphere (IPCC 2007). There are potentially other important knock on effects of reductions in the snow and ice fields. Fifteen to twenty percent of the world's population live in river basins that are supplied by glacier or snowmelt water. Both glaciers and groundwater provide long-term storage of freshwater. The water in streams and rivers originating from mountainous areas come from three main sources, rainfall, snow melt and glacier melt. Snow cover in mountain regions has provided, through spring snow melt, much of the streamflow water. Large areas of the world such as the semi-arid American West and Central Asia largely depend on snow melt either directly or indirectly to supply agriculture and cities. These sources of water have always been seasonal but fairly regular. With global climate warming, it is less likely that winter precipitation will fall as snow and that which does fall will melt earlier in the spring (Barnett et al. 2005). Changes in the amount of precipitation will also potentially affect the amount of snow accumulation, which typically occurs towards the end of the winter. Snow melt and subsequent run-off will tend to be earlier which, with less snow, will be reduced as will summer rainfall if temperatures increase. Even if temperature rise is kept to the lower end of targets, e.g. 1.5 °C, the impact on snowmelt and subsequent freshwater resources will be large (Barnett and Pennell 2004). Water availability problems through lack of winter snow and snowmelt are already being felt in some regions, for example in California. The largest impact predicted on the hydrological cycle from the point of view of snow cover is in the mid-latitudes to low latitudes (Nijssen et al. 2001). In the Western United States, reduced snowfall and spring snowmelt will reduce reservoir volumes, potentially causing stress to all users including hydroelectric schemes (Barnett et al. 2004). Problems will also arise for regions dependent upon glacial meltwater for their spring and summer supplies. This is because in a warmer world snowfall will be reduced and water lost from a glacier will not be replaced (Barnett et al. 2005). Glaciers reached their maximum sizes towards the end of the Little Ice Age (the early fourteenth to mid-nineteenth century). Much of the water stored in glaciers can be regarded as 'fossil' water as much of it has originated over hundreds or thousands of years and cannot be replaced in the short term. Glacier melt water tends to reach a maximum in the summer unlike the spring peak for snow melt water. Globally, most glaciers are shrinking although there are some exceptions, e.g. in the Karakoram (Hewitt 2005). Generally, a warmer atmosphere leads to less snowfall and greater melting, which results in glaciers retreating. UNEP (2007) reported that 30 glaciers used for reference purposes, as they have had almost continuous measurements made on them for the past 40 years, showed an annual loss in mass averaging 0.58 m water equivalent for the decade 1996–2005. This was nearly double that for the previous decade. As glaciers flowing directly into the oceans melt, they directly contribute to sea level rise. European glaciers, which are mainly medium sized, are likely to retreat with rises in global temperatures, whereas larger type glaciers in Alaska and Patagonia are more likely to get thinner in depth rather than retreating. When inland glaciers melt, the water they discharge can be hazardous causing avalanches and flooding. These effects can be restricted and local or have widespread impacts and do have an economic cost estimated to be, on average, in the order of several hundred million US$ annually with the most devastating disaster killing more than 20 000 people (Kääb et al. 2005). There are a number of potential hazards from glacial and permafrost melting. These include breaching of moraine dams, glacial outbursts, displacement waves caused by avalanches, changes in glacial run-off, rock and ice fall avalanches, landslides, thaw settlement and soil heave if permafrost thaws, destabilisation of frozen slopes leading to landslip and increased flows of debris from permafrost. These are discussed in detail in Kääb et al. (2006). Glacier and Permafrost hazards are a greater problem in high mountain areas. Examples include rock-ice avalanches in Peru (1970) and problems of debris flows in Columbia (1985). Even if global warming is limited to 1.5 or 2.0 °C, there will still be considerable effects on glaciers, snowfall and permafrost. Any sustainable ice, snowfields and glacial lakes could be disrupted. Permafrost thaw would allow different species of plants to grow. There could also effects on socio-economic development. In many mountain areas, human activities are increasing as is the demand for land which is encroaching upon potentially unstable areas. In developing countries, e.g. the Andes, Central Asia and the Himalayas, where this is occurring, the countries do not have the resources to cope with these events. They also lack the capability to monitor potential hazards, so problems cannot be dealt with prior to an event. These types of hazard do occur in developed countries, e.g. the European Alps, but there the funds are more likely to be available to build expensive protective measures.

Groundwater

Groundwaters provide a more stable and reliable source of freshwater as long as they are used sustainably and not overexploited. They are ultimately fed from surface waters and precipitation, so there is interdependency between them. They may, in turn, discharge to surface waters as springs, the oceans, or evaporate to the atmosphere. They act as a store of water and also play an important role in freshwater circulation by feeding streams and rivers. Although it is difficult to precisely estimate the volume of water stored in groundwaters, the amount is quite large and can vary geographically. Shiklomanov and Rodda (2003) estimate that the volume is approximately 96% of all of the earth's unfrozen freshwater. Döll and Fiedler (2007) estimated that there is a mean renewable freshwater volume in aquifers of 2091 m^3 per person per year potentially making it an important contributor to general water resources. Many groundwaters are not subject to short-term fluctuations unlike many surface and atmospheric water sources. Some aquifers are replenished annually, but others, for example in arid areas, replenishment may only be intermittent or, in terms of human lifespans, be never. In some aquifers where there is a great depth of water, the upper layers may be recharges annually, but the lower ones may be unchanged for a very long time. With these and many others where the replacement time is measured in hundreds or even thousands of years, the waters may be regarded as non-renewable. The distribution of groundwaters is related to the geology of an area as well as the behaviour of the hydrological cycle in that region. Shiklomanov and Rodda (2003) distinguish the following three zones of groundwater storage and movement.

1) A zone near to the soil surface where water exchanges actively with the surface and atmospheric components of the hydrosphere. The quality of the water is closely related to the geology of the rocks and overlying soils.
2) Below the first zone is one that has less active water exchange. It is only affected by large rivers with deep channels. Some aquifers of this type lie below the sea and may discharge into the sea. Even though there is less water movement in these aquifers, they tend to have only weakly mineralised water.
3) These deeper aquifers lie below the second one and may be down to a depth of 2000 m. Only the upper part of these aquifers contain freshwater, and the lower parts tend to be saline.

Korzun (1974) estimated that there was 23.4 million km^3 of water stored in these three zones. By far the largest abstractors are India, China and the United States. Interestingly, Saudi Arabia and Syria are both in the top 15 abstractors. The amount of water abstracted for drinking supplies has substantially increased for urban populations (+63.1%) but decreased for rural populations (−4.0%). Groundwaters are largely abstracted for agricultural use. Very few countries use groundwaters predominantly for industry, the main one being Japan. The global increase in the size of cities, especially megacities, has led to the growth of groundwater use for human domestic purposes. If current trends continue, this sector is likely to increase significantly in the future.

Although groundwaters are largely unseen, they play a vital role in supporting the biosphere. Without them, the surface of the earth would be quite different. They support wetland habitats, streams and many deeper rooted trees.

Groundwaters have been exploited for many thousands of years, but they have only been extensively utilised for a little over the last 100 years. During this period, they have been used to meet the needs of agriculture, ecosystems and human communities (Llamas and Martinez-Santos 2005). Currently, approximately 22% of the global population rely upon groundwater for their water supply. Unfortunately, about 10% of that consumption is unsustainable, i.e. it is being used faster than it is being replenished. Groundwaters are being increasingly used to supply human needs in many countries, e.g. Jamaica, Saudi Arabia,

Libya and Lithuania, where they supply over three quarters of the freshwater needs. The increasing global population and ever-increasing demand for food has driven this increase in use. Groundwaters are now essential for domestic and food security for over 1.7 billion people, mainly in Africa and Asia. This increase in use of groundwater has been made possible through advances in technology that has enabled aquifers to be exploited to a greater degree. This improved access to groundwater has greatly helped farmers in many regions as it can often be obtained on demand and used more efficiently (UNESCO World Water Assessment 2003). The distribution of groundwater greatly varies globally (Vrba 2004). Shallow aquifers where the water table is near to the surface are usually closely integrated with surface water behaviour, whereas deep groundwaters are often linked with historic hydrological regimes, for example some aquifers were created mainly during wet Pleistocene periods. Groundwater has also been viewed as being more reliable throughout the year as compared to surface waters and rainfall. This gives greater more constant yields and hence better economic returns. In some countries, farmers and municipalities have made their own investments in groundwater technology, e.g. in Bangladesh. However, when many small farmers and communities exploit an aquifer, it can lead to management problems as abstraction rates need to be regulated to prevent over abstraction. Groundwater pumping does involve direct costs for the user, so there is an incentive to use water as efficiently as possible. This is good except where energy costs are subsidised by the state when overuse can occur and accelerated depletion can result, e.g. in India and Pakistan. The water quality in groundwaters varies with the geology of the region as well as changes that can be brought on by human influences, for example increased nitrates from agricultural fertilisers. Parameters such as total dissolved solids and chemical composition vary not only regionally but also with depth and along the flow path of the water within the aquifer. Although many chemicals in the water may be harmless, some may be present in higher than acceptable concentrations. Examples include iron, manganese, fluoride and arsenic (see Table 2.3).

Overabstraction of an aquifer can result in wells having to be sunk much deeper, and the pumping costs become much higher. This could take their use beyond the reach of poorer

Table 2.3 Potential chemical problems and concentrations in groundwaters from various countries.

Country	Origin	Chemical	Conc. (mg/l)	WHO DW standard
Bangladesh	Minerals and rocks	Arsenic	1.1–5.0	0.01
Czech Republic and India	Minerals and rocks	Fluoride	4.5–10	1.5
China	Sea water	Chloride	8818	250
Australia	Salinisation	Chloride	4718	250
USA	Salt containing strata	Sodium	121 000	200
USA	Marine argillaceous rocks	Aluminium	64	0.2
China and India	Acid mine drainage	pH	2.0–4.0	6.5–8.5

Source: Data adapted from Vrba (2004). These are some examples, but it should be noted that these problems occur in other aquifers and countries than those listed above.

farmers and communities. Consequently, it is vital that groundwaters are protected from both contamination and overabstraction. Groundwaters already contribute between 20% and 50% of municipal water supplies as well as being vital for agriculture in many regions.

In both the lifetime of an individual and for future generations, we rely upon and are affected by the movement of water within the hydrological cycle. Sometimes it is disadvantageous, e.g. causing floods and droughts, but generally humans and the entire biosphere depend upon it. What we must understand in its management in the knowledge that not everything is short term. Indeed, some aspects of the hydrological cycle could have originated 3.5–4.0 million years ago (Klige et al. 1998). As far as humans are concerned, this water is irreplaceable. Some aquifers are isolated from the ground surface by an impermeable geological layer. These are called confined aquifers (Figure 2.1). In arid regions, these confined aquifers can give rise to an oasis. Water from a confined aquifer is called artesian water and may be very old. It can be accessed through an artesian well. The water in the well can rise, and the level it reaches is called the potentiometric level. This can reach the surface and produce an artesian well with flowing water. For the pressure to be maintained, the water flowing from the well must be replaced by infiltrated water to recharge the aquifer. Examples of artesian springs are oases in the desert where the recharge area is often from a mountain area some distance away. An example is the Nubian sandstone that underlies a large part of northern Africa. Another example is the confined aquifer that underlies London. The deepest part is under central London feeding the fountains in Trafalgar Square.

Rivers, Lakes and Reservoirs

These are three of the main freshwater resources used by humans. Lakes and rivers are naturally occurring, but reservoirs are generally man-made and constructed to store water and sometimes for hydroelectric power and to regulate water flow in a river.

Rivers

Apart from atmospheric water, the main way that surface water moves across the land is along rivers. These do not only provide a way of draining the land but also play an important role in shaping it. In turn, the landscape and geology determine the path and flow of the river and play an important role in the water quality. Although the majority of rivers (about 80%) ultimately drain into the sea (exorheic basins), some do not. These latter ones have closed basins and are termed endorheic and amount to 20% of all rivers. In England, nearly 20% of the land surface does not initially drain into the sea. Most other endorheic river basins occur in Asia and Africa where about one-third of rivers drain internally. In Asia, some rivers drain into a sea as opposed to the ocean. Examples include the Amu Darya and Syr Darya that drain into the Aral Sea and the Volga and Kura that drain into the Caspian Sea. In Africa almost one-third of the rivers drain internally for example into the Sahara and Kalahari deserts. The North American Great Salt Lake is endorheic. Other examples include Lake Eyre in Australia and the Middle Eastern Dead Sea, the lowest location on the land surface, fed by the River Jordan. These inland lakes and seas tend to be more saline because of evaporation that concentrates dissolved minerals. If the river flow into an endorheic basin is low, then there may not be a permanent standing body of water but a white salt pan deposit. The total area of inland run-off, according to Korzun (1974), is

30.2 million km^2 of which 2.2 million km^2 is in Europe, 12.3 million km^2 is in Asia, 9.6 million km^2 is in Africa, 3.9 million km^2 is in Australia, 1.4 million km^2 is in South America and 0.88 million km^2 is in North America. Most rivers are part of a network of feeder streams that together form a river basin. The length of river, volume of flow and the area drained determine their classification into very large, large medium, small and very small (Shiklomanov and Rodda 2003). River basin areas can vary from the Amazon which has 6 160 000 km^2 to a few fields on a hillside. Rivers can be fed by springs, groundwater seepage or by precipitation run-off, although the latter can vary from near to 100% to almost nothing. The amount can be determined by local topography, geology and the land use. It is important to understand the way in which water flows across a river basin in order to optimise its management and, if required, the river flow. If the water is able to move through the soil or rocks, it is termed throughflow. This throughflow can help maintain river flows more evenly. The rate at which this flow reaches the river depends upon the route it takes. Management of flow is sometimes essential to control flooding. The channel along which the river flows is not static but is subject to continuous changes caused by the flow itself. These changes can be rapid and dynamic or very slow, again depending upon the geology and topography of the area. Erosion of the river banks also depends upon the river flow, volume and geological nature of the banks and substrata. Erosion will lead to higher sediment loads in the river. This sediment may be deposited elsewhere in the basin and could even change the course of the river or it may be carried to the river outlet and be deposited in the estuary or delta. Precipitation landing in the catchment can follow a number of routes. It may be taken up by plants and eventually be transpired to the atmosphere or it can be directly evaporated into the air or it can infiltrate into the soil depending how porous it is, and if these routes are saturated, it can flow across the surface forming overland flow. This latter can particularly occur during very heavy precipitation when infiltration rates and the soil are saturated and cannot meet the needs of drainage. This type of overland flow is common where the soil depth is shallow and the underlying rocks are impermeable or when rain falls on a sloping hillside. Overland flow can also lead to soil erosion and local flooding. River and catchment management has been based on the control of these dynamic changes and keeping the water within a defined channel. It must be recognised that these management changes invariably affect the ecology of the river. Different components contributing to river flow usually act at different rates and therefore result in peaks in the flow hydrograph for that river depending upon the catchment. Unfortunately, the flow in many rivers is not even but subject to large peaks and troughs, causing potential difficulties for the rivers used by humans. Knowing the amount of water flowing in a river is called its discharge. This is important in water resource management, but unfortunately it is not always available. Example of selected river run-offs for the main rivers in Asia are given in Table 2.4.

Knowing the amount of water flowing in a river, its discharge and how this may vary is important in water resource management. Unfortunately, in many countries this data is not available. The discharge of a river is the total volume of water flowing along its channel at a given point and is usually measured in cubic metres per second (cumecs). Drainage basin discharge equals precipitation minus evapotranspiration plus (or minus) changes in storage. Measurement is usually made at a gauging station. These are often sited at weirs. Modern technological methods include ultrasonic discharge gauges and Doppler probes. Unfortunately, measuring discharges is decreasing because of cost (Vorosmarty et al. 2000). Over a short period, a storm or flood can be tracked on a hydrograph, but these only

Table 2.4 Run-offs of selected rivers in Asia (units in km^3/yr).

River	Average	Minimum	Maximum
Amur (Russia)	267	135	422
Yangtze (China)	734	451	1009
Ganges (India)	380	244	543
Brahmaputra (Bangladesh)	574	396	881
Lena (Russia)	225	159	331
Niger basin (Africa)	302	163	482
Nile basin(Africa)	161	95	248
Murray (Australia)	320 (m^3/yr)	1.16	129
Loire (France)	33.6	19.5	33.9
Rhine (Europe)	85.9	61.2	86.1

The figures for European rivers are estimates for comparison. *Source:* Adapted from Shiklomanov and Rodda (2003).

represent what is happening over a few hours, and their purpose is to give information on the relationship between rainfall and discharge. Each rainfall/storm event will vary from one catchment and whole river basin to another. There are a number of factors affecting these short-term storm hydrographs, including the size of the drainage basin. Larger basins have potentially a greater area for precipitation to fall, so they can register a higher peak discharge compared with smaller basins. Steep-sided catchments drain water more quickly, so they have shorted lag times between an event and it showing in the river. This is also true for catchments with impermeable rocks underlying them. If, on the other hand, there is a large amount of vegetation that intercepts the rainfall and slows the movement of the water, this will reduce the peak discharge. Deforestation can increase soil erosion but also reduce rain and surface flow interception, thus increasing flood risk. Agriculture can also have a direct impact. If soils are ploughed, they allow more infiltration that in turn reduces peak discharges and increases lag times. Drainage ditches and pipes, however, speed up the water flow from the field into the stream or river increasing peak discharges and reducing lag times. Moreover, when the crop in a field is harvested, intercepting is reduced and overland flow is encouraged. If there is a large amount of human activity, this can also affect discharge. For example, if artificial drainage systems have been constructed or a large number of impermeable surfaces have been created, they will lead to shorter lag times and/or increased peak sizes. River discharges will vary from year to year in response to climate, temperature, evapotranspiration, and the characteristics of the basin. In the United Kingdom, the peak flows are often in the winter although this is not always the case, especially with a changing climate. The most northerly rivers have a more pronounced winter peak than those in the drier south-east such as East Anglia.

Lakes

Shiklomanov and Rodda (2003) estimate that there are about 15 million lakes worldwide with a total area of 2 million km^2. This amounts to 1.5% of the total land area. Table 2.5 lists the most important lakes in various countries. Although most are freshwater, some are

Table 2.5 Examples of major lakes and their volumes.

Lake	Country	Maximum depth (m)	Volume (km^3)
Ladoga	Russia	230	908
Onega	Russia	120	295
Baikal	Russia	1741	23 000
Aral Sea	Kazakhstan, Uzbekistan	68	1064
Balkhash	Kazakhstan	25.6	106
Sevan	Armenia	86	56
Poyang Hu	China	20	—
Balaton	Hungary	12.2	1.9
Ohrid	Macedonia	256	61
Vattern	Sweden	122	74
Malaren	Sweden	61	14.3
Victoria	Tanzania, Kenya, Uganda	84	2750
Tanganyika	Tanzania, Zaire, Zambia	1471	17 800
Superior	USA, Canada	406	11 600
Michigan	USA	281	4680
Manitoba	Canada	28	17
Geneva	Switzerland, France	310	88.9

Source: Saline. Adapted from Shiklomanov and Rodda (2003).

saline. These latter are defined as lakes with >5000 ppm dissolved solutes. These are more common in arid and semi-arid regions. They may also occur in closed drainage basins. Freshwater lakes and reservoirs vary greatly in size (see Table 2.5) and distribution across continents. Korzun (1974) recognises 145 large lakes distributed over the continents, 19 of which have a surface area of greater than 10 000 km^2.

Lakes and reservoirs provide medium- to long-term storage of water. Lakes have an inflow and most have an outflow. They also lose water by evaporation from their surface although their levels are mainly influenced by river flows and precipitation as well as abstraction for human use. Depending upon location and surface area, evaporation can amount to a considerable water loss. Lake levels could also be affected by a warming climate that could also affect their thermal behaviour. The majority of lakes occur in the northern hemisphere and are associated with the retreating ice from the last glaciation, for example the Great Lakes of North America. Finland has a number of lakes separated from the Gulf of Finland by a large group of terminal moraines known as Salpa-Uselka. Others are located where there were movements of the earth's crust such as tectonic depressions in mountainous areas, e.g. Baikal, the African Rift Valleys and Titicaca. In the Pamirs and the Altai, many are formed by rock falls caused by an earthquake blocking a valley some of which are relatively recent. Both the Caspian Sea and the Dead Sea were formed in depressions, which were below sea level (−27 and −392 m). Some smaller lakes are even formed when a river flow is slow and the gradient is less. The main channel then meanders, and sediments are deposited on bends isolating a loop of the river to form a shallow lake. Examples of this type of lake can be found along the Amazon, Mekong and Niger rivers. Lakes generally have an inflow and an outflow so, depending upon their size, depth and volume, the water is in theory being changed over a period of time. The rate of flow through a lake will depend upon size and number of inflows as well as the size of the outflow. Three of the largest lakes and lake complexes are the Laurentian Great Lakes of North America, Lake Baikal in Russia and Lake Victoria in Africa. The Laurentian Great Lakes together form the largest freshwater body on earth although Lake Baikal, as an individual lake, has the largest single volume of freshwater. The area surrounding the Great Lakes has some large cities such as Chicago, Detroit and Toronto. In industrial cities, there are also some important industrial areas as well as important agricultural ones. Both industry and agriculture have contributed to a gradual deterioration in water quality in the area.

The lakes were formed about 10 000–15 000 years ago by glacial action and ice melt. Their drainage area is relatively small for such large lakes, so the inflow is correspondingly small resulting in a slow turnover rate. Lakes Superior and Huron do not have many cities or much agriculture along their banks, so they are relatively clean. Lakes Erie, Ontario and Michigan have more inputs because of the greater number of towns and cities around them. The area of the Great Lakes is so large that it significantly influences the climate of the area. With the possible exception of Lake Erie, none of the lakes completely freeze over in winter and they warm up only slowly in summer. They also generate cool breezes in summer. Each of the lakes is situated at a different altitude and this results in a flow of water from the highest, Superior, to the lower, Huron. Lakes Huron and Michigan flow into Erie and then Ontario. From there the outflow is the St. Lawrence River. This provides access for international shipping from the Atlantic and is a major route for international movement of goods, especially agricultural products, from both the United States and Canada. Modern shipping has increased both in size of ships and volume of trade, and to accommodate this, the Welland Canal was constructed in 1833 alongside the river avoiding the Niagara falls and connecting the lower lakes to Lake Superior. The canal was upgraded several times until the mid-twentieth century to accommodate both larger ships and increased trade. The distance between the Atlantic Ocean and the western shore of Lake Superior is 3700 km. The connection to the Atlantic and between all of the lakes has also provided a route for invasive animal and plant species to migrate into areas where they did not otherwise occur. All of the lakes suffer from contamination from both human and industrial waste. Lake Erie is the worst affected as it is the shallowest (mean depth 19 m) and has the largest amount of development along its shoreline.

As an individual lake, Baikal in the south eastern part of Siberia is the deepest and oldest being 20–25 million years old and has a maximum depth of 1620 m (5 315 ft), making it the deepest in the world. It also holds the largest volume of ice-free freshwater. It has a volume of 23 000 km^3, one-tenth of the world's liquid freshwater. Baikal is 636 km long with an average width of 48 km. There are more than 330 rivers and streams feeding into the lake, the largest being the Selenga, Barguzin, Upper Angara, Chikoy and Uda. There is only one major outflow, the Angara River, located at the south eastern end. The lake is in a tectonic fault surrounded by mountains. It was well known for the purity of its water and the many unique species of organisms that live there. There are many islands, most uninhabited. The largest is Olkhon that is inhabited. The lake usually freezes in winter thawing in early summer. The ice can be so clear that it is possible to see through it like glass. When not frozen, the visibility depth can be as much as 40 m. Industries around the lake include mining for mica and marble, cellulose and paper manufacturing, shipbuilding, fisheries and timber. A pulp and paper mill was built at the south end in 1966, which was controversial as its poorly treated waste effluent was badly polluting the lake. As a result, a governmental decree was issued to protect it from polluting emissions but concerns still remain about its impact. In the second half of the twentieth century, there has been a considerable increase in tourism, especially in the south eastern region. During the first decades of the twenty-first century, tourism increased threefold from 500 000 to over 1.5 million. These tourists required a larger infrastructure to support them, so the number of hotels and campsites increased dramatically. Unfortunately, the infrastructure to deal with their wastes did not increase in proportion. Previously, in 1916, the Barguzinsky nature reserve was created. In the 1970s and 1980s, more nature reserves were created. The lake and its environment were declared a UNESCO World Heritage site in 1966. Climate change is now having an effect, especially on the feeder rivers that have reduced the spring run-off into the lake causing a decrease in the lake's water level. However, the winter run-off and inflow increased and by controlling the outflow through the Angara River lake water levels have now been controlled and maintained (Sinukovitch and Chernyshov 2019). Most rivers in Russia have undergone changes regarding their flow and regulation. A number of reservoirs have been constructed in the Angara river catchment, often containing hydropower plants, and these have transformed natural flows. The lake outflow contributes about 47% of the total Angara flow. The upper reaches of the river vary with the flow from the lake, but the lower reaches depend upon the tributaries that are fed by a combination of snow and rain. The Ilim river, for example, is characterised by spring floods. The flow of some of the rivers is regulated by reservoirs built for hydropower plants. This in turn regulates the outflow from the lake and has an impact on its biology.

Land use changes also generally impact both streamflow and lake inflows and outflows. Rising temperatures will also reduce the amount and period of ice cover of northern and far southern lakes. Of the precipitation that falls upon the land globally, approximately one-third drains back into the sea along streams and rivers. As they flow, they erode rocks and soils and transport sediments. The size and length of a river are partly dependent upon the size of the river basin and the climate of that area. Human populations have relied upon rivers for their development and food for thousands of years. Very large rivers, such as the Amazon, Congo, Mississippi and Ob, drain large areas of land. There is a problem however when trying to assess the volumes of water discharged by rivers as there is a lack of gauging stations on most of them, so accurate measurements are not available. A further problem is that most rivers have large seasonal variations in flow. Very few rivers are still in their original natural state as they have been modified by human activities by building dams, reservoirs and altering the river channel. Europe has probably the largest density of gauging stations of any of the continents. The total amount of run-off for Europe during the period 1921–1985 was estimated at 2900 km^3/year. Over time human use has modified the hydrology of most river systems, especially in developed countries, so it is difficult to obtain regional data although some data for larger rivers is available. Some examples of discharges from selected major rivers by continent are given in Table 2.6.

Table 2.6 Examples of discharges of some major rivers by continent.

River and continent	Catchment area (km^2 × 10^3)	Average discharge (m^3/s)
Africa		
Congo	3680	41 200
Nile	2870	3750
Niger	2090	5589
Zambesi	1330	7070
Orange	1020	[a]
Asia including China and India		
Ob	2990	12475
Yenisi	2850	19600
Lena	2490	16871
Amur	1855	11400
Yangtze	1808	30166
Ganges[b]	1746	38129 (16648)
Indus	960	6600
Mekong	795	14800
North America		
Mississippi	2980	16782
St Lawrence	1926	16800
Colorado	637	637
Frazer	233	3475
South America		
Amazon	6915	209000
La Plata	3100	22000
Orinoco	1000	36000
Europe		
Volga	1380	8060
Danube	817	7130
Neva	281	2445
Rhine	252	2900
Loire	120	835
Rhone	98	1710
Australia		
Murray	1072	3490
Cooper Creek	285	73
Fly	64	6000
Sepic	81	3804

Publications consulted: Rivers of the Earth (http://home.comcast.net/-igpl/Rivers.html).

[a] No data available.

[b] The first figure is for Ganges only, and the second includes the Brahamaputra and Maghna rivers.

Source: Adapted from Gupta (2007), Kumar et al. (2005), Benke and Cushing (2005) and Nezhilovsky (1981).

The average total river flows, allowing for flood flows, as long as suitable holding measures are in place, such as dams, and there is usually enough for both humans and wildlife. This also assumes good water quality and proper management of any abstractions. The distribution of precipitation run-off and large rivers is very uneven across the continents. It greatly exceeds globally the amount of water currently being withdrawn for human use. Regions of highest run-off occur in the Amazon basin, parts of Indonesia and the southern part of the Himalayas down through Malaysia. Other regions, such as northern Siberia, parts of central Asia, central and eastern Africa (the Congo area), eastern and a narrow north-eastern coastal strip of North America, currently have adequate run-off. Areas of least run-off are in sub-tropical regions, parts of central Asia, most of Australia and western North America. There are seasonal variations in all of these areas. In marginal areas where there can be scarcity, it is essential to have correct water management and conservation. Although the size of rivers is frequently measured by their length from the point of view of water resources, the average discharge is frequently more relevant as it gives a better indication of the amount of water available. The Amazon is one of the longest together with the Yangtze, but the former has approximately six times the discharge of the latter. The Ganges is a little over half the length of the Mekong but has twice the discharge volume. Rivers have been used since ancient times for activities other than drinking and cooking. Apart from irrigation, rivers have been, and still are, used for navigation and a source of power. Large amounts are used for cooling waters and some for feeding boilers. Water is used for many types of hydropower. Many rivers cross national boundaries. Quite often lower basin countries do not have control of the river's headwaters, so agreements between countries on their management are essential.

Lakes and Reservoirs

Lakes and reservoirs are important reserves of freshwater. Examples of the principal lakes in different countries are given in Table 2.7.

Table 2.7 Some of the principal reservoirs in the world.

Reservoir	Country	Basin	Volume (km^3)	Main use
Bratskoya	Russia	Angara	169.3	HNTWFR
Nasser	Egypt	Nile	160	IHANF
Volta	Ghana	Volta	148	HNIF
La Grande 2	Canada	La Grande	61.7	H
Ataturk	Turkey	Euphrates	48.7	HI
Sanmenxia	China	Huang He	35.4	HISN
Itaipu	Brazil, Paraguay	Parana	29.0	H
Mica	Canada	Columbia	24.7	HA
Fort Peck	USA	Missouri	23.0	SHIN

Key to symbols: H, hydropower; N, navigation; W, water supply; I, irrigation; F, fishery; T, timber rating; R, recreation; A, accumulation; S, flood control.

Reservoirs, which are man-made, have long been used for controlling flows in rivers. They are often filled at times of high river flows and used for augmenting the flow in drier times. Although reservoirs have been constructed for thousands of years, their rate of construction has increased greatly during the last century. These have been both large and small. Avakyan (1987) estimated that there are about 30 000 reservoirs worldwide with a volume in excess of 1×10^6 m^3. Where suitable rivers and location are available, a dam has been created across a valley to hold the river water in a reservoir. If there is not a suitable valley with river to create a dam, an available flat area of land can be used to build a waterproof wall and base to enclose a piece of land and water pumped in from the non-tidal part of a river to fill it. These are called pump storage reservoirs and are used, for example, to the west of London in the River Thames valley to supply parts of the City. More than half the world's rivers have one or more dams across them and in excess of 7000 km^3 of water is stored in them. As much as 40% of the water that would have flowed from continental land masses directly to the sea is intercepted by dams, and they hold back more water than is present in natural river channels (Vorosmarty et al. 2003, 2005). Dams have become a major tool in water management and storage. As many as 800 000 small dams exist in rural communities used for both domestic and industrial supplies as well as irrigation (Vorosmarty et al. 2005). The overall storage of water has dramatically changed in both volume and distribution of water on the continents. Currently, the largest valley reservoir is that formed by the Three Gorges Dam in China, but others, including the Bratskoya Reservoir on the Angara River in Russia and the Volta Dam in Ghana, are also very large (see Table 2.7).

Some lakes have had dams constructed across their exits in order to raise the existing lake level. An example of this is Lake Victoria in Africa. Reservoirs, as can be seen in Table 2.7, can have a number of uses apart from storage of water for use at drier times of the year. The most common use is for providing irrigation water, but energy production through hydropower is very common. They may also be used for fisheries producing food. Flood control and prevention is an important function as is navigation and in more recent times recreation. In recent years, especially during the period of massive dam construction in China and India in the second half of the twentieth century, many objections to their construction have been raised. The downsides of dam construction include flooding of large areas of land, displacement of properties including whole villages, inundation of useful agricultural land and changing the river ecology both upstream and downstream of the reservoir. Silt deposition can affect the life of the dam, which reduces the silt loads downstream including estuaries. Reservoirs and lakes are both subject to the same chemical and ecological problems. Increases in nitrates and phosphates from both flooded land and basin drainage can lead to eutrophication and algal blooms. All dams are not large-scale projects. Where there are marked seasonal differences in flows in smaller rivers and communities, the scale of construction will be much smaller, for example building sand dams in parts of Africa. Reservoirs are continuing to be built as rising human populations require more water both for themselves and for food production. Water use in industry is also increasing.

In both the lifetime of an individual and for future generations and as well many ecosystems, all rely upon and are affected by the movement of water within the hydrological cycle. Sometimes it may be disadvantageous, e.g. causing floods and droughts, but generally humans and most of the biosphere depend upon it either directly or indirectly for their water availability. What must be recognised is how those parts impinging upon our societies can be sustainably managed. Not everything in the cycle is short term. Indeed, we must thing in the long term and acknowledge that some of our water sources are millions of years old (Klige et al. 1998). As far as humans are concerned, such water is irreplaceable and must be used sparingly. Water demand will continue to rise in the future with the increasing population, food demand and improved standards of living. A proper understanding of the behaviour of water in the environment and its sustainable, i.e. non-wasteful, use made aware to all users, including agriculture, industry and the public as well as protection of ecosystems.

References

Ananthaswarmy, A. (2012). *The Great Thaw*. New Scientist No. 2889, November 3, 2012.

Avakyan, A.B., Saltankin, V.P., and Sharapov, V.A. (1987). *Water Reservoirs*. Moscow: Mysl Publication. (in Russian).

Barlow, M., Solstein, D., and Cullen, H. (2005). Hydrologic extremes in Central Southwest Asia, meeting report. *EOS Transactions AGU, 218–221*.

Barnett, T.P. and Pennell, W. (eds.) (2004). Impact of global warming on Western US water supplies. *Climate Change*, 62 (Special) 3.

Barnett, T.P., Malone, R., Pennell, W. et al. (2004). The effects of climate change on water resources in the west: introduction and overview. *Climate Change* 62: 1–11.

Barnett, T.P., Adam, J.C., and Lettenmaier, D.P. (2005). Potential impacts of a warming climate on water availability in snow dominated regions. *Nature* 438 (23): 734.

Bates, T.S., Lamb, B.R., Guenther, A. et al. (1992). Sulphur emissions to the atmosphere from natural sources. *Journal of Atmospheric Chemistry* 14: 315–337.

Benke, A.C. and Cushing, C.E. (2005). *Rivers of North America*. Academic Press.

Benn, D.I. and Evans, D.J.A. (1998). *Glaciers and Glaciation*, 734. Arnold.

Bowers, R.M., McLetchie, S., Knight, R., and Fierer, N. (2011). Spatial variability in airborne bacterial communities across land-use types and their relationship to bacterial communities of potential source environments. *The ISME Journal* 5: 601–612.

Burrows, S.M., Elbert, W., Lawrence, M.G., and Poschi, U. (2009a). Bacteria in the global atmosphere – Part 1: review and synthesis of literature data for different ecosystems. *Atmospheric Chemistry and Physics* 9: 9263–9280.

Burrows, S.M., Butler, T., Jockel, P. et al. (2009b). Bacteria in the global atmosphere – Part 2: modeling of emissions between different ecosystems. *Atmospheric Chemistry and Physics* 9: 9281–9297.

Church, T.M. (1996). An underground route for the water cycle. *Nature* 380: 579–580.

Cohen, D., Person, M., Wang, P. et al. (2010). Origin and extent of fresh paleowaters on the Atlantic continental shelf USA. *Groundwater* 48 (1): 143–158.

Conrad, R. (2009). The global methane cycle: recent advances in understanding the microbial processes involved. *Environmental Microbiology Reports* 1: 285–292.

Darwin, C. (2003). *The Voyage of H.M.S. Beagle*, 518pp. The Folio Society.

DeLeon-Rodriguez, N., Lathem, T.L., Rodriguez, L.M. et al. (2013). Micrbiome of the upper troposphere: species composition and prevalence, effects of tropical storms, and atmospheric implications. *PNAS* 110 (7): 2575–2580.

Döll, P. and Fiedler, K. (2007). *Global-Scale Modelling of Groundwater Recharge*. Institute of Physical Geography.

FAO (2014). AQUASTAT Global Information System on Water and Agriculture. https://www.fao.org/aquastat/en.

Fierer, N., Liu, Z., Rodriguez-Hernandez, M. et al. (2008). Short-term temporal variability in airborne bacterial and fungal populations. *Applied Environmental Microbiology* 74: 200–207.

Gupta, A. (2007). *Large Rivers: Geomorphology and Management*. Wiley.

Gupta, S.K. (2011). *Modern Hydrology*. Chichester, UK: Wiley Blackwell.

Hewitt, K. (2005). "The Karakoram Anomoly". Glacier expansion and the elevation effect; Karakoram Himalaya. *Mountain Research and Development* 25 (4): 332–340.

Holden, J. (2014). *Water Resources*. London & New York: Routledge.

Hughes, K., Mc Cartney, H., Lachlan-Cope, T., and Pearce, D. (2004). A preliminary study of airborne microbial biodiversity over peninsular Antarctica. *Cellular and Molecular Biology* 50: 537–542.

Intergovernmental Panel on Climate Change (IPCC). (2007). *Climate Change 2007. Impacts, Adaptation and Vulnerability*. Contribution to the Fourth Assessment, Working Group II, UNEP. UK: Cambridge University Press.

Jones, J.A.A. (2010). *Water Sustainability; A Global Perspective*. London: Hodder Education.

Kellogg, C.A. and Griffin, D.W. (2006). Aerobiology and the global transport of desert dust. *Trends in Ecology and Evolution* 21: 638–644.

Klige, R.K., Danilov, I.D., and Konishshev, V.N. (1998). *History of the Hydrosphere*. Moscow: Scientific World Publication (in Russian).

Kohout, F.A., Hathaway, J.C., Folger, D.W. et al. (1977). Fresh groundwater stored in aquifers under the continental shelf: implications from a deep test, Nantucket Island. *Journal of the American Water Resources Association* 13 (2): 252–254.

Korzun, V.I. (1974). *World Water Balance and Water Resources of Earth*. Leningrad Hydrometeorizdat (in Russian).

Kumar, R., Singh, R.D., and Sharma, K.D. (2005). Water Resources of India (PDF). *Current Science (Bangalore)* 89 (5): 794–811.

Kääb, A., Reynolds, J.M., and Haeberli, W. (2005). Glaciers and permafrost hazards in high mountains. In: *Global Change and Mountain Regions (A State of Knowledge Review)* (ed. U.M. Huber, H.K.M. Bugmann, and M.A. Reasoner). Dordrecht: Springer.

Kääb, A., Huggel, C. and Fischer, L. (2006). Remote sensing Technologies for Monitoring Climate Change Impacts on Glacier and Perafrost related Hazards. *Engineering Conferences International, Proceedings on Geohazards*.

Lavers, D.A., Allan, R.P., Wood, E.F. et al. (2011). Winter floods in Britain are connected to atmospheric rivers. *Geophysical Research Letters* 38 (23).

Llamas, M.R. and Martinez-Santos, P. (2005). Intensive groundwater use. A silent revolution that cannot be ignored. *Water Science and Technology Series* 51 (8): 167–174.

Margat (2008). *Les Eaux Souterraines: Une Richesse Mondial*. Paris: United Nations Educational Scientific and Cultural Organization.

Moore, W.S. (1996). Large groundwater inputs to coastal waters revealed in radium 226 enrichments. *Nature* 380: 612–614.

Morris, B.L., Lawrence, A.R., Chilton, B. et al. (2003). Groundwater and its susceptibility to degradation: A global assessment of the problem and options for management. *Early warning and assessment series, RS. O3*. United Nations Environmental Programme, Nairobi, Kenya.

Nezhilovsky, R.A. (1981). *Neva River and Neva Bay*. Gidrometeroizdat.

Nijssen, B., O'Donnell, G.M., Hamlet, A.F., and Lettermaier, D.P. (2001). Hydrologic sensitivity of global rivers to climate change. *Climate Change* 50: 143–175.

Oki, T. and Kanae, S. (2006). Global hydrological cycles and world water resources. *Science* 313: 1068–1072.

Shiklomanov, I.A. and Rodda, J.C. (ed.) (2003). *World Water Resources at the Beginning of the 21st Century*. Cambridge University Press.

Sinukovitch, V.N. and Chernyshov, M.S. (2019). Water regime of Lake Baikal under conditions of climate change and anthropogenic influence. *Quaternary International* 524: 93–101.

Strakhoiv, N.M. (1963). *Types of Lithogenesis and Their Evolution in the Earth's History*. Moscow: Geoltekhizdat Publication. (In Russian).

UNEP (2008). *World Glacier Monitoring Service (WGNS). Global Glacier Changes: Facts and Figures*. Nairobi, Kenya.

UNEP United Nations Environmental Programme (2007). *Global Outlook for Ice and Snow*. Nairobi Earthprint Publications.

Vorosmarty, C.J., Gren, P., Salisbury, J., and Lammers, R.B. (2000). Global water resources: vulnerability from climate change and population growth. *Science* 289: 284.

Vorosmarty, C.J., Meybeck, M., Fekete, B. et al. (2003). Anthropogenic sediment retention: major global impact from registered river impoundments. *Global Planetary Change* 39 (1–2): 169–190.

Vorosmarty, C.J., Douglass, E.M., Green, P., and Revenga, C. (2005). Geospatial indicators of emerging water stress: an application to Africa. *Ambio* 14 (2): 230–236.

Vrba, J. (2004). The World's groundwater resources. International Groundwater Resources Centre (IHRAC). *Report Nr. IP 2004-1*.

Womack, A.M., Bohannan, B.J.M., and Green, J. (2010). Biodiversity and biogeography of the atmosphere. *Philosophical Transactions of the Royal Society B* 365: 3645–3653.

World Water Assessment Programme (2003). *Water for People. Water for Life*. United Nations Educational, Scientific and Cultural Organization (UNESCO).

Zhu, Y., and Newell, R.E. (1998). A proposed algorithm for moisture fluxes from atmospheric rivers. *Monthly Weather Review Journal. Amelsoc.org*.

3

Human Needs and Water Demands. How Much Water Do We Need?

Water is the thread that draws all lives, including human activities, together. This thread runs through all living things as well as all human activities, including food production and preparation, health, energy, industry and all aspects of our economy as well as being essential to living things and the ecosystems in which they live, especially those that provide earth's life support systems. Unfortunately, there is a finite amount of freshwater on this planet, and this is under considerable strain with increasing demands. Although global climate change has come into prominence in recent years, it is its impacts on the hydrological cycle that will have the greatest immediate impact on human well-being. Management plans that are developed must not only consider these needs, but also how the resources can be used sustainably both for humans and ecosystems. This has to involve an integrated approach from governments, water authorities, industry and the public. Because of the multiplicity of water uses and needs and the complexity of its behaviour in the environment, a properly funded and structured management plan is needed. This plan must be flexible enough to adapt to the change and uncertainty as well as a growing population and climate variability. Because water movement transcends regional, national and even continental boundaries, there needs to be more than just local management. Integrated water resource management (IWRM) must occur at all levels and needs including health, food, economic development and recreation as well as sustaining healthy ecosystems. Water provision for human health, agriculture, etc. must not only be treated as merely meeting certain specific sectoral requirements but must be based upon a holistic picture. A better, indeed essential way, if a sustainable future is to be achieved, is to plan all human activities in such a way as to optimise water resource use in a sustainable way taking into account possible impacts on other users. In particular, water must not be treated as an unlimited resource that will always be renewable. Some parts of the water cycle are more vulnerable than others and must be treated as non-renewable. This particularly applies to some ancient groundwater aquifers. Estimates of the volumes of freshwater withdrawn for either direct or indirect human use are provided in Table 3.1. As can be seen, there are large differences between different countries as well as for sectors within countries.

Water: Our Sustainable and Unsustainable Use, First Edition. Edward G. Bellinger.

Table 3.1 The estimated volumes of freshwater withdrawn by selected countries.

Country	Total withdrawal (km^3/year)	Per capita withdrawal (m^3/year)	Domestic withdrawal (%)	Industrial withdrawal (%)	Agricultural withdrawal (%)
India	645.84	585	8	5	86
China	549.76	415	7	26	68
United States of America	477	1600	13	46	41
Japan	88.43	690	20	18	62
Indonesia	82.78	372	8	1	91
Russia	76.68	535	19	63	18
Pakistan	71.39	847	8	24	68
Canada	44.72	1386	20	69	12
Italy	41.98	723	18	37	45
Germany	38.01	460	12	68	20
Spain	37.22	864	13	19	68
France	33.16	548	16	74	10
Australia	24.06	1193	15	10	75
Saudi Arabia	17.32	705	10	1	89
United Kingdom	11.75	197	22	75	3
The Netherlands	8.86	544	6	60	34
Libya	4.27	730	14	3	83
Namibia	2.95	977	30	4	66
Sweden	2.4	519	23	67	10
Denmark	0.67	123	32	26	42
Malta	0.02	50	74	1	25

For full list of 170 countries, see 'List of countries by freshwater withdrawal, Wikipedia accessed June 2021'.

Our Basic Water Needs

Gleick (1996) attempted to define basic water requirements (BWRs) not just of individuals but also for society. He pointed out that water is used for removing and diluting wastes, growing food crops, producing and using energy and as a medium for and method of transport among other things. The amount of water needed for each of these sectors will vary from one country to another. All water for human use is not available as a piped supply, and the type of access available makes a difference to the rate of consumption (see Table 3.2). The harder and more time consuming obtaining the water, more frugally it is used. Easy access encourages greater use, sometimes unnecessarily.

The very basic needs for a human will depend upon not only their daily activities but also the climate of that area. Dry warm conditions can double the amount needed depending on vigorous activities a person is undertaking.

Table 3.2 Domestic water use relative to distance to source of supply.

Source of supply	Water use (l/person/day)
Public standpipe more than 1 km away	<10
Public standpipe closer than 1 km distance	20
Connection to house with flush toilet	60–100
Connection to house with gardens	150–400

Source: Adapted from Gleick (1996).

Our Right to Water

Although some progress has been made over the past decades, the problems concerning the provision of safe drinking water and proper sanitation still affect millions of people. The Universal Declaration of Human Rights (UHDR) of the United Nations in 1948 stresses the right for an adequate standard of living and health for all of us (UN 1948). Article 3 of this declaration states that 'Everyone has the right to life, liberty and security of person'. Article 25 states that 'Everyone has the right to a standard of living adequate for the health and well-being of himself and of his family, including food, clothing, housing and medical care and social services'. Although not explicitly stated in order to achieve most of these rights, access to safe drinking water and sanitation is required. There have been a number of covenants over the years that either directly or indirectly imply the right to an adequate supply of water for everyone. These include the 1948 UHDR, the 1966 International Covenant on Economic, Social and Cultural Rights, the Inter-American Convention on Human Rights and the European Convention on Human Rights. The UN General Assembly and the Human Rights Council in 2010 recognised and resolved an individual's right to water (Winkler 2012). Although not unanimously accepted by all countries, none voted against it but a few did abstain. It was thus accepted by consensus. Following on from this resolution, the right to safe drinking water also implies an adequate standard of living and a high standard of physical and mental health and the right to life and human dignity (for others see Gleick 2007). Although the International Bill of Rights does not specifically mention water, it is generally agreed that water availability is crucial to aspects of many rights but many people still do not have proper access to it. Rights to water are guaranteed for states that have ratified the social and civil covenants under the rights for housing, health and life. Other subsequent treaties do explicitly mention the right to water, e.g. the African Charter on the Rights and Welfare of the Child and the Arab Charter on Human Rights regarding the right to health. In addition, the Convention on the Elimination of All Forms of Discrimination against Women (UN, CEDAW 1979) requires the right to an adequate standard of living. The original document has had many reservations (alterations) included predominantly by Muslim countries. For some countries, the convention is aspirational and not legally binding. This convention was developed through a number of conventions on women's rights dating from 1953 (Cole 2016). To provide this and requirements of other declarations and conventions, states are obliged to respect the human right to water by protecting it from pollution and also not preventing proper access or endangering human health. Water must be both accessible and affordable although not necessarily free from a reasonable charge. These rights should be achieved as quickly as possible which has not always happened. No single use of water can be granted unconditional priority for use over the basic right of human needs which must come first. The right to food implies, however, that some allocation of water for food production is required and this, apart from human health, must be given some priority. As water is a finite resource and is required by both humans and ecosystems, it must be managed both efficiently, equitably and most of all sustainably. With the increasing human population, the alternatives are either to tap into more of the resource, if it can be found, or each person and user must have a smaller allocation. Although the 1948 UHDR was not binding upon member states, in subsequent years the UN produced a covenant that was binding stating that all member states should recognise the right of individuals to have an adequate standard of living, which includes food, clothing and housing. They should also strive to continuously improve living conditions and control disease. Although climate change has come into prominence in recent years, the problems of drinking water supplies and adequate sanitation must be close behind or, for some countries and international organisations, at the very top. Although some progress has been made, there are nearly three million people without access to proper sanitation. Such facilities are fundamental to a proper life and existence. As previously stated, safe drinking water and sanitation are essential to humans and all life on this planet so that resources that are available must be shared between humans and ecosystems as the latter are essential for providing the biosphere in which we live. The right to life was explicitly stated in the International Covenant for Civil and Political Rights (Tomuschat 2008). The UN expanded upon this pointing out that the term 'inherent right to life' must not be too narrowly interpreted and that states should adopt positive measures to ensure this right. Over the past few decades, there have been a number of conferences where human needs and the right to water have been discussed. Perhaps the most significant one was in 1977 at Mar del Plata (Argentina) where the right of access to drinking water of adequate quality and quantity for human needs was reinforced. A review of subsequent international meetings reinforcing different aspects of water including IWRM can be found in Mitchell (2005) and Thomas and Durham (2003). All of the available water is neither accessible nor suitable for human use, and these factors can be exacerbated by its uneven distribution across the planet. In some regions, especially with the increasing human population, it is conceivable that not everyone could have unlimited amounts, and there could even be rationing. Every person should, however, have access to an 'adequate amount'. The WHO (1993, 2002) defines domestic water as 'water used for all usual domestic purposes including consumption, bathing and food preparation'. The question then arises as to what is adequate or what is the minimum requirement? No single use of water can be granted unconditional priority, but when there are a number of different types of users, the basic rights of humans come first. The main other user given priority is food production, but this is still second to basic human individual needs. In order to provide adequate amounts to all peoples, it must be managed efficiently, equitably and most of all sustainably and not forgetting ecosystem needs. No two population groups are the same. There are large differences in climate, diets, nature of work, culture and religion that result in differences in individual requirements. If these requirements cannot be met, water scarcity is said to exist. Rijsberman (2006) and IWMI-CA (2007) define scarcity as 'when an individual does not have access to safe affordable water to satisfy her or his needs for drinking, washing or their livelihoods we can call that person water insecure. When a large number of people in an area are water insecure for a significant period then we can call that area water scarce'. UN-Water (2006) and UN-Water and FAO (2007) define scarcity as 'Water scarcity is the point at which the aggregate impact of all users impinges on the supply or quality of water under prevailing institutional arrangements to the extent that the demand by all sectors, including the environment, cannot be satisfied fully'. While the first definition covers much, it does not specifically state whether the scarcity is because of mismanagement or just not enough water in that area. This gives rise to two other definitions, namely (i) *economic water scarcity* which is 'when investments needed to keep up with growing demand are constrained by financial, human or institutional capacity' and (ii) *physical water scarcity* that occurs 'when available water resources are insufficient to meet all demands'. From these two definitions, it can be seen that theoretically an area can have plenty of water but still be water scarce as the water is not available to people for practical reasons such as poor infrastructure, bad institutional arrangements or some other factor (Lautze 2014). This is an important distinction. It is reasonable to assume that mere survival is not enough as that would require less than 5 l/day. The Mar del Plata conference stated that basic needs were that 'all peoples whatever their stage of development and social and economic conditions, have the right to have access to drinking water in quantities and of a quality equal to their basic needs'. This was expanded at the Earth Summit in Rio de Janeiro to encompass ecological/ecosystem needs: 'In developing and using water resources, priority has to be given to the satisfaction of basic needs and the safeguarding of ecosystems'. Four basic human needs were also defined. These were (i) drinking water for survival, (ii) water for human hygiene, (iii) water for sanitation services and (iv) modest household needs for food preparation. Implicit in this last need was that water would also be needed for growing food. Unfortunately, many nations still do not provide enough water for these needs either some or all of the time. The WHO/UNICEF (2000) indicated that 'the availability of at least 20 l/person/day from a source within one kilometre of the users dwelling' was a reasonable target. They were, however, mainly discussing access rather than the total amount of water needed and attempted to define BWRs not just for individual humans but also for society. It was pointed out that water is used for removing and diluting wastes, growing food, producing and using energy and transport among other things. The true amount of water needed for each of these will vary from one country to another, but the basic minimum (in litres per person per day) is 5 for drinking, 20 for sanitation, 15 for bathing and 10 for cooking and food preparation making a total of 50. Not all water for human use is available as piped supply. The type of access to the water makes a difference to the consumption (see Table 3.2). This illustrates the point that the easier the access the greater the consumption. The very basic water needs for a human will depend upon not only their daily activities but also the climate. Dry warm conditions can double the amount of water needed as can the needs for a person engaged in vigorous activities. The water drank is to replace that lost through the skin and lungs as well as excretion. Estimates of daily water requirements range from 2.0 l/day (USEPA 1976) to 5.0 l/day (Saunders and Warford 1976). For an average human in a temperate climate the requirement for fluid replacement is about 3.0 l/day, but in a hot climate a 70 kg person will need more as water is lost through sweat and breathing so they will need 4–6 l/day. The minimum water requirement must include sanitation services but not the water used in

Table 3.3 Water requirement estimates from different authors.

Author	Daily intake (l/capita/day)
White et al. (1972)	1.8–3.0
US EPA (1976) and NAS (1977)	2.0
Vinograd (1966), Roth (1968) and WHO (1971)	2.5
US DANRCS (1997)	0.23–3.78
NRC–NAS (1989)	2.0–4.5
Saunders and Warford (1976)	5.0

Source: Adapted from Inocencio et al. (n. d.)

Table 3.4 Estimates of water use in different countries that use less than 50 l/person/day.

Country	Population (millions)	Domestic water use (l/person/day)	Domestic use as a % of 50 l/day
Somalia	7.5	8.9	18
Uganda	18.79	9.3	19
Ethiopia	49.24	13.3	27
Nepal	19.14	17.0	34
Bangladesh	115.59	17.3	35
Nigeria	108.54	28.4	57
Afghanistan	16.56	39.3	79
Kenya	24.03	46.0	92

Source: Adapted from Gleick (1993), FAO (1995) and Gleick (1996).

producing any food consumed. Estimates of water requirements for various uses from different authors are presented in Table 3.3. Sanitation requirements are about 20 l/day with another 20 l/day needed for food preparation and dishwashing. Estimates of water use in different countries are provided in Table 3.4.

The United States is a relatively high water consumer although it must be stressed that not every state in the United States of America is the same. There the overall use is nearly three times that of the Netherlands and 50% more than Sweden. All of these countries are highly industrialised and have well-developed economies, but the situation in less-developed countries will be quite different. The water use per household greatly varies depending upon the income of the occupant. In low-income properties, such as tenements where there may be a shared WC, the water consumption is about 55–70 l/capita/day; for a middle-income property where there will be at least 1 WC and two taps, the consumption is 110–160 l/capita/day and in high-income properties, such as detached houses and luxury apartments that have at least 2 WCs and a minimum of three taps per household, the consumption is about 150–260 l/capita/day. The average daily average consumption per person for the European Union is 144 l, which is well above that required for basic human needs.

This is divided into showering, brushing teeth, each toilet flush, dishwasher and 60 l for each wash in a washing machine (all of these figures are for using efficient modern equipment). If the equipment is of older types, the amount could rise to 315 l but could be even higher for large households. The WHO indicates that a bare minimum survival level of 20 l/day/person for temperate climates but more in hot dry climates. If all other factors such as sanitation and other health requirements are taken into account, the quantity rises to 50–100 l/day/person. Although these figures are estimates, they do indicate that where the economic status is rising, as is an aspiration in most economies, the demand for water will rise.

Estimating the water demand for a whole region or country with a view to calculating water needs and management strategies is difficult because of the complexity and requirements of different users as well as the potential variability of different resources and changing climate. One way of attempting to overcome this problem is to reduce the overall water needs to a simple index. In the 1990s, the term water security and its concept became more widely and discussed. Initially, there was a range of approaches depending upon the particular interests of the people concerned. Cook and Bakker (2012) reviewed the published literature on the topics covered. In many articles, the term water security also covered water scarcity, water stress and water sustainability. Depending upon the authors' background, the aspects emphasised ranged from atmospheric sciences to water resources, the latter being the most widely used approach among the 198 articles surveyed which also included environment, engineering, geosciences, agriculture, geography, public health, social science, natural sciences and meteorology. The use of an index to bring together a range of aspects gained favour with many government and other agencies because of the complexity of the problem and the wider implications of the topic. There are, however, a number of different ways of calculating water indices. Various scales can be used, for example national, state, watershed or village, depending upon the particular interest of the author. The amount of water available in the region also determined which factors were important in their case. Some included assessment of hazards if flooding and/or drought was a problem.

Basic Water Requirements

Ban Ki-moon, the then UN Secretary General, stated that 'Safe drinking Water and adequate sanitation are crucial for poverty reduction, crucial for sustainable development and crucial for achieving any and every one of the Millennium Development Goals'. This was a number of years ago, but still at least 884 million people do not have access to safe drinking water and 2.6 billion lack access to basic sanitation. This amounts to about 40% of the world's population. This was reinforced through the UN General Assembly Resolution A/RES/64/29 on 28 July 2010, which declared that safe drinking water and sanitation were a human right essential to the full enjoyment of life and all other human rights. There were a number of discussions in the UN in subsequent years confirming these rights, and in 2011 the Council adopted another resolution (No. 16/2) stating that access to safe drinking water and sanitation was a human right affecting the right to life and human dignity. They stressed that the water supply must not only be sufficient but also continuous and should not only contain enough for drinking but also for personal sanitation, washing clothes, food preparation and personal and household hygiene. The amount required for these is 50–100 l/person/day. A number of studies have been made to more precisely quantify human needs, and in order to achieve these aims, it is necessary to know the basic needs for water which can vary.

Water requirements are not just a matter of quantity but also good quality. This will vary with the country, culture, gender and their lifecycle and privacy. This includes constructing public and school toilets that are separate for males and females. The Mal del Plata conference and the subsequent Earth Summit in Rio (1992) (Gubb et al. 2019) strongly emphasised the concept of basic water requirements for both humans and ecosystems. Water of sufficient quantity and quality must be provided to meet four basic needs: drinking water for survival, water for hygiene, water for sanitation and water for food preparation and other modest household needs Gleick (1996). White et al. (1972) identified the following three categories of applications in studies of domestic water use in East Africa: (i) consumption (drinking and cooking), (ii) hygiene (including basic needs for normal personal and domestic cleanliness) and (iii) amenity use (lawn watering and car washing). A fourth category was later added by Thompson et al. (2001) who included 'productive use'. This included animal watering, construction, brewing and small-scale horticulture. There should also be consideration for growing food and ecosystems. The fact that also has to be considered is that different groups of people living in different areas probably use water in different ways including removing waste, manufacturing and energy production. These, as well as climate change, affect water needs.

Factors Driving Water Demand

In many countries, an important factor is the type of access to the water. This combined with the climate in that area has a significant impact. If there is a public standpipe that is more than 1 km away, the daily use is less than 10 l/day. At the other end of the scale, if there is a household connection, especially if there is a garden as well, the use could increase to 400 l/person/day. The situation is worse in dry hot climates where consumption could double compared with a cooler humid climate. Gleick (1996) does not use the term 'water use' as there are many types of use but prefers the term 'withdrawal'. He argues that not all water that is withdrawn is consumed. Some users will return the water directly to its source, e.g. cooling water. He also uses the term 'consumptive use' where the water cannot easily or directly be reused. Examples include plant transpiration, contaminating of the water or incorporation of the water into a finished product. The question still arises as to what is the absolute minimum requirement for a human? This is the minimum amount required for survival. Again this can vary with climate and also the physiological state of the person. A person can lose up to 1% of their body fluid and just feel thirsty, but if they lose 10% they are in danger of dying. Water loss in humans can occur through the skin, by excretion and though the lungs. Taking into account these losses, estimates for minimum water intake range from 1.8 to 5.0 l/day (average 3 l/day). This is for a 70 kg person in an average climate, but in a hot climate this would increase to 4–6 l and even more for a person doing strenuous work. In addition to these physiological needs, there is also a need for sanitation. Proper sanitation will help protect against disease, so proper disposal of human waste is essential.

Waterborne diseases are one of the main causes of death, especially in young children. Not only human waste should be carried away for treatment, but there should also be a facility for handwashing at the household or location. There are some methods of disposal that do not require water. Flush toilets, especially in villages, towns and cities, are essential, and these should be linked to a piped transport system to the proper wastewater treatment facility. Older flush toilets that are not very efficient use up to 75 l/person/day, whereas more modern ones use as little as 5.0 l/person/day. Dietary differences in different communities will use different amounts of water in food preparation. These differences are reflected in the amounts of water needed to grow the food (see also virtual water).

Water Withdrawals and Use in the United States

Examples of the overall average amounts of water used in different parts of the United States of America are presented in Tables 3.5 and 3.6. In general terms, the three largest categories were thermoelectric power generation, irrigation and public supply, which cumulatively added up to 90% of the national total. Saline water withdrawals, which are

Table 3.5 Total water withdrawals for selected states in the United States of America by source for 2010.

		Withdrawals (million gallons per day)		
State	**Population in thousands**	**Groundwater**	**Surface water**	**Total**
Alaska	710	622	472	1090
Arizona	6390	2550	3540	6090
California	37 300	12 700	25 300	38 000
Florida	18 800	4120	10 800	14 900
Kansas	2850	3200	800	4000
Michigan	9880	694	10 100	10 800
Montana	989	286	7360	7650
New York	19 400	704	9870	10 600
Oklahoma	3750	2030	1140	3170
Pennsylvania	12 700	657	7480	8130
Texas	25 100	7710	17 100	24 800
Vermont	626	41.6	389	431
Wyoming	564	617	4080	4700
US total[a]	313 000	79 300	275 000	355 000

NB The withdrawals given in the above table for groundwater and surface water are freshwater withdrawals only and do not include any saline water withdrawals. The totals for each state do include both freshwater and saline water withdrawals.

[a] The US total includes all states as well as Puerto Rico and US Virgin Islands.

Source: Data from the USGS Circular 1405, US Geological Survey and US Department of the Interior (2014).

Table 3.6 Total water withdrawals by water-use category, 2010, in million gallons per day.

State	Public supply	Self-supplied domestic	Irrigation	Livestock	Aqua-culture	Self-supplied industrial	Mining	Thermo-electric power	Total[a]
Alaska	790	14.8	1.59	0.25	684	7.78	24.1	58	1090
Arizona	1210	27.2	4570	27.0	47.3	12.9	86.6	104	6090
California	6300	172	23 100	188.0	973	400	36.4	65.4	38 000
Florida	2270	214	2920	21.3	1.86	213	113	613	14 900
Kansas	391	14.9	3040	114.0	12.9	40.3	13.3	377	4000
Michigan	1090	231.0	209	19.6	82.7	612	76.2	8520	10 800
Montana	138	22.2	7160	41.8	18.9	66.4	27.9	151	7650
New York	2260	152	70.4	22.6	40.2	352	72.4	2760	10 600
Oklahoma	657	26.8	564	88.8	10.7	20.8	18.0	385	3170
Pennsylvania	1420	201.0	27.1	52.3	108.0	866	62.0	5390	8130
Texas	3990	259	6830	259.0	31.4	680	203	10 500	24 800
Vermont	43.1	13.6	2.5	5.6	10.9	5.7	3.9	345	431
Wyoming	99.0	8.6	4370	16.5	20.8	6.7	50.1	63.4	4700
US total[b]	42 000	3600	115 000	2000	9420	15 000	2250	117 000	355 000

[a] The US total is all states including Puerto Rica and US Virgin Islands.
[b] The state total column includes both freshwater and saline water withdrawals. Individual volumes of each are not listed in the table.
Source: Data from USGS (2014).

not listed here separately, were mostly used in thermoelectric power generation, but most water used in this industry was freshwater. Overall surface water withdrawals accounted for about 78% of the total, and of these about 84% were freshwater. Irrigation water accounted for approximately 38% of all freshwater withdrawals and over two and a half times that withdrawn for public supplies. California had the largest withdrawals accounting for 11% of the total. Saline groundwater withdrawals were greatest in Oklahoma and Texas, and these were used mostly in mining. The largest surface freshwater withdrawals were in California (28 800 million gallons per day) followed by Texas (24 800 million gallons per day), and then Idaho, Illinois, Minnesota and North Carolina all with just over 10 000 million gallons withdrawn daily. Not all people in the United States rely on public supplies. About 14% of the population provided their own water for domestic use in 2010, most of which (98%) were from groundwaters. Irrigation water was all classed as freshwater and included not only agriculture but also golf courses, parks, nurseries, turf farms, cemeteries and other landscape watering. Farm irrigation includes frost protection, crop cooling, application of chemicals and weed control. Three types of irrigation are used in the estimates: sprinkler, micro-irrigation and surface flood systems. The largest amounts of irrigation water were used in the western states, in particular California, although usage has declined in some states.

Different parts of America are below the national average, e.g. Massachusetts, but others are much higher, e.g. California. The US Geological Survey (USGS) has carried out estimates of water use every five years since 1950. From 1950 to 2010, the US population grew from 150.7 to 313 million. Public supply has risen by 300%. Irrigation withdrawals have increased by a little under 40%, and thermoelectric power water demand has quadrupled. Fresh groundwater use has more than doubled as has fresh surface water withdrawals. During the five-year period 2005–2010, most sectors experienced small to modest reductions in withdrawals. This probably reflects greater efficiencies in many sectors and more reuse and recycling in many industries. Aquaculture and mining, however, both showed increases. Although the overall use in America is considerably larger than that in Sweden and the Netherlands, the five yearly figures for 2010–2015 showed that the daily total water use in the United States in 2015 was an estimated 322 billion gallons, which was 9% less than in 2010. Fresh surface water withdrawals were 14% less than 2010. Among different categories of uses, only irrigation showed increase in water use (USGS 2015). Most of the freshwater withdrawn in the United States is from surface waters.

The use of water extends to a range of industrial and commercial requirements as well as power generation and cooling. All of these could be classified as 'wants' rather than basic needs (Gleick 1996). When considering water requirements and needs, the requirements of natural ecosystems must also be included. It is essential to meet ecosystem needs as these provide not only the overall life support systems of the planet but also provide important food sources for fish and other aquatic organisms, in many countries. This problem extends beyond the amounts of water and importantly includes water quality (see Chapter 6). The United Nations produced a Manual on the Right to Water and Sanitation that outlined different water needs (COHRE, AAAS, SDC and UN-Habitat 2007). This states that all humans have the right to (i) sufficient water, (ii) clean water, (iii) accessible water and sanitation, (iv) affordable water and sanitation, (v) non-discrimination and inclusion of vulnerable and marginalised groups, and (vi) access to information and participation. They also point out several misconceptions regarding the right to water and sanitation. These are as follows:

The right entitles people to free water	These services need to be affordable, but people are expected to contribute financially but only within their means
The right allows unlimited use of water	Everyone should have sufficient water for personal and domestic use, but this must be sustainable and is not unlimited
Everyone is entitled to a household connection	Water and sanitation need to be within a the immediate vicinity of the house
The right entitles people to use resources in other countries	You cannot claim water from another country
A country is in violation of the right if all of its people do not have access to water and sanitation	A state must be taking reasonable steps to maximise availability or progressively move towards that right

Source: Taken from COHRE, AAAS, SDC, and UN-Habitat.

Climate change is arguably the most important environmental factor affecting both the hydrological cycle and human populations now and in the future. Climate change will have a marked effect on seasonal temperatures, weather conditions and water distribution. This latter, it must be remembered, is not because the total amount of water on this planet is changing for, as was noted earlier, there is a finite and constant amount here. One impact of climate change is to significantly alter its geographical and temporal distribution. A further important driver affecting water availability for humans is population growth. About 10 000 years ago, the human population of this planet was approximately five million. Over the next 8000 years, it grew to about 200 million with a growth rate averaging 0.05% (World Population Clock-Worldometers). By 1900 it had grown to 1.6 billion, and by 1999 had reached 6 billion. Currently, it stands at about 9.7. billion. By the end of this century, depending upon the prediction used, the population could reach 10.9 billion. The growth rate at the start of this millennium is about 1.3% per annum. Although the growth rate peak may have passed, there will still be about 90 million more people per year. The result is that, purely on the basis of increasing numbers, the demand for resources, in particular water, will increase. If the targets set out in the Millennium Declaration are met, the demand will increase further as not only will the needs of the increased population have to be met but also there will have to be large improvements to the large numbers of people who already do not receive enough. Just over 10 000 years ago, the human population were hunter gatherers living in small groups and using local sources of water in small amounts. With the onset of settled agriculture food supplies, water became easier even though demand increased. Eventually, as the population grew, settlements grew into towns that again required more water. The requirement for more organised waste disposal also arose. The population really began to mushroom in the eighteenth century when the industrial revolution began. In northern Europe, there was a large migration of workers from rural areas into towns and cities to provide labour for the newly developed factories (UN Urbanisation Prospect 2009 revision). This again produced extra demands on water resources and sanitation. Population growth, urbanisation, industrial growth, energy to drive machines as well as water for human consumption all contributed, and much of this has continued until the present day. The potential problem of population growth and the resulting increase in demands were recognised as early as 1798 when Thomas Malthus published his Essay on the Principle of Population in which he argued that birth rates were too high and needed to be checked if hunger and famine were to be avoided. Although Malthus could not anticipate the huge advances that have occurred in both technology and agriculture, both of which to an extent countered his arguments, he was right in flagging up these potential problems and in showing concern that food production was not keeping pace with the population's needs. At this stage, we should ask why humans, among all animals on this planet, have been so successful. Perhaps it is because humans are very adaptable (see Tickell 1993) giving them the ability to switch from one resource base to another and exploit it as required. Success is one thing, but out of control success is perhaps not so good! Certainly, one resource we cannot switch from is water. Efforts to solve the food problem by increasing crop yields in the Green Revolution only worked by using more fertilizers and, in particular, water.

An overview of the changing world's population is shown in Figure 3.1. From the point of view of water demand, a rising population will, if everyone is to be supplied with an adequate quantity, continue to place large strains on both supplies and the infrastructure to meet the demand. There are many factors involved, including geographical differences in resources, climate, economic capacity of the country and rural vs. city populations. During the past few decades, there has been a remarkable growth in large cities. Those with a population of 10 million or more are referred to as 'mega-cities'. Table 3.7 indicates some of the mega-cities.

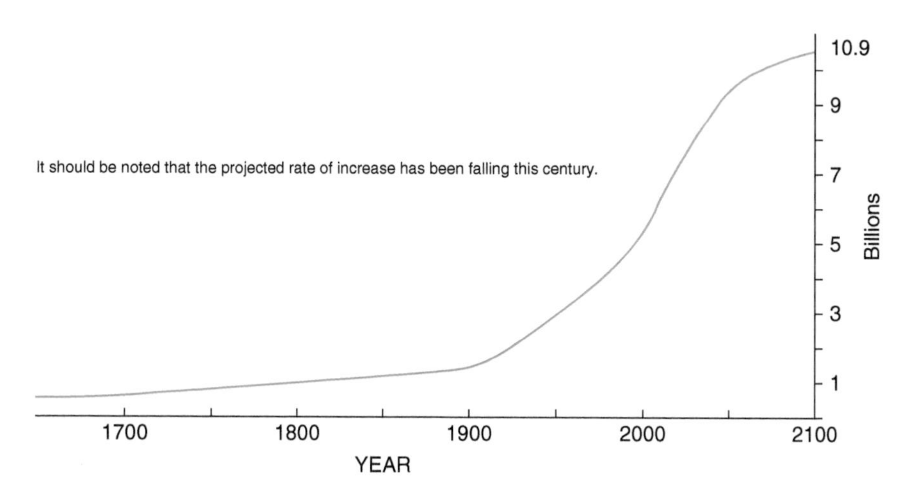

Figure 3.1 World population growth from 1700 and projected numbers for 2100. It should be noted that the projected rate of increase has been falling this century.

Table 3.7 Seven of the world's largest cities where water shortages are likely.

City	Population (million)	Report of water shortage
Mumbai, India	19.8	Mumbai faces water shortage 2011 http://news.bbc.co.uk/2/hi/8138273.stm
Karachi, Pakistan	15.4	Karachi facing shortage of 70 million gallons of water 2010: www.dailytimes.com.pk/default.asp?page=2010%5C09%5C26%5Cstory26-9-20_pg7_22
Delhi, India	28.5	Water shortage in South Delhi, May 2010. http://www.hindustantimes.com/Watershortage-in-South-Delhi/Article1–545037.aspx
Manila, Philippines	13.4	Water shortage hits 50% or Metro Manila, 2010. http://newsinfo.inquirer.net/topstories/view/20100720-282193/Water-shortage-hits-50-of-Metro-Manila0DPWH-Chief
Mexico City, Mexico	21.6	Dry taps in Mexico City: A Water Crisis Gets Worse, April 2009 http://www.time.com/time/world/article/0.8599,1890623.00.html
London, UK	5.59	Leaks May Cause Water Shortages, March 2005. http://news.bbc.co.uk/2/hi/uk_news/England/London/4330721.stm
Beijing, China.	19.8	Beijing Water Supplier Faces Severe Water Shortage, March 2009. www.chinadaily.com.cn/china/2009-03/21/content_7603383.htm

Source: Adapted from World Economic Forum (2011). Water Security. The water–food–energy–climate nexus and various sources. Population figures are for urban areas not just the main city.

Such large concentrated populations place an excessive demand upon local resources in particular drinking water and sanitation. Cities such as Lima, Lagos, Delhi, Calcutta, Mumbai and Beijing may all suffer from water scarcity in the future. Many large cities are afflicted by hazards other than water scarcity. Several are situated in flood prone areas. Of the 633 largest cities listed by the UNDESA (2007) in 2012, at least 233 are located near to high-risk flood zones. This could potentially affect 600 million people. Not all of these cities are coastal. The regions most at risk are in Asia, Latin America, parts of North America and the Caribbean. Flooding is the most frequent hazard with drought the second. Currently, the shift from rural to city dwelling is continuing and with it the problem of providing adequate water supplies and sanitation.

The worldwide increase in population to date, and as projected for the future, has without human society realising it, the power to seriously interfere with the life support systems of the planet. Tackling this problem is extremely complex as it impinges on every person's freedom of choice as well as religion, economic status, etc. All of the food we eat comes either directly or indirectly from photosynthesis, i.e. primary production of plant material (with the minor exception of artificially produced protein). All animals either directly or indirectly depend upon plants. We either consume plants directly or use them to feed animals that we then consume. The problem will also get more severe with increasing affluence creating a greater demand for protein in their diets.

Sectoral Water Demand and Consumption

Agriculture

All agricultural plant crops and animal livestock require water to live and grow. About 70% of abstracted water globally is used for agriculture, and this may reach 90% in some countries. Agricultural use of water is often not very efficient, so it is important to understand where and how it is used in order to manage it and, if possible, reduce the demand because future increases in freshwater abstraction could lead to a global scarcity in freshwater. The Green Revolution did increase crop productivity but also considerably increased water demand and abstractions. Food security, an objective of all governments, depends upon adequate water availability. Although globally there is possibly enough freshwater for our present and even foreseeable needs, if used sustainably and managed properly, this does not take into account the huge geographical variations in supply and the many areas where water supplies are scarce. In recent centuries, the demand for meat protein has increased dramatically partly because of population growth and also because a growing middle class will probably have different dietary preferences. To meet these demands, it is quite likely that food production may have to double. For example, cereal demand is projected to grow from 585 million tonnes currently to 828 million tonnes by 2025 (World Economic Forum 2011). Not only will more production be required but also large savings of post-harvest losses as well. In 2008, about 3350 million hectares were used for meadows or permanent grassland for livestock grazing. This is more than twice that used for arable and permanent crops (UNESCO 2012). This increase in livestock production, which is still continuing, already accounts for 40%, in value, of the global agricultural output. The result of greater numbers of livestock is greater demands for water, cereals for feed and also land, which is being met in some countries, e.g. Brazil, by deforestation. Meat requires approximately 10 times as much water per calorie as plants. More meat consumed equals more water required. For example, the average daily diet in California requires about 6000 l of water in agriculture, whereas in Egypt and Tunisia it only requires 3000 l (International Water Management Institute 2007). This then has a knock on effect for not only other forms of agriculture but other sectors as well. Although livestock production officially includes aquaculture for fish, this does not consume significant volumes of water compared with cattle, pigs, sheep, etc. Overall cereal crops are by far the most important source of total food production, as measured in calories, in developing countries, and the main meat consumers are in the developed world. Total world food consumption, in kcal/person/day, has risen from just less than 2358 in 1965 to a projected 3050 in 2020 (UN World Development Report 2003). Food security concerns are growing in many countries, and one measure to combat this is that meat consumption may have to be reduced. Irrigated crop yields generally exceed those of rain-fed crops by as much as threefold. The area of irrigated crop land is steadily increasing. It is estimated that it has risen by over sixfold over the last century from about 40 million hectares in 1900 to more than 260 million hectares by the end of the last century (Chartzoulakis and Bertaki 2015; Postel 1999). Almost 40% of the world's food comes from 18% of the cropland that is irrigated. In Greece, agriculture uses about 80% of the water abstracted. Currently, the irrigation efficiency is low. Only an estimated 55% of the water is used by the crop. This will increase the demand for water unless large improvements are made to irrigation efficiency. This is a particular problem in areas such as the Mediterranean and similar areas. The United Nations Food and Agricultural Organisation (FAO) estimates that there is still the potential for an increase in farmland area although this would certainly involve using poorer or more marginal soils. The growth in demand for meat in our diets and increasing numbers of animals can also result in more land pollution from their wastes which can be leached from the soil affecting ecosystems and impact on waters causing eutrophication. Intensive livestock production, if not properly managed, can be a major source of environmental pollution. We must also remember that existing farmland is being lost annually through land degradation (e.g. through salt build up) and urbanisation. The increase in population will result in a greater need for food. According to the United Nations Environmental Programme (UNEP 2009), the world cereal demand will grow from 585 to 828 million tons by 2025, and to meet this target, the water requirement will also have to grow. Water scarcity in several countries will mean that they cannot meet these food demands, and the only way to meet this demand is to increase world food trade. At the present time most food trade is between a relatively small group of countries, but to meet future demand this will have to widen. In 2001, 60% of world food exports was from the United States of America, the EU and Canada, and 60% of agricultural import was from the United States of America, EU and Japan which is counter to the world's needs (World Economic Forum 2011). This will result in an increase in population growth also means that the amount of farmland per person has decreased from 0.4 ha/person in 1961 to 0.2 ha/person in 2005. In addition, the scarcity of water is likely to reduce yields in several countries. This could lead to the growth of protectionism and price rises to the detriment of poorer countries. This is at a time when the global demand for food is increasing and will do so into the future. The projected rise in agricultural water abstraction is likely to be about 8515 km^3 by 2050 (Comprehensive Assessment of Water Management in Agriculture 2007). In addition to agricultural food crops, there has been an increase in recent years of using land for biofuel production. This not only takes up valuable agricultural land, but also uses significant amounts of water. All other water sectors are likely to be competing for this finite resource including industry and energy. This emphasises the need for proper planning and management, including the use of marginal waters such as drainage, reclaimed and even mildly saline for irrigation and more efficient water application to the crop. Irrigation is only a sustainable option for food production if enough reliable water supplies are available. This might involve using crops or varieties that do not require as much water. Pressure on land from other users will also increase in the future.

One interesting growth area within the agricultural sector is the growth in the cut flower trade. Many of the cut flowers in florists and supermarkets are imported from Africa and in particular from Kenya. Kenya has developed a large and lucrative flower export industry which is generally regarded as a significant economic success. The main flower cultivation area is around Lake Naivasha. This lake, in the Kenya rift valley, has important Ramsar designated wetlands around its shores which are being threatened as the lake level drops Everard and Harper (2002). The human population in the lake basin is growing and poses another threat, and the area requires wise management. It has contributed an average of $141 million annually between 1996 and 2005 (Mekonnen and Hoekstra 2010). The industry has grown to produce $352 million in 2005 alone. In addition, foreign exchange for the country the flower industry provides employment helps finance infrastructure such as schools and hospitals for many people in the Lake Naivasha basin. Flower cultivation uses considerable amounts of water, much of which is obtained either directly or indirectly from the lake, groundwater and rivers flowing to the lake. These rivers need to be monitored properly and so does the abstractions from groundwater and the lake itself in order to construct a proper viable water balance. Algal blooms in the lake are increasing due to nutrient pollution making the water less useful for irrigating the flower crops, especially the sensitive species, without expensive treatment. This water use has resulted in the lake level gradually reducing, which in turn has adverse effects both on the lake and surrounding wetlands biodiversity. One of the main flowers grown is the rose. The water footprint of a rose flower is about 7–13 l. The amount varies for other types of flower, but overall the virtual water export from them is 16 Mm^3/year, which is comprised of 45% blue water, 33% grey water and 22% green water. The flower farms around the lake have thus significantly contributed to the drop in the lake level. Some farms are also responsible for polluting the lake with nutrients fed to the flowers which then seeps into the lake. If the flower farms are to continue, it is essential that they adopt sustainable water management. Reducing water usage could be through pricing the water at its full cost in addition to other regulatory measures. Although full-cost pricing of the water has been agreed, in principle, it has been difficult to implement it in the agricultural sector. Kenya is unlikely to be able to introduce full-cost pricing as farmers resist even modest price rises, and it is unlikely that the authorities could enforce them. One alternative has been to try to introduce more sustainable water use in flower farming, but this also needs agreement with all major agents in the global flower market along the cut-flower supply chain as a premium would need to be added to the final product by the retailer who could also promote the idea that the premium would go towards promoting water sustainability which could be more acceptable to the public. The problem with the farming industry is how to have sustainable development and still allow economic development. Businesses will have to achieve a zero impact approach as it would be catastrophic to lose the unique ecosystems. The lake and other waters must not be regarded as 'common property' and managed in a sustainably manner. This will take time and require a proper information base, but it must be done. Managing water sustainability will take a whole catchment approach. To achieve this, it will certainly require external aid and a unified regional plan and that plan must not only protect the ecology of the region but also its social and economic well-being. The ability to achieve sustainability will become more difficult with time and must be started as soon as possible. If there is any chance of meeting the aspirations of local societies and needs, this must be initiated now. If started, the social and economic aspirations of local people can be achieved and this applies globally. Action along these lines is a pressing priority if the long-term well-being of the local communities is to be protected. If the problems of Lake Naivasha are not solved, the area may well have a bleak long-term future, and this applies to many other areas around the world.

Combining all of these probable trends, i.e. increasing population, increasing water scarcity, possible food shortages in some countries and climate change in the coming decade over half of the world's population, will become reliant upon food imports for their survival. This can only work if there are changes to the way that world trade in food is currently working and cash poor countries will suffer. The richer countries will continue to alter the geopolitical landscape by buying farmlands in other countries to grow their crops using other countries water. The benefit is that it does create employment in those countries but may not solver either their water or food problems.

Energy

Probably, one of the most used engineering structure for water management is the creation of a dam. These were initially used to store water in reservoirs and control river flows both to prevent flooding downstream and to use the controlled flow to drive machines such as watermills. In this way, the water flow could be turned into an asset rather than a possible hazard. Dams have been used by beavers for thousands of years both to create a home and to control the water for food. Water power can be directly used to create energy for human use, but it is also needed as an integral part of other forms of energy production. These include cooling, cultivation of biofuel crops and extraction of energy-producing raw materials. Water and energy also have a circular relationship, and energy is often required to move, pump and treat water, especially with the use of membrane technologies such as desalination. There is a projected rise in energy consumption until 2050. Non-OECD (Organization for Economic Cooperation and Development) countries are expected to show both the greatest rise and the greatest overall consumption of water associated with their projected increase in economic status and rise in gross domestic product (GDP). Energy sources requiring water can be divided into primary and secondary sources. Primary energy sources are extracted, captured or cultivated and include oil, coal, biomass and geothermal, whereas secondary sources have to be transformed into hydrocarbons and electricity generated from thermal processes using fossil fuels, geothermal, nuclear and hydropower, solar, photovoltaic and wind (Overaard 2008). Overgaard defines energy sources as

> Primary energy is energy embodied in sources that involve human-induced extraction or capture, that may include separation from contiguous material, cleaning or grading, to make the energy available for trade, use or transformation.
>
> Secondary energy is energy embodied in commodities which comes from human-induced energy transformation.

The interaction of water and energy is often indirect. For example, wind power may be used to pump water from one place to another and the water may then be used to provide energy for another purpose. This would be classed as non-thermal energy. In any case, water and energy are inextricably entwined and interdependent. The production of primary sources is expected to increase steadily over the next decades, mainly through the use of biofuels, natural gas, wind and possibly photovoltaic with perhaps some nuclear but less use of coal and oil. Electricity demand and consumption is also projected to rise, especially with its increasing use for heating and transport. Water is required in the production of traditional primary fuels such as coal, oil and natural gas as well as uranium ore. More water is used, however, in generating stations using traditional fuels in creating energy. Water is not only needed for traditional transport using boats but also in producing fuels for transporting materials, and this, for example crude oil, needs water during the drilling stage, in pumping, refinement and treatment. Thermal power plants generate electricity using fossil fuels or nuclear by heating water that is then run though turbines to drive electricity generators. The water in this cycle is then usually cooled and recycled. These types of processes account for over 70% of global energy production (US EIA 2010). The amount of water consumed is not very large as cooling and recycling are frequently used, and water is returned to the environment. However, its temperature may be considerably higher than the receiving stream, river or lake, so have a marked effect on the native biota. If the system is a closed loop and the water is virtually all recycled, such a discharge will not happen. Losses during the cooling cycle are usually from cooling towers, where some is lost to the atmosphere through steam and the rest is cooled for either reuse or discharge. Hydropower electricity generation is often the largest source of renewable energy for a region (15% of global production in 2007). There is still scope for expanding this energy source in the future. Hydropower uses gravity to encourage the water to flow downwards through turbines that generate the electricity. The water is not consumed or polluted during this action, but the hydrology of the system may be altered and there can be ecological impacts both upstream and downstream of the installation. Wind, solar and photovoltaic electricity production use virtually no water but use merely small amounts for cleaning the installations, especially if located in dusty or polluted areas. Large-scale solar power plants can be used to heat water creating a steam cycle similar to that used in thermal power stations and would thus have a similar water use.

Energy is needed for many aspects of water use. EPRI (2002) estimates that between 2% and 4% of the total US electricity consumption is used for water provision in both drinking and wastewater treatment plants. Including end users, the US energy consumption for water is about 10%. Groundwater pumping generally takes 30% more energy than surface water, and this is likely to increase as water tables fall, often through overabstraction. Energy is needed in the treatment of water to render it safe for drinking (World Economic Forum 2011). Desalination using membrane technology requires quite large amounts of energy, usually electrical, to provide the required pressure to pass the water through the membrane. Generally, all membrane technology requires the pressure of some description. The result is that the cost of water produced by this method is closely linked to the cost of energy. In addition, brine disposal from desalination plants can be a severe problem for the receiving ecosystems. The overall demand for energy will increase in the future as will the demand for water if the ‘business as usual’ scenario is followed. Gleick (2009) suggested several ways in which the energy demands on water might be helped. These include (i) reducing the use of cooling water used in power stations by changing to air cooling or other technologies and (ii) using renewable energy, e.g. solar, wave and wind, to reduce demands on freshwater. Solar power and other renewable forms of energy are already being used in both drinking water and wastewater treatment plants as well as some smaller service reservoirs being covered with solar panels to provide energy for the works. A closer integration of all aspects of energy production and use with water is needed to produce yet more savings.

Industry

The diversity of industrial processes means that only general statements about water requirements and amounts can be made. Broadly speaking, industry does not use a large amount of water when compared with other sectors, but there are high requirements for accessible, reliable and sustainable supplies. About 20% of global water abstraction/use can be attributed to industry, but there are large variations between different countries. It is not difficult to obtain water use estimates from large industries, but consumption figures for the myriad of small firms are quite difficult to obtain.

The portion of water used by industry is often proportional to the per capita income level and can range from 5% of total withdrawals in lower income countries to over 40% in some high-income countries. Industrial expansion and demands can be as big an influence on water consumption in developing countries as population growth. In the past, and also to some extent today, water management in industry has not always been very good. Generally, much more was withdrawn than was actually used. Of course, water requirements will vary from one industry and processes within an industry to another. For some industries recycled or reclaimed water is adequate, but for others the quality standards required may be much higher than even those for drinking water, e.g. the food industry and electronics industry. Water withdrawal = water consumption + effluent discharge. Depending upon the process, effluent discharges can have a significant impact on the receiving water. Economic growth and development are often the main drivers of government policy. The result is that industrial expansion and increased building on 'green belt' areas is often encouraged at the expense of other factors. Industry may be given preferential use of water compared with other users, and regulations regarding effluent wastewater quality are not always enforced. This can give rise to antagonism with the local community and degradation to the environment. International trade also has an important role. To meet export requirements of their customers, the developing countries may have to meet many standards both for their products and the environment, e.g. ISO Certification, EMS Environmental Management Systems and Corporate Social Responsibilities. It is hoped that such requirements would prevent the development of 'pollution havens' in countries with lax regulations and poor enforcement capabilities. Water availability for industry is now often a major concern when deciding where to locate a new enterprise. The amount, sustainability of supply, quality and costs all have to be considered.

Humans and Human Settlements

Estimates of the global population over the last 1000 years (Figure 3.1) show that the overall numbers did not change dramatically until the early seventeenth century. From then, they slowly increased but then showed a dramatic increase from the early nineteenth century onwards. Since the Second World War, the global population has been increasing by approximately a billion people every 15 years, resulting in a doubling of the population from 1970 to the present day. This increase will obviously put much greater demands upon all global resources. Between 2009 and 2050, the global population is expected to increase by 2.3 billion to at least 9.1 billion. Over the same period, urban populations are predicted to grow by 2.0 billion to 6.3 billion, i.e. 60% of the total population, and by 2050 this will reach 70%. Most of these increases will be in developing countries. Population growth will largely be an urban problem as a significant part of the urban increase will be migration from rural areas. China has over 100 cities with more than one million inhabitants and India has 35 cities. There are currently 24 megacities. These are cities with more than 10 million people. Cities are not just places to live but are a complex of interactions covering all aspects of human activities and are also always changing. Currently, cities are usually growing and placing greater demands upon resources and as such require careful planning, not haphazard growth. All of these cities will not only house people but also associated industries and businesses. These will not only take up more land but also demand more energy, food and water. These cities will have a cross section of society including a growing middle class. These will demand more food, possibly with a higher meat content in their diets, and a greater consumption of water as they will have more baths, showers, dishwashers and washing machines. Their water footprints will be higher than average. There will also be a large proportion of the population living in poverty. For example, 360 million urban dwellers in India only have access to drinking water of varying quality for an average of less than three hours a day, and 780 million people in South Asia are forced to defecate in the open not only polluting the environment but also spreading disease. Up to 90% of waste from open defecation find its way into rivers and streams some of which are used for drinking and washing. Unfortunately, the water and wastewater infrastructure has not always kept pace with this growth, and slum development on the edges of cities is growing. The question now arises as to how city planners and policymakers harmonise these diverse interests and the inherent contradictions that arise (UN Habitat 2008). A city cannot be harmonious if large sections of its population are lacking in basic needs while large parts of its residents live in opulence. In addition, it must be recognised by both planners and policymakers that cities, and in particular the growth of cities, have an adverse impact on the environment. Harmony in a city must be based upon sustainability and protecting the earth's natural assets. This must be achieved in parallel to the recognition that cities frequently drive a nation's economy. The growth of cities is higher in in the developing world which is responsible for 95% of the world's urban population growth. The growth rate of these cities was between 2% and 4% per annum. In the developed world, the growth rate has slowed and some are even experiencing slight declines. Planning for the needs of the developing world's urban populations is to provide basic drinking water and sanitation needs. This is not easy and requires considerable financial inputs. Fourteen of the world's 19 largest cities are located near to large water bodies that can be used as links to regional and global trade as well as possibly supplying water if they are freshwaters not saline (UN Habitat 2008).

Most city dwellers and most of the largest cities are concentrated in the world's largest economies (Satterthwaite 2002). These cities also develop the most highly productive skills among the population that help drive development, but this also places increased demands upon resources, especially water. To meet the demands for freshwater, both groundwaters and surface waters are being overexploited, a trend which is unsustainable. About 1.2 billion urban dwellers rely on groundwaters, and a further 1.8 billion rely on surface waters. They are at a disadvantage as they have to compete for that water with agriculture and other, often more affluent, rural areas. Many in the urban poor populations have only intermittent water supplies, and many have no sanitation and cannot afford the price of treated drinking water. The net result is widespread disease and ill health, inhibiting their productivity and ability to contribute to society. It is not always easy for the authorities to construct the required infrastructure to meet all drinking water and sanitation requirements. Table 3.7 shows examples of seven of the world's largest cities facing water shortages.

Water and energy are both becoming a problem in Beijing. The water–energy nexus is a problem in many cities, not only in developing countries. The growth of cities is, in many cases, outstripping the fundamental resources, water and energy, which has boosted their economies in the past. The growth has also outplaced the ability to provide both the above fundamental resources and acceptable provision of sanitation for all. Some cities are attempting to solve the problem by 'importing' both energy and water in one form or another. The United Nations in 2008 (UN 2008) estimated that 75% of the global energy production and 10% of the water consumption would be by cities by 2050 (UN 2008; WWAP 2014). With supply problems, the obvious initial response is to save energy use and manage the water supply in a sustainable way. Moreover, if cities are going to thrive, in addition to water and energy, proper sanitation must be provided (Liu et al. 2019). Traditionally, in past years different government departments have been responsible for different resources including water and energy. This invariably results in different, and sometimes uncoordinated, strategies being used to increase efficiency. Sometimes the needs of each department directly conflict others. This can result in attempts in progress being neutralised because one department does not speak to the other. The result is that, in order to develop, self-sufficiency is sacrificed for sustainability. The way out being sought is to go to external resources and import what is needed. Unfortunately, external water sources require a lot of energy either to obtain them or to transport them. Among the water sources being used in Beijing is recycled water. This covers both encouraging industry to recycle and reuse as much of their water as possible and also, for a range of purposes, reuse treated wastewater. Unfortunately, most of these alternatives require large amounts of energy to produce, treat and move. Although many plans have been drawn up to address these problems, they are frequently large and require large investments to construct trunk sewers, storm water drainage systems, new drinking water and new wastewater treatment works only to falter because the money is not available. However, these plans have merit many improvements if could be done on a smaller scale to improve the situation. This is not to say that the problems are easy to solve, but they are not. Figure 3.2, adapted from Khatri and Vairavamoorthy (2007), Kelay et al. (2006), Segrave (2007) and Zuleeg (2006), illustrates the number of drivers and their complex interactions with water management in the future. The key is how to manage various interactions of these and how to obtain the finances required. Of the drivers, climate change, population growth, urbanisation and ageing infrastructure are the main challenges. In addition to these challenges, it is essential to have accurate maps of the existing system and the water quality throughout the drinking water distribution system. Water scarcity can not only affect people's health but also the quality of water they receive. Indeed, in some cases it may be more efficient to introduce home treatment systems (Rheingans and Moe 2006). These could, especially for home sanitation, reduce the spread of disease and contamination of water. Dry sanitation can be a

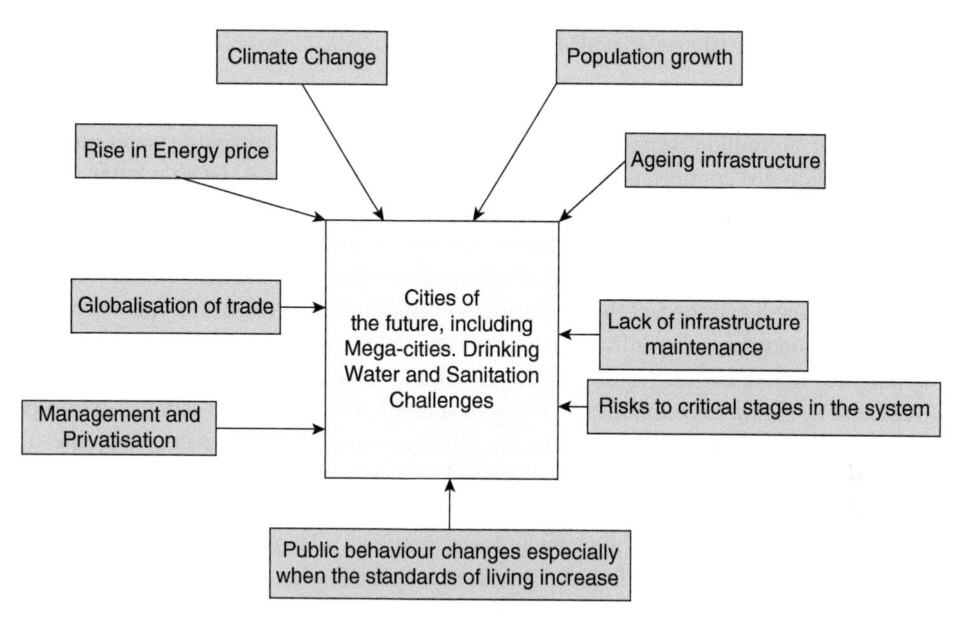

Figure 3.2 The number of potential driving forces/influences that affect water management in cities. *Source:* Adapted from Khartri and Vairavamoorthy (2007).

satisfactory option, especially where water is scarce. The containment of human excreta would help minimise environmental contamination and, after suitable treatment, could be used as fertilizer for domestic garden crops. Composting toilets are available and are often based upon long-term storage to destroy any pathogens (Winblad et al. 2004). Recycling of the nutrients contained in human wastes (faeces and urine) has been used in some regions for many years although not widely in large cities. As the growth in some megacities has resulted in large slum developments, it is hoped that permanent small treatment systems for both waste and drinking water could greatly help until proper development can take place.

Problems will vary with location and country, but Rheingans and Moe (2006) identify four universal barriers to progress. These are (i) inadequate investment in water and sanitation infrastructure and (ii) lack of political will. Often effort is devoted on economic development rather than water and sanitation. (iii) Developments tend to avoid new technological approaches and rely upon conventional methods. In addition, there is often no community involvement in and proposed developments, so local community needs are not considered. (iv) Monitoring the benefits and evaluating the performance of drinking water and sanitation programmes are rarely carried out, especially regarding water quality and health benefits in the community. However, relying on technological advances is not the only solution and should not be relied upon. Although public–private partnerships are increasing in number, they are not always the solution. Private investors wish to not only recoup their investments but also make a profit, and if the water charging in those cities or areas is not yielding an acceptable return, then they may fail to complete their side of the scheme and withdraw from the agreement.

Water Abstractions and Use in Europe

Most parts, but not all, of Europe would be considered to have a reasonable standard of living. About 499 million people are estimated to be connected to a drinking water network and 450 million are connected to a wastewater collection network. Of this latter group, 435 million are connected to a wastewater treatment plant (EurEau 2017). People connected to their individual waste treatment systems are not included in most data. Their waste still must be treated before being discharged into the environment. Because of both population growth and the fact that there are many old/ancient cities in Europe where a lot of the water and wastewater infrastructure was created many years ago, there is a continuing need for maintenance, repair and upgrading of the network. This requires constant investment by the water companies and/or government. Annually, water services in Europe invest approximately 45 billion euros in the infrastructure which, on average, equates to 93.5 euros per person each year. Switzerland, Germany, Denmark, Ireland, Norway, Slovenia and the United Kingdom invest the most. The investment is mainly financed through water bills (tariffs). The water industry employs a large number of people. It is estimated that about 476 000 full time equivalent people equating to 0.1% of the population are employed. The type of management of the water sector varies with country and includes local government, publicly owned companies, private operators where the infrastructure is publicly owned but is run by a private operator, full privatisation where the structure is also owned privately and joint private–public arrangement.

Whichever arrangement is being used, operating and maintaining the infrastructure is a large commitment. For example, there are an estimated 4 225 527 km of pipes in the drinking water network, and this amounts to between 4.92 and 19.55 m/inhabitant. This difference is largely due to population density. The amount of drinking water delivered is about 44.7 billion m^3/annum. This includes non-revenue water that is water lost in the system by leakage and does not actually reach the consumer. Two units are used to express this volume, percentage and volume per kilometre of pipe. Non-revenue losses include many aspects other than just leakage, for example maintenance, firefighting, street cleaning and ornamental fountains in parks. The mean values for losses are 23% and 2171 m^3/km/year in EurEua member countries. Where does all of this water come from? Most of this water comes from surface resources than groundwater, but there is a small but growing use of desalination.

Similar data is available for wastewater services. The mean value of the length of sewer network per connected inhabitant is 7.3 m of sewer/person. The total length of sewer network is approximately three million kilometres. European Union countries and associate countries are required to treat wastewater in accordance with the Urban Wastewater Treatment Directive (91/271/EEC [UWWTD]). If authorised by local authorities, sewage works can also take industrial waste as well as domestic. In Europe, 3.1% of the wastewater is treated at primary level, 28.5% at secondary level and 68.4% at tertiary level. The level of treatment does depend upon the sensitivity of the receiving water to the treated effluent. One of the problems of wastewater treatment plants is the production of sewage sludge that must be disposed of. The final destination, depending upon its composition, could be agriculture 49.2%, incineration 24.9%, land reclamation 12.4%, landfill 8.7% and other 4.9%.

The European Union has acknowledged the problem of resource shortage with freshwater and, within their policy of environmental and resource efficiency, has promoted sustainable water use. Part of this is efficient use of water, but another part is looking at water use as a circular action involving preventing wastewater, preventing its contamination and promoting reuse. This includes reclaiming and reuse of water rather than using new supplies, preventing waste, loss in the system and greater efficiency in its use. Although this is aimed at EU countries, the principles apply globally. These problems apply not only to the present in some countries but also for the future for all. To address this problem, the EU has passed a number of directives, including the Water Framework Directive, the Groundwater Directive, the Environmental Quality Standards Directive and the Floods Directive. The object of these directives is, in general, to guarantee that urban wastewaters are treated and disposed of with minimum impact on the environment and promote water reuse. This latter is an important feature of a circular water economy in which as much as possible water is used within the human system and reused without needing to tap external resources. The *Circular Economy Action Plan* published by the European Commission in March 2020 promotes water reuse, highlighting the recently adopted *Water Reuse Regulation (WRR)* that was aimed at being a key tool encouraging agriculture to reuse water as much as possible. The Action Plan is also expected to be extended to some industrial processes with an overall aim of reducing pressure on freshwater resources. The WRR will come into force from June 2023. Before this regulation comes into force, the European Commission estimates that about 1 billion m^3 of treated water (this is water that has received at least tertiary or advanced treatment) is reused annually. This equates to 2.4% of urban wastewaters but is less than 0.5% of annual freshwater withdrawals in the EU. This figure should grow to about 6 billion m^3 with the new rules. The reuse of water in different sectors after tertiary treatment in the EU is estimated as agriculture, 32.0%; industrial, 19.3%; landscape irrigation, 20.0%; non-potable urban uses, 8.3%; environmental enhancements, 8.0%; recreational, 6.4%; indirect potable use, 2.3%; groundwater recharge, 2.1%, and others, 1.5%. These figures are from the European Commission 2018. The key to these rules of reuse in the EU is that the reused water should be properly treated and thus it is safe and helpful in areas that have suffered shortages and drought in the past.

Canada is generally assumed to have a wealth of water, but there are regions where demand exceeds supply. In order to meet any shortfall, the reuse of water is practiced. Water from treated municipal effluents has been used as non-potable sources that do not require high-quality water. A further benefit of reuse and recycling of treated effluents reduces effluent discharge volumes to the environment. In the assessments of potential health risks with these types of water, it must be known that microbial health risks and chemical contaminants, such as endocrine-disrupting chemicals and pharmaceuticals, may be present. Health and environmental risk assessments are essential steps in safe water reuse (Exall et al. 2004). Table 3.8 shows examples of general characteristics of water reuse guidelines. Water reuse is a viable option both for saving water and costs by not using normal municipal supplies. Health guidelines must be followed, but there can still be a degree of possible public resistance. Both authorities and users must have the knowledge of treatment technologies concerning wastewater reuse.

Table 3.8 Examples of general characteristics of reuse guidelines and regulations.

Agency or province	Coliform limit for unrestricted	References
WHO guidelines	<200FC	WHO (1989)
Texas regulations	≤20FC (geometric mean)	
	≤75FC (single sample)	State of Texas (1997)
Alberta guidelines	≤200FC (geometric mean)	
	≤1000TC (geometric mean)	
	Golf courses and parks only.	Alberta Environment (2000)
British Columbia guidelines	≤2.2FC (median)	
	≤14FC (single sample)	B.C. MELP (1999)

All coliform numbers apply to regulations for two categories of water quality where category 1 is unrestricted public access and category 2 is restricted public access.

Figure 3.2 indicates the number of possible driving forces/influences that affect water management in cities. As can be seen, there are multiple factors all interacting and must be dealt with in the future.

If these guidelines are followed, reused water is perfectly safe and provides a significant saving of resources.

References

Alberta Environment (2000). *Guidelines for Municipal Wastewater Irrigation*. Alberta: Edmonton.

British Columbia MELP. (1999). *British Columbia Ministry of Environment, Lands and Parks*. Water Management Act – Municipal Sewage Regulation. Victoria, B.C.

Chartzoulakis, K. and Bertaki, M. (2015). *Sustainable Water Management in Agriculture under Climate Change*, 88–98. Chaniac, Greece: Agriculture and Agricultural Science Procedia, 4.

COHRE, AAAS, SDC and UN-Habitat. (2007). Manual on the Right to Water and Sanitation.

Cole, W.M. (2016). Convention on the Elimination of All Forms of Discrimination Against Women (CEDAW). In: *The Wiley Encyclopedia of Gender and Sexuality Studies*, 3. Wiley.

Comprehensive Assessment of Water Management in Agriculture (2007). *Water for Food, Water for Life: A Comprehensive Assessment of Water Management in Agriculture: Summary*, 1–95. London, U.K.: Colombo, Sri Lanka; Earthscan.

Cook, C. and Bakker, K. (2012). Water Security: Debating an emerging paradigm. *Global Environmental Change* 22: 94–102.

EPRI. (2002). Electric Power Research Institute, Inc. Water and Sustainability (Volume 4); U.S. Electricity Consumption for Water Supply and Treatment – The Next Half Century. *Technical Report, 1006787*, pp. 1–50.

EurEau (2017). Europe's water in figures. In: *An Overview of the European Drinking Water and Wastewater Sectors*. Brussels, Belgium: European Federation of National Associations of Water Services.

Everard, M. and Harper, D.M. (2002). Towards sustainability of the Lake Naivasha Ramsar site and its catchment. *Hydrobiologia* 488: 191–203.

Exall, K., Marsalek, J., and Schaefer, K. (2004). A review of water reuse and recycling, with reference to Canadian practice and potential: 1. Incentives and implementation. *Water Quality Research Journal of Canada* 39 (1): 1–12.

Food and Agricultural Organization (FAO) (1995). Irrigation in Africa an Figures. *Extract from Water Report No. 7*. FAO, United Nations, Rome, Italy.

Gleick, P.H. (ed.) (1993). *Water in Crisis: A Guide to the World's Fresh Water Resources*, 1–24. New York: Oxford University Press.

Gleick, P. (1996). Basic water requirements for human activities: meeting basic needs. *Water International* 21: 83–92.

Gleick, P.H. (2007). The human right to water. *Water Policy* 1 (5): 487–503.

Gleick, P.H. (2009). N World Economic Forum Water Initiative, *The Bubble is Close to Bursting*.

Gubb, B., Koch, M., Thompson, K. et al. (2019). *The Earth Summit Agreements. A Guide and Assessment: An Analysis of the Rio 92 UN Conference on Environment and Development*. Routledge.

Inocencio, A.B., Padilla, J.E., and Javier, E.P. *Determination of Basic Household Water Requirements (Revised)*, PIDS Discussion Paper Series No. 1999-02. Makati City: Philippine Institute for Development Studies (PIDS).

International Water Management Institute 2007, Water for Food, Water for Life. A Comprehensive Assessment of Water Management in Agriculture.

IWMI-CA (2007). *Water for Food, Water for Life; a Comprehensive Assessment of Water Management in Agriculture*. London: Colombo International Water Management Institute. Earthscan.

Kelay, T., Chenoweth, J., and Fife-Schwa, C. (2006). Trend Report on Consumer Trends, Cross-cutting issues across Europe, *TECHNEAU*, pp. 1–46

Khatri, K.B. and Vairavamoorthy, K. (2007). Challenges for urban water supply and sanitation in the Developing Countries. *UNESCO-IHE, Institute for Water Education, Discussion Draft Paper*, Delft, The Netherlands, pp. 93–112.

Lautze, J. (2014). *Key Concepts in Water Resource Management. A Review and Critical Evaluation*. London, UK: Earthscan, Routledge.

Liu, J., Li, X., Yang, H. et al. (2019). The water–energy Nexus of megacities extends beyond geographic boundaries; a case of Beijing. *Environmental Engineering Science* 36 (7): 778–788.

Mekonnen, M.M., and Hoekstra, A.Y. (2010). Mitigating the water footprint of export cut flowers from the Lake Naivasha Basin, Kenya. *Value of Water Research Report Series No. 45*. UNESCO-IHE, Delft, The Netherlands, pp. 1–105.

Mitchell, B. (2005). Integrated water resource management, institutional arrangements and land – use planning. *Environment and Planning A* 37: 1335–1352.

National Acadamy of Sciences (1977). *Drinking Water and Health*, vol. 1. Washington, DC, USA: National Academy Press.

National Research Council (1989). *Recommended Dietary Allowances*, 10e, 1–290. Washington, DC, USA: National Academy Press.

Overaard, S. (2008). *Definition of primary and secondary energy*. Oslo Group on Energy Statistics. http://www.og.ssb.no.

Postel, S. (1999). *Pillar of Sand*, 1–313. London, U.K.: W.W.Norton & Company.

Rheingans, R.D. and Moe, C.L. (2006). Global challenges in water, sanitation and health. *The Journal of Infectious Diseases* 4 (S): 41–57.
Rijsbermann, F.R. (2006). Water scarcity: fact or fiction? *Agricultural Water Management* 80: 5–22.
Roth, E.M. (1968). *"Water", Compendium of Human Responses to the Aerospace Environment* (ed. E.M. Roth) Ch 15. Albuquerque, NM, USA: Lovelace Foundation for Medical Education and Research.
Satterthwaite, D. (2002). *Coping with Rapid Urban Growth.* London: RICS International Paper Series, Royal Institution of Chartered Surveyors.
Saunders, R.J. and Warford, J.J. (1976). *The Goal of Improved Health. Village Water Supply. Economics and Policy in the Developing World*, 279pp. Baltimore, USA: World Bank/John Hopkins University Press.
Segrave, A.J. (2007). Report on trends in the Netherlands. *TECHNEAU*, p. 113.
Texas, State of. (1997). *Use of reclaimed water.* Texas Administrative Code, Title 30, Part 1, Chapter 210. http://www.tnrcc.state.tx.us/oprd/rules/indxpdf3.html#210
Thomas, J.S. and Durham, B. (2003). Integrated water resource management: looking at the whole picture. *Desalination* 156: 21–28.
Thompson, J., Porras, I.T., Tumwin, J.K. et al. (2001). *Drawers of Water II: 30 Years of Change in Domestic Water Use and Environmental Health in East Africa.* London, UK: HED.
Tickell, C. (1993). The human species: a suicidal success? *The Geographical Journal* 159 (2): 219–226.
Tomuschat, C. (2008). *International Covenant on Civil and Political Rights. United Nations Audiovisual Library of International Law.* UN.
UN Water and FAO. (2007). *Coping with water scarcity.* Challenge of the 21st century. http://www.fao.org/nr/water/docs/escarcity.pdf.
UN Water. (2006). *Coping with water scarcity.* UN Water Thematic Initiatives. http://www.unwater.org/downloads/waterscarcity.pdf.
UNESCO (2012). Managing Water under Uncertainty and Risk. *The United Nations World Water Development Report 4*, Volumes 1 & 2. Paris, France.
United Nations (1948). *Universal Declaration of Human Rights*, 14–25. UN General Assembly.
United Nations (UNDESA) (2007). *UN Dept. of Economic and Social Affairs.* World Economic and Social Survey. New York.
United Nations Environment Programme (UNEP) (2009). *The Environmental Food Crisis.*
United Nations Human Settlement Programme (UN-HABITAT) (2008). *State of the World's Cities 2008–2009*, 1–223. London: Earthscan Publishing.
United Nations World Water Development Report. (2003). Water for People Water for Life World Water Assessment Programme.
United Nations (1979). Convention on All Forms of Discrimination against Women (CEDAW).
United Nations (2008). *State of the World's Cities 2008/2009. Harmonious Cities.* New York, NY: United Nations Human Settlements Programme.
United States Environmental Protection Agency (USEPA) (1976). EPA-570/9-76-003.
US EIA. United States Energy Information Administration. (2010). *EIA's Outlook through 2035 from the Annual Energy Outlook, 2010.* Diane Kearney, EIA, U.S. Department of Energy, Washington, DC.
US Geological Survey (USGS) (2014). *Estimated Use of Water in the United States in 2010. Circular 1405.* U.S. Department of the Interior & U.S. Geological Survey. By M.A. Maupin, J.F. Kenny, S.S. Hutson, Lovelace, J.K., Barber, N.L, & Linsey, K.S.

US Geological Survey (USGS) (2015). *Total water use in the United States.*
Vinograd, S.P. (1966). *Medical Aspects of a "Orbiting Research Laboratory", Space Medicine Advisory Group Study, NASA-SP-86*, 1–144. Washington, DC, USA: National Aeronautics and Space Administration.
White, G.F., Bradley, D.J., and White, A.U. (1972). *Drawers of Water: Domestic Water Use in East Africa*, 63–73. Chicago, USA: University of Chicago Press.
WHO and UNICEF (2000). *Global Water Supply and Sanitation Assessment 2000 Report.* Geneva/New York: WHO/UNICEF.
Winblad, U., Simpson-Hebert, M., Calvert, P. et al. (2004). *Ecological Sanitation: Revised and Enlarged Edition.* Stockholm, Sweden: Stockholm Environmental Institute.
Winkler, I.T. (2012). *The Human Right to Water*, 242–254. Oxford, UK: Hart Publishing Ltd.
World Economic Forum (2011). *Water Security: The Water-Food-Energy-Climate Nexus.* World Economic Forum, Island Press.
World Health Organisation (WHO) (1989). *Health Guidelines for the Use of Wastewater in Agriculture and Aquaculture.* Geneva, Switzerland: WHO.
World Health Organisation (WHO) (1993). *Guidelines for Drinking Water Quality,* Recommendations, 2e, vol. 1. Geneva: WHO.
World Health Organisation (WHO) (2002). World Health Report 2002. Geneva.
WWAP (United Nations World Water Assessment Programme) (2014). *The United Nations World Water Development Report 2014: Water and Energy.* Paris, France: UNESCO.
Zuleeg, S. (2006). *Trends in Central Europe*, 83. Germany/ Switzerland: TECHNEAU.

4

Water Resources

Generally, water is a relatively common compound on this planet covering about 73% of the earth's surface. Most is saltwater that dominates the type of water with 97% coverage. Freshwater, which is required by all living things including humans, covers only about 3% and is very unevenly distributed. Table 4.1 presents the volumes of the main stocks of water. The amount and availability of freshwater also varies geographically because of the earth's natural geology and climate system variations. Freshwater systems are basically confined to the continents although both evaporation and precipitation take place over the oceans as well and are exclusively controlled by the factors driving fundamental Earth Systems such as climate, lithology, topography and ecology. This natural variability is now added to by anthropogenic factors, such as climate change, land use changes, population change, changing food demands and improvements in economic status. In many regions, these anthropogenic drivers have become more important than the Earth System drives. With population growth and the increased demand for building land, there is an increased demand for controls over some of the adverse effects of the hydrological cycle, e.g. floods and droughts. However, any modifications to any part of the water cycle must take into account wider implications to change for other stages in the cycle, including effects on ecosystems.

For thousands of years before rapid population growth, water was plentiful in most areas. It was largely free, but now the situation is changing, especially in semi-arid and arid regions where water availability, or lack of it, is the greatest threat to food production, human health and ecosystem survival (Seckler et al. 1999). International organisations have, in recent decades, come to the conclusion that improving water availability in many parts of the world is a key factor in both relieving starvation and lifting communities out of poverty. Seckler et al. estimated that nearly 1.4 billion people (about 20–25% of the world's population or one-third of the people living in developing countries) will live in regions of severe water scarcity within the next 20 years. A further one billion live in areas of absolute water scarcity. The result will be that they cannot be self- sufficient for food and to survive will need to import more food. Meeting this aspiration will certainly lead to greater demands on this finite resource, so efficient and sustainable use is imperative if scarcity is to be avoided.

Of the freshwater, over 50% is in the form of ice and permanent snow cover. Table 4.2 shows estimates of renewable water supplies accessible to humans and their communities in different regions of the world. Freshwater systems are basically confined to the continents and the atmosphere and in the past were exclusively governed by the factors driving

Water: Our Sustainable and Unsustainable Use, First Edition. Edward G. Bellinger.

Table 4.1 The main stocks of water.

	Volume ($km^3 \times 10^3$)	% of total water	% of freshwater
Salt water			
Oceans and seas	1 338 000	96.54	
Saline/brackish groundwater	12 870	0.93	
Salt water lakes	85	0.006	
Inland waters			
Glaciers and permanent snowfields	24 064	1.74	68.7
Fresh groundwater	10 530	0.76	30.06
Permafrost, ground ice	300	0.022	0.86
Freshwater lakes	91	0.007	0.26
Soil moisture	16.5	0.001	0.05
Atmospheric water vapour	12.9	0.001	0.04
Marshes and wetlands	11.5	0.001	0.03
Rivers	2.12	0.0002	0.006
Incorporated in biota	1.12	0.0001	0.003
Total water	1 386 000	100	
Total freshwater	35 029	100	

Marshes, wetlands and biota may contain some salt water.
Source: Adapted from Shiklomanov (1990).

Table 4.2 Estimates of renewable water supplies accessible to humans in different regions.

Region	Area ($km^3 \times 10^3$)	Total accessible blue water supply	Population served billions	% of world population
Asia	20.9	9.3	2.56	42
Soviet Union	1.8	0.27	4	21.9
Latin America	8.7	0.43	7	20.7
North Africa including				
Middle East	11.8	0.24	0.22	4
Sub-Saharan				
Africa	24.3	4.1	0.57	9
OECD	33.8	5.6	0.87	14

Source: Adapted from UN World Water Development Report (WWAP) 4 (2), (2012).

fundamental Earth Systems, such as climate, lithosphere, topography and ecology. Table 4.2 clearly shows not only the large geographical variations in accessible water between regions but also their population differences. This natural variability is now added to by anthropogenic factors such as land use changes, population changes both in numbers and where they are living, polluting discharges, changing food demands and improving in economic status. In many regions, these anthropogenic drivers have become more important than the original Earth System drivers. With population growth, there is an increased demand for building houses, other properties and the growth of cities all of which take up more land. This results in a need to instigate more controls over some of the adverse impacts on the hydrological cycle, such as floods and droughts. However, any modifications to one part of the water cycle must take into account the wider implications to change for other parts of the cycle, including ecosystem needs being brought about by climate change.

Water resources differ. It is normal practice to divide them into two separate categories. These are renewable waters, which are surface and sub-surface that are part of the short-term hydrological cycle and are replaced within a short period of time. Non-renewable water resources are generally deep aquifers that have a negligible rate of recharge in terms of human timescales, and thus if used are not replaced (FAO 2015). Not all water, be it either renewable or non-renewable, is available for use. In human terms, the water resources that we are able to use are termed exploitable. Exploitable resources include those that have a regular flow, including both surface and groundwater (although some with irregular flows may be included as long as they are recharged within a lifetime), that can also be used for storage behind dams, can be used for artificial aquifer recharge or may be used directly. These are sometimes referred to as being mined. The type and amount of use, including environmental needs, will depend upon local conditions. Of the renewable water resources, some will be internal, i.e. derived from inside the boundary of the region, and some may be external, i.e. flowing from outside the region's boundary, including transboundary lakes, rivers and aquifers. Some rivers and lakes may actually form a political boundary between two or more countries. In that case, there are usually agreements sharing the resource, for example the Ganges and Rhine rivers, Lakes Victoria, Kariba, Itaipu, Mourzouk, India River Plain and Cuenca Baja del Rio aquifers.

With increasing pressure on water availability, other sources, such as desalinated sea water, are becoming more widely used to supplement natural freshwater supplies, which is discussed in more detail in Chapter 7. Assessments of the world's water resources vary between different studies and authors. There are a number of reasons for this, including lack of comprehensive monitoring in some regions, incomplete data provided by some governments and problems of how to account for regions and countries that cross river basin boundaries. Examples of some of these different estimates are provided in Table 4.3.

Table 4.3 Estimates of renewable water resources availability and population numbers in different continents.

Continent	Population ×10^8	Water resources (km^3/year)	Water availability/ person (m^3/year)
Europe	685	2900	4.24
North America	453	7870	17.4
South America	315	12030	38.3
Africa	708	4047	5.72
Asia	3345	13510	3.92
Australia and Oceania	28.7	2400	83.60

Source: Adapted from Shiklomanov and Rodda (2003).

Table 4.4 Examples of exploitable criteria for freshwater resources in some Mediterranean countries.

Country	Total water resource (km^3/year)	Exploitable water resource (km^3/year)	Exploitable criteria
Algeria	14.32	13.00	Technical–economic
Cyprus	0.78	0.54	Technical–economic
France	203.70	100.00	Technical–economic and environmental
Gaza strip	0.06	0.06	Physical and technical–economic
Italy	191.30	123.00	Technical–economic and dam capacity
Malta	0.05	0.02	Physical criteria Saltwater balance in main aquifer
Portugal	77.40	68.70	Technical–economic and environmental
Tunisia	4.56	3.63	Technical–economic
West Bank (Palestinian Authority)	0.75	0.71	Physical and technical–economic

Source: Adapted from FAO Water reports, 2015.

An additional problem is which countries are included in each region. For example, Gleick treated each on its own, but this was not done by most other authors. Whether the Former Soviet Union is placed in Europe or part in Asia, for example, skews the data.

Estimates of the regional water resources per capita in those regions is shown in Table 4.4. Nine countries, Brazil, Russian Federation, Canada, Indonesia, China, Colombia, United States of America, Peru and India, have in their internal water resources 60% of the world's natural freshwater (FAO 2003, 2015; Barnett and Pennell 2004; United Kingdom Environment Agency 2010). Arid areas and some islands, on the other hand, are some of the world's water poor countries. These include Israel, Jordon, Libya Cape Verde, Qatar, Malta, Gaza Strip, Bahrain and Kuwait. Thirty-three countries depend upon other countries for over 50% of their renewable water resources. These are Argentina, Azerbaijan, Bahrain, Bangladesh, Benin, Bolivia, Botswana, Cambodia, Chad, Congo, Djibouti, Egypt, Eritrea, Gambia, Iraq, Israel, Kuwait, Latvia, Mauritania, Mozambique, Namibia, Netherlands, Niger, Pakistan, Paraguay, Portugal, Moldova, Romania, Senegal, Somalia, Sudan, Syria, Turkmenistan, Ukraine, Uruguay, Uzbekistan, Viet Nam and Yugoslavia (FAO Aquastat). Only part of the natural freshwater resource can be utilised, especially where there are difficult considerations of water quality including salinisation. There may also be technical difficulties in accessing the water resource and these, together with the economic status of the country, can restrict access. This can happen even in the Mediterranean region (see Table 4.4).

Basin management is essential although not always adequately practiced. Any sustainable management of freshwater must be based upon a full understanding of its spatial and temporal distribution, especially if scarcity, which increasingly could become a problem in some locations, is to be avoided. There is also an additional problem to water availability, which is the possibility of some of the changes in the hydrological cycle causing geological hazards. Some examples of these are given later. There is increasing evidence from many river basins that significant reductions are occurring in streamflow (IPCC 2007). Unfortunately, some of the river basins with the greatest reductions include those with the highest population densities and hence the highest demand. Some river basins in higher latitudes may have increases in streamflow in the future, but these are not usually those with high population densities. Over half of all potable water used is abstracted from rivers either directly or via reservoirs. Any changes in discharge from rivers is likely to have a major impact on human and ecosystem use. Scarcity was not perceived as a problem in developed countries as they became industrialised, but in many developing countries water availability, especially those in the northern hemisphere, has been either a seasonal or continuous problem. Even in some developed countries, water quantity and availability has become a major issue. Water, including freshwater, is in a state of flux undergoing movement from one physical state to another over different timescales.

Water resources both regionally and locally can change for many reasons. There are a range of geophysical factors, including

- changes due to natural climate variability
- climate change causing increased floods and droughts
- increased evapotranspiration due to increased temperatures
- changes in seasonal timing of events, e.g. snowmelt
- changes in flows from glaciers due to their retreating
- groundwater depletion and changes in soil moisture

In addition to these factors, the amount of water available also depends upon how much we abstract from the resource and how efficiently we use it. Some of the socio-economic factors include

- population growth
- economic development
- changes in diet
- hydro engineering
- efficiency of governance, political environment and culture

Because of these multiple factors, it is not possible to accurately predict future water behaviour and use based upon historical data. An understanding of water as a compound and the global geo-physical environment is needed.

Water has several special properties shown by only a few other substances. It can exist as either a solid (ice), a liquid or a gas (vapour). It is a very good, naturally occurring solvent that allows a vast range of substances to be carried in solution, some in large amounts and some in very small amounts. This availability of a large range of elements and compounds in solution is vital to living organisms. Water has a number of physical properties that affect its behaviour in the environment and greatly determine how organisms behave in relation to the aquatic environment. Freshwater is at its densest at 4 °C. Water at this temperature will sink to the bottom of a lake or pond, whereas colder water or ice, being less dense, will float on top. This means that although the water body may freeze over, there may, unless conditions are extremely cold, be liquid water in the lower parts allowing organisms such as fish and many insects to survive. As water increases in temperature above 4 °C, it expands. Although this expansion is only small, from 4 to 10 °C, it is less than 0.03%, and it is an important factor in sea level rise caused by global climate change. The IPCC (2007) estimates that about half of the 0.44 m increase in sea level during the twenty-first century will be due to thermal expansion. It should be noted that salt water does not behave like freshwater because of its high salt content, so its density is not at its maximum at 4 °C. It behaves more like a normal liquid (Berner and Berner 1987). The specific heat of water, i.e. the amount of energy required to raise the temperature of 1 kg of water by 1 °C, is more than for any liquid other than ammonia at 4200 J. The same amount of energy is released when cooling water. Water provides a very stable thermal environment for organisms living there. Although the temperature of the air above a lake may fluctuate widely from day to night or with daily weather changes, the water temperature, in the short term, will change very little except on a seasonal basis. The effect of this high specific heat is that water takes a long time to both heat up and cool down. Water also has one of the highest latent heat of vaporisations that is known. This means that it takes a lot of energy (2.3 million Joules) to evaporate 1 kg of water without any change of its temperature. Water vapour is the most mobile form of water and may travel great distances before condensing into water droplets in, for example, a cloud. This releases some of the energy into the atmosphere, so in this way water vapour transports energy from one part of the planet to another, e.g. from the tropics to higher and lower latitudes. Water as a low thermal conductivity, which means that warm or hot water does not pass its heat to cool water very quickly. The result is that waters of different temperatures do no mix very easily unless physical mixing is applied. Under a relatively calm condition, waters may become thermally stratified as their different temperatures result in different densities that are quite stable (see idealised diagrams of a thermally stratified lake in Figure 4.1). Water has a high surface tension that allows insects, such as pond skaters, to skim across the surface without sinking in. Adding

Figure 4.1 A thermally stratified lake showing the main layers and a typical summer temperature depth profile. E, Epilimnion; T, Thermocline; H, Hypolimnion.

detergents to the water, as for example may occur when a wastewater is discharged to a lake or river, reduces the surface tension so that the water skater (an insect) would not be held up at the surface. Water, as noted previously, is an excellent universal solvent so that many compounds will dissolve in water. This property enables minerals to be eroded and dissolved from geological deposits by leaching and then distributed widely by the flowing water. The minerals may then also be made available for uptake by living organisms. Water is vital to ecosystems.

The Driving Forces Behind the Global Climate

The mobility of water, especially on the surface of this planet, is a major factor determining its abundance or scarcity. The temporal and geographical variations in the hydrological cycle are driven by global phenomena originating in the oceans and the atmosphere. These include the jet stream in the atmosphere, the interactions between the atmosphere and oceans exchanges of heat and the circulation of waters in the ocean currents. A general picture of the main ocean currents is shown in Figure 4.2. These currents transport cold water to warm areas, e.g. the Benguela current off the south west coast of Africa, or warm water to cooler regions, such as the Gulf Stream from the Caribbean to north-west Europe. The main general ocean circulations are the El Nino Southern Oscillation, the North Atlantic Oscillation, the Atlantic Multidecadal Oscillation and the Pacific Decadal Oscillation. A recent understanding of these drivers has allowed more efficient prediction and planning of resources. An understanding of these major drivers of climate can give not only an understanding of current conditions but also greatly enhance the ability to predict

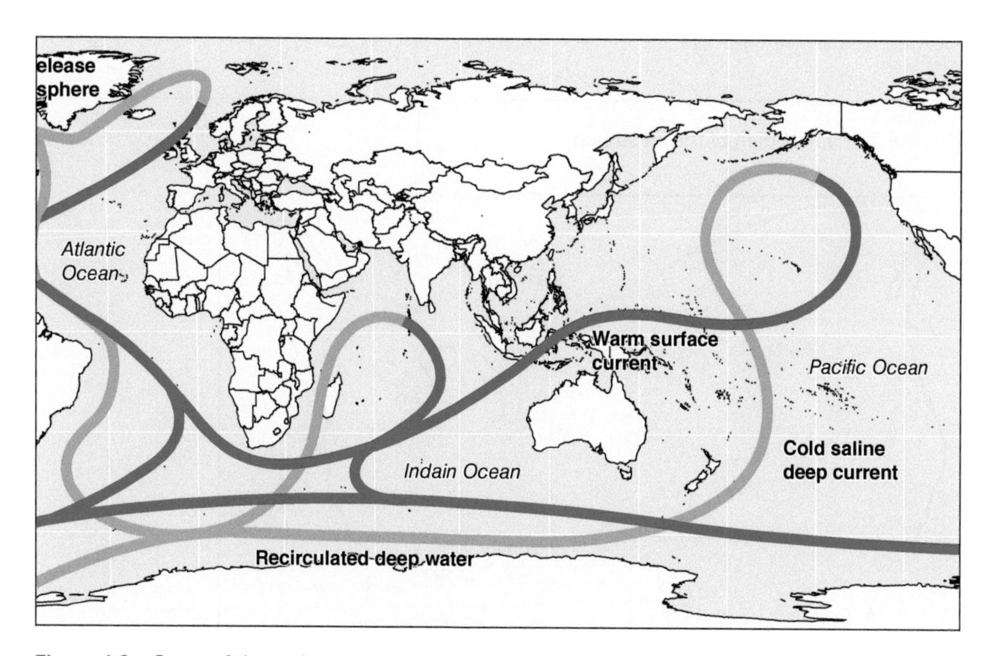

Figure 4.2 Some of the main ocean currents and their path around the globe.

future resources. Although these natural cycles originate over the oceans of the world, they do have serious effects on the land-based hydrological cycle. This in turn affects the economics of many countries (Brunner 1998; Tol 2009). A further factor that has received attention in the literature and media recently has been the jet stream. The jet stream is a string of very strong winds circulating the globe at about 10–15k above the land surface just below the tropopause. There are two main jet streams, the polar and the subtropical Jets (see Figures 4.3 and 4.4). The wind speeds can reach up to 200 mph and move weather systems from location to location. The jet streams affect North America in a similar way with the polar jet usually impacting Canada and the Great Lakes in the winter and in the summer sweeping down from around Montana to the far south (see Figure 4.4). The subtropical jet affects the south-eastern United States, especially during winters when there is a strong El Nino in the Pacific. The jet stream arises because of temperature differences between tropical air masses and polar air masses. The jet stream does not follow a

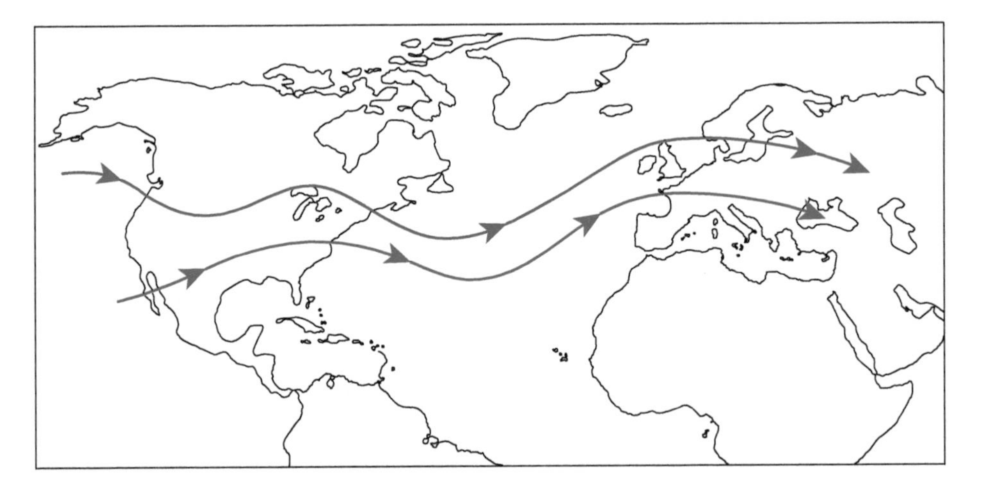

Figure 4.3 Typical path of the jet stream.

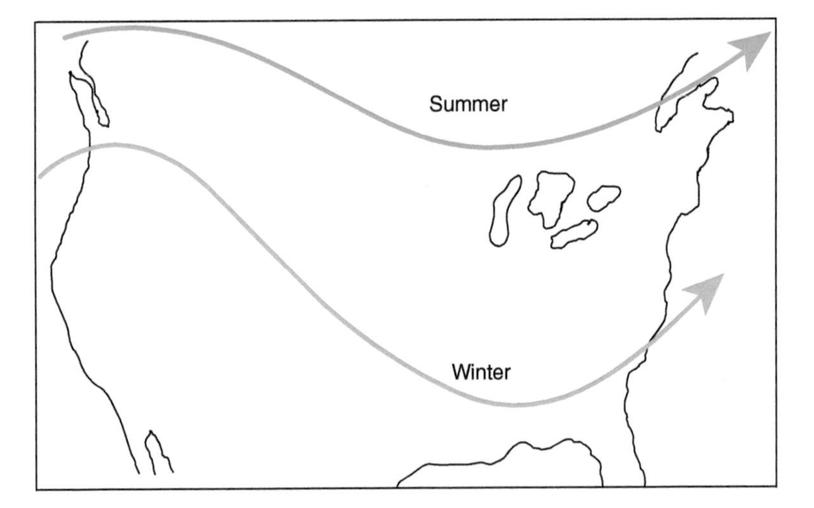

Figure 4.4 The jet stream over North America.

pre-defined pathway but has ripples or wobbles that may cause deepening of Atlantic depressions or high-pressure zones that then affect Europe. They can affect precipitation and temperature and thus affect water resources over the land masses.

In general environmental terms, water can be thought of as either green or blue water. Green water refers to rainwater that is stored in the soil which is either then incorporated into plants and other organisms or evaporates into the atmosphere. Blue water is renewable surface water run-off and water recharging groundwater. Green water is the main source of water used by ecosystems and rainfed agriculture that produces about 60% of the world's food. Blue water is the main source for human withdrawals and is the main type of water targeted in water management schemes. Shiklomanov (2000) estimated that there was about 42 780 km^3 of blue water available and of this 3800 km^3 was, in 1995, withdrawn for human use and 2100 km^3 was classed as consumed. Consumed in this context means that it has been incorporated into products or organisms or has evaporated so that it becomes unavailable to other users (Cosgrove and Rijsberman 2000). Different authors have slightly used different methods of computing resource figures, but there is general agreement. FAO (Aquastat), averaging a number of sources, indicates a total resource of 43 764.3 km^3/year, which is between the figure of Shiklomanov and 44 540 km^3/year indicated by Gleick (2000). Of course, only a proportion of the total resource can be utilised by humans. There are also inequalities between nations and regions regarding not only the total amount of water resource but also how much can be utilised. These inequalities are magnified when factors such as economic and technical development, and basin management and political regimes are taken into account. The effort required also depends on the regularity of recharge and the accessibility of the particular resource. Countries such as Israel, the Gaza Strip and the Palestine Authority that utilise mainly groundwater have a ratio close to 100% for use of exploitable water to total renewable water resource, whereas in countries such as Turkey and Greece that rely quite heavily on surface waters have a ratio of less than 70% (FAO 2015).

Humans and natural ecosystems all require water to survive so that any strategy to manage water must include the needs of both, and the amount of water on this planet is finite and that freshwater is only a small proportion of the pressure on freshwater resources will grow. The world oceans hold by far the largest proportion of water on the planet, and because of its large area, it takes up almost twice as much solar energy as the land, a large proportion of which is used in evaporating over 500 000 km^3/year which provides global water exchange (Shiklomanov and Rodda 2003). Of course, the ocean water is salty so has only recently, and to a limited scale, been available for use by humans by using modern technologies (see Chapter 7). The most likely consequence of climate change is an increase in average global temperatures. As a result of this, there will be increased evaporation and transpiration. This is likely to impact on the amount of surface run-off. The largest store of freshwater is in the planet's glaciers and permanent ice fields. The largest ice fields are in Antarctica where 21 600 000 km^3 of water is stored as compared with Greenland that has 2.9 km^3 stored water. With the growing world population, water resource management is becoming critical. In order to achieve a sustainable management strategy, we need to know how much there is in different components of the hydrological cycle and how it behaves. There are uncertainties about the amounts of water in these different segments of the hydrological cycle, but it is possible to make reasonable estimates of both quantities and, what is also important in resource management, residence times in each sector.

Evaporation and Precipitation

The global water cycle is largely driven by energy from the sun. This energy is the main driving force that transfers water from liquid form in oceans, lakes, etc. into the atmosphere. Most evaporation occurs over the oceans and is estimated to amount to over $430 \times 10^3 \text{km}^3$/year. The amount evaporated over land is somewhat less and is estimated at about $65 \times 10^3 \text{km}^3$/year and includes transpiration losses to the atmosphere from vegetation. This land-based transfer is termed evapotranspiration. Not all of the water associated with vegetation has been taken up by the plants. Some precipitation is intercepted by leaves and lays on their surfaces from where it evaporates directly. Transpiration is the passage of water from the soil or substrate, through the plant tissues to aerial parts and from there it is evaporated into the atmosphere. Most of the water that evaporates from the oceans falls back as rain onto their surface, but once in the atmosphere water vapour represents the most mobile phase of the hydrological cycle and can move from over the sea to land. Of the water evaporated from the oceans, approximately 10% falls as precipitation on land. Amounts of precipitation vary across the globe. In Mediterranean climates most precipitation occurs in the winter, whereas in countries with a monsoon climate most occurs in the summer. Table 4.5 shows the average precipitation in the main regions of the world.

Table 4.5 Average precipitation in the continental regions.

Region	Precipitation depth (mm/year)	Precipitation volume (km^3/year)
Africa – Northern	96	550
Sub-Saharan	3304	16048
Total	3400	16598
Americas – Northern	1383	13881
Central and Caribbean	5279	1515
Southern	6022	29012
Total	12684	44048
Asia – Middle East	1409	1422
Central Asia	273	1271
Southern and Eastern Asia	6316	24162
Total	1409	26855
Europe – Western and Central	3266	4099
Eastern	1045	8462
Total	4311	12561
Oceania – Australia and N. Zealand	4598	574
Pacific Islands	2055	4733
Total	2629	9331

Source: Adapted from FAO (2014).

Glacier and Snowmelt

Glaciers and snow are major storage components of freshwater. High mountain chains can be regarded as 'water towers' because of their contribution to river flows from ice and snowmelt including several major rivers (see Figure 4.5). The growth and shrinkage of ice sheets and glaciers over many thousands of years have profoundly affected the global water cycle. If glaciers continue to shrink and even disappear, the meltwater streams, both large and small, could potentially dry up completely during hot summers. This could impact both humans and ecosystems. Fifteen to twenty percent of the world's population live in river basins that are supplied by glacier or snowmelt water. Both glaciers and groundwater can provide long-term storage of freshwater. The water in streams and rivers originating from mountainous areas come from three main sources, rainfall, snowmelt and glacier melt. Snow cover in mountain regions has provided, through spring snowmelt, much of the stream-flow water. Large areas of the world such as the arid American West and Central Asia largely depend on snowmelt to supply cities and agriculture. These sources of water have always been seasonal but fairly regular. When the global climate warms, it is likely that less winter precipitation will fall as snow and that which falls will melt earlier in the spring (Barnett et al. 2005). Changes in the amount of precipitation will also potentially affect the amount of snow accumulation which typically occurs towards the end of winter. Snowmelt and subsequent run-off will tend to be earlier which, with less snow, will be reduced as will summer rainfall if temperatures rise. Even if global temperature rises are kept to the lower end of suggested targets, e.g. 1.5 °C, the impact on snowmelt and subsequent freshwater resources will be large (Barnett and Pennell 2004). Water availability problems caused by a lack of winter snow and snowmelt are already being felt in some regions (see Gleick for

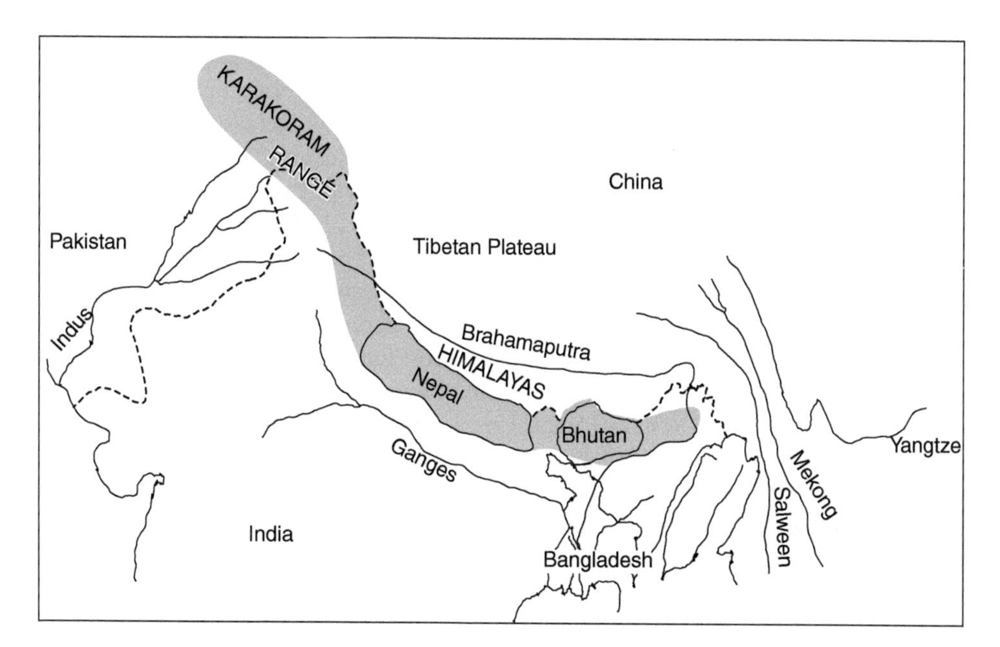

Figure 4.5 The Himalayan water towers of the Karakoram range of mountains.

California), but in addition changes in glaciers and melting permafrost can cause geohazards. The largest impact predicted on the hydrological cycle from the point of view of snow cover is in the mid-latitudes to low latitudes (Nijssen et al. 2001). In the western United States, reduced snowfall and spring snowmelt will reduce reservoir volumes potentially causing stress to all users, including hydroelectric schemes (Barnett and Pennell 2004). Problems will also arise for regions dependent upon glacial meltwaters for their spring and summer supplies. This is because in a warmer world snowfall will be reduced and water lost from the glacier will not be replaced (Barnett et al. 2005). Glaciers, in recent times, reached their maximum sizes towards the end of the Little Ice Age (the early fourteenth to mid-nineteenth century). Much of the water stored in glaciers can be regarded as 'fossil' water as it has originated over many hundreds or even thousands of years and cannot be replaced in the short term. Glacier meltwater tends to reach a maximum in the summer, unlike the spring peak in snowmelt water. Globally, most glaciers are now shrinking although there are some exceptions, e.g. in the Karakoram. Generally, a warmer atmosphere leads to less snowfall and greater melting that results in the glacier retreating. UNEP (2007) reported that 30 glaciers used for reference purposes as they had almost continuous measurements made on them for the past six decades showed an annual loss in mass of between 20% and 25%. As glaciers flowing directly into the oceans melt they contribute directly to sea level rise. European glaciers, which are mainly medium sized, are likely to retreat with global temperature rise, whereas larger type glaciers in Alaska and Patagonia are more likely to get thinner in depth before retreating. When inland glaciers melt the water, the discharge can cause hazards such as avalanches and flooding. These effects can be restricted and local or have widespread impacts and do have economic costs estimated to be, on average, in the order of several hundred million US dollars annually with the most devastating disaster killing more than 20 000 people (Kääb et al. 2005). There are a number of potential hazards from glacial and permafrost premature melting. These include breaching of moraine dams, glacial outbursts, displacement waves caused by avalanches, changes in glacial run-off, rock and ice fall avalanches, landslides, thaw settlement and soil heave if permafrost thaws and the destabilisation of frozen slopes leading to landslip and increased flows of debris from permafrost. These are discussed in more detail in Kääb et al. (2006). Glacier and permafrost hazards are a greater problem in high mountain areas. Examples include rock-ice avalanches in Peru (1970) and the problems of debris flows in Columbia (1985). Even if global warming is limited to 1.5 or 2.0 °C, there will still be a considerable effect on glaciers, snowfall and permafrost. The effects are likely to be greater in mountain regions. Any unstable ice and snow fields and glacial lakes could be disrupted. Steep frozen slopes could become unstable. Rockfalls could increase. Permafrost thickness is likely to decrease, and the thickness of the active layer, that layer which thaws in the summer allowing greenhouse gases to be released as well as more plants to grow, would increase. Another major factor is the impact of socio-economic development. In many mountain areas, human activities are increasing and demand for land has meant that these activities are increasingly encroaching upon potentially unstable and hazardous areas. Where this is occurring in developing countries, e.g. the Andes, Central Asia and the Himalayas, the countries do not have the resources to mitigate for these events. They often also lack the capability to monitor potential hazards so problems cannot be dealt with before a disaster occurs. These types of hazard do also occur in developed countries, e.g. the European Alps but here the funds are more likely to be available to build expensive protective structures.

Groundwater

Groundwaters are fed from precipitation and surface waters and so there is considerable interdependency between them. They may, in turn, discharge into surface waters as springs, and the oceans, or evaporate to the atmosphere. They can act as storage for water and also play an important role in freshwater circulation by feeding streams and river. Although it is difficult to precisely estimate the volume of water stored in groundwaters, the amount is quite large but can also vary geographically. Shiklomanov and Rodda (2003) estimate that the volume is approximately 96% of all the earth's unfrozen freshwater, a huge amount. Some aquifers are replenished annually but in others, for example in arid areas, this may only be intermittent or, in human terms of human lifespans, be never. In some aquifers where there is a great depth of water, the upper layers may be recharged annually but the lower ones may be unchanged for very long periods of time. With these and any others where the replacement time is measured in hundreds or even thousands of years, the waters may be regarded as non-renewable. Table 4.6 shows estimates of groundwater resources in different continents.

The distribution of groundwaters is related to the geology of an area as well as the behaviour of the hydrological cycle in that region. Shiklomanov and Rodda (2003) distinguish three zones of groundwater storage and movement, which are as follows:

1) A zone near to the soil surface where water is exchanged actively with the surface and atmospheric components of the hydrosphere. The quality of the water is closely related to the geology of the rocks and the overlying soils.
2) Below the first zone is one that has less active water exchange. It is only affected by large rivers with deep channels. Some aquifers of this type lie below the sea and may discharge into the sea. Even though there is less water movement in these aquifers, they tend to have only weakly mineralised water.
3) These deeper aquifers lie below the second one and may be down to a depth of 2000 m. Only the upper part of these aquifers contain freshwater, and the lower parts tend to be saline.

Korzun (1974) estimated that there was 23.4 million km^3 water stored in these three zones. Examples of countries with the largest groundwater abstractions are shown in Table 4.7.

Table 4.6 Groundwater resources by continent.

Continent	Total groundwater resource (km^3 × 10^6)
Europe	1.6
Asia	7.8
Africa	5.5
North America	4.3
South America	3.0
Australia and Oceania	1.2

Source: Adapted from Shiklomanov and Rodda (2003).

Table 4.7 Major aquifers and current extraction rates.

Country	Aquifer	Exploitable reserve (mm^3)	Extraction rate (mm^3/year)
Egypt, Libya, Chad,	Nubian Sandstone	6 500 000	1600
Algeria, Libya, Tunisia	North Western Sahara	1 280 000	2500
Botswana	Central Kalahari Karroo Sandstones	86 000	Low
Australia	Great Artesian Basin	170 000	600

A percentage of the abstracted water will be used for irrigation as well as other purposes. By far, the largest abstractors from wells are India, China and the United States. Interestingly, Saudi Arabia and Syria are both in the top 15. The amount of water abstracted for drinking supplies globally has increased substantially for urban populations (+63.1%) but decreased for rural ones (−4.0%). Groundwaters are largely abstracted for agricultural use. Very few countries use groundwaters predominantly for industry, the main one being Japan. The growth of cities globally, especially megacities, has led to the growth of groundwater use for human domestic purposes. Following current trends, this sector is likely to increase in the future.

Groundwaters are largely unseen, but they play a vital role in the biosphere. Without them, the surface of the earth would be quite different. They support many wetland habitats, streams and many deeper rooted trees.

Although surface waters have been exploited for many thousands of years, groundwaters have only been extensively utilised for a little over the past 100 years. During this past century, they have been exploited in many countries and now play an important role in meeting the needs of agriculture, municipal needs and ecosystems (Llamas and Martinez-Santos 2005).

The United States of America has one of the world's largest aquifers situated beneath the high plains in mid-west of the country. This is the Ogallala or high plains aquifer. It spans parts of eight states, South Dakota, Nebraska, Wyoming, Colorado, Kansas, Oklahoma, New Mexico and Texas. It is a relatively shallow water aquifer in unconsolidated, poorly sorted clay, silt, sand and gravel. The pores or spaces between these particles are filled with water. It was formed about 3.8 million years ago (in the Tertiary period) from streams that flowed from the Rocky Mountains in an easterly direction (http://www.waterencyclopedia.com/Oc-Po/Ogallala-Aquifer.html). It is classed as an unconfined aquifer and is the largest in the United States of America. Its location and area covered are given in Figure 4.6. It covers almost 175 000 square miles (453 250 km^2) and contains about 4000 km^3 of water (Peck 2007). The thickness of the aquifer can vary from 30 cm (1 ft) at its edges to over 400 m (1300 ft) in the central part of Nebraska, but most is in the range 15–90 m (50–300 ft). The saturated thickness varies from 60–300 m (200–1000 ft). The greatest saturated thickness as well as the largest area is in Nebraska, about two-thirds of the aquifer's water is found in that state. Most of the water pumped from the Ogallala aquifer is used for irrigation, and without it agriculture in the arid high plains and Texas Panhandle could not exist and the regional population would be restricted. The high plains

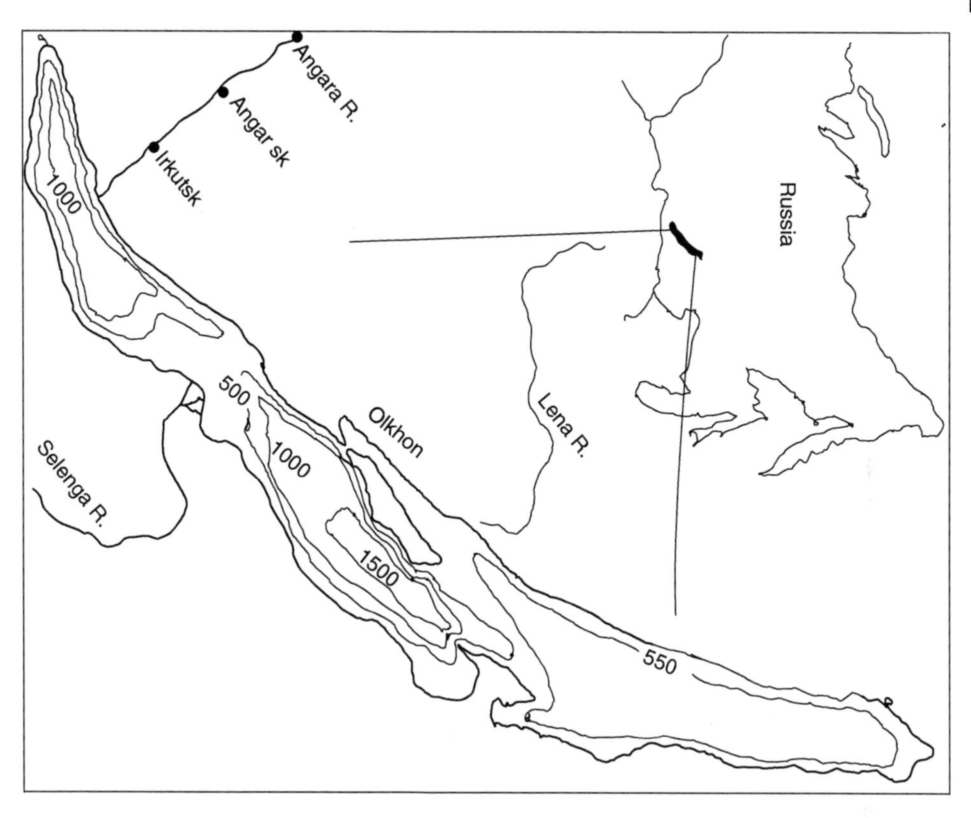

Figure 4.6 Lake Baikal, the largest body of freshwater. Also shown are the depth contours in metres.

area is one of the leading irrigation areas in the western hemisphere, but unfortunately this has resulted in groundwater mining where more is being withdrawn that is being replaced by natural recharging. This has resulted in declines in water level and volume in some areas and also pollution. Declines in levels of over 30 m (100 ft) have occurred in parts of Kansas, New Mexico, Oklahoma and Texas. Smaller declines have occurred in South Dakota, Nebraska, parts of Kansas and Colorado. The expansion in agriculture and hence demand for water accelerated greatly after the Second World War with the introduction of centre-pivot sprinklers. Currently, more water use efficiency is being introduced, which is slowing down the rate of depletion. The US Geological Survey in collaboration with numerous state and federal agencies monitored water levels in over 7000 wells in 1988 to chart the changes in water levels. This was repeated in 2005 when 9068 wells were monitored and in 2007 when 9340 were monitored. Overall, the results showed that over this period levels rose by 84 ft in Nebraska but declined by 234 ft in Texas. There was an overall decline in level for the whole aquifer for the entire period of 14 ft (McGuire 2009). It is generally agreed by the high plains states that the future economy of the region depends upon the future sustainability of the Ogallala aquifer. It needs to be managed for water sustainability and prevention of its contamination.

Unfortunately, calculations of how long the water in the aquifer will last which are needed in order to regulate pumping withdrawal rates cannot just be based upon recharge amounts vs. pumping withdrawals. Devlin and Sophocleous (2005) pointed out that several other factors were important. The amount of recharge of an aquifers affected not only by simple climate in the amount of rain falling on an area and its geology but also the land use. As much aquifer water is used in irrigation, some of the irrigation water will percolate through the soil back to the aquifer adding to the recharge. There will also be some induced recharge from adjacent surface waters such as lakes and streams. If an aquifer is overpumped, then its rate of discharge will naturally decrease so water levels in wells will also decrease. Few, if any, aquifer systems are truly fixed directional systems where water only enters through prescribed pathways and exits through other defined pathways. This would be like a kitchen sink with water only coming in through the tap and only leaving through the drain hole (Devlin and Sophocleous 2005). A knowledge of recharge rates is important in managing the water in an aquifer but to manage the system sustainably a broader concept needs to be understood. Sustainability depends upon an entire, and sometimes complex, system including forests, fish and aquatic ecosystems, wetlands and all plants and animals that need water in that area (Sophocleous 2000; Sphhocleous and Devlin 2004). It is always important to take a holistic view if a truly sustainable outcome is to be achieved. In the case of the Ogallala, agriculture dominates the area, and water is essential for current and future development. The Ogallala Aquifer Summit report of 2018 recognised that there were both similar and in some cases varying strategies for the management of their part of the aquifer depending upon their individual priorities. Discussions with all states involved recognised that in some areas the aquifer was being overabstracted. There was general agreement that more collaboration between the states could benefit the Ogallala states. Several suggestions were agreed upon. These included looking at a wider range of water-use efficient crops and to provide more funding for this and other conservation measures. They also advocated the use of social media and other outreach efforts to increase overall visibility and understanding not only of the value of the water being produced but also that everyone should participate in water efficiency actions. There should also be a shift in research, policy, education and outreach to maximise yields per drop of water used. This would not only save water but cut production costs. There was also a need for better data collection. There must be better interstate communication and an effort to 'normalize' conservation practices and technology across all eight states. It is also important to establish a region-wide vocabulary to aid communication within states to be adopted by all stakeholders, including crop advisors, academics, local and Federal agencies, multinational companies, landowners and any others. There needs to be an increase in education and dissemination of information on the latest irrigation technologies as well as providing on-farm demonstrations. Many, if not all, of these things can only be achieved by multistate collaboration.

Currently, approximately 22% of the global population rely upon groundwater for their water supply. About 10% of that consumption is unsustainable, i.e. it is being used faster than it is being replaced. Groundwaters are being increasingly used to supply human needs in many countries, e.g. Jamaica, Saudi Arabia, Libya and Lithuania. They supply over three-quarters of the needs for freshwater. The increasing global population and the ever-increasing demand for food has driven this increase in use. The use of groundwater is now essential for domestic and food security for over 1.7 billion people, mainly in Africa and Asia. This increased use of groundwater has been made possible through advances in technology that enabled aquifers to be accessed to a greater degree. This improved access to groundwater has greatly helped farmers in many regions as it can often be obtained on demand and used more efficiently (World Water Assessment Programme 2003). The distribution of groundwaters is very variable globally and largely depends upon the geology of the region (see Vrba and van der Gun 2004). Shallow aquifers where the water table is near to the surface are usually closely integrated with surface water behaviour, whereas deep groundwaters are often linked with historic climate and hydrological regimes. In arid regions, for example, some aquifers were created, mainly during wet Pleistocene periods. Groundwater has also been viewed as being a more reliable source when compared with surface water supplies as it is regarded as being less susceptible to seasonal variations. This gives greater yields and hence better economic returns. In some countries, farmers and communities have made their own investments in groundwater technology, e.g. in Bangladesh. However, when many small farmers and communities exploit an aquifer, this can lead to management problems as abstraction rates need to be regulated to prevent overabstraction that is more difficult with multiple wells. Groundwater pumping does involve direct costs for the user so there is an incentive to use water as efficiently as possible. This is good except where energy costs are subsidised by the state when overuse can occur and accelerated depletion can result, e.g. in India and Pakistan. The water quality in groundwaters varies with the geology of the region as well as changes caused by human influences (see section on water quality). Parameters such as total dissolved solids and chemical composition vary not only regionally but also with depth and along the flow path of the water within the aquifer. Although many chemicals in the water may be harmless, some may be present in higher than acceptable concentrations. Examples include iron, manganese, fluoride and arsenic (see Table 4.8).

Table 4.8 Potential chemical problems and concentrations in groundwaters from various countries.

Country and origin	Chemical	Concentration (mg/l)	Drinking water standard (mg/l)
Bangladesh and arsenic-rich minerals and rocks and India	Arsenic	1.1–5.0	0.01
Czech Republic minerals and rocks and India	Fluoride	4.5–10.0	1.5
China sea water intrusion	Chloride	8818	250
Australia soil salinization	Chloride	4718	250
USA salt-containing strata	Sodium	121 000	200
USA marine argillaceous rocks	Aluminium	64	0.2
China and India acid mine drainage	pH	2–4	6.5–8.5

Source: Adapted from Vrba and van der Gun (2004).

It should be noted that these problems occur in other aquifers and countries than those cited above, but these are some of the main ones.

Overuse can lower the water table to such an extent that the wells to abstract have to be much deeper and pumping costs much higher, which could take their use beyond the reach of poorer farmers and communities. Management of aquifers may not always be straightforward. Groundwaters often cross national or administrative boundaries, so any control over abstraction rates requires international agreement. Traditionally, groundwaters have a legal status that regards them as part of land property so that the competing interests of users wishing to withdraw water from a common property may clash. The use of the land above an aquifer can, if not regulated, have a considerable and deleterious effect on water quality. The use of pesticides, herbicides and fertilisers can all contaminate groundwater as they percolate through the soil. There has been a dramatic increase in the use of groundwater during the second half of the twentieth century, and by far the greatest proportion of the water abstracted is used in agriculture for irrigation. The growth in the world population has inevitably meant that there has been a corresponding increase in the demand for food, irrespective of possible dietary changes due to improvements in economic status. Table 4.9 shows the proportion of groundwater used for irrigation, industry and domestic supplies in some of the nations with the largest rates of groundwater abstraction.

The case of transboundary aquifers (TBAs) can be quite problematical. Apart from agreeing how much can be abstracted by each country, it could be that the aquifer is recharged in one country but largely abstracted in another. Compared to surface waters, groundwaters have received relatively little attention both from the point of view of legislation and policy development. There are many different types of aquifers depending largely upon the geology containing them (for more details, see Cech 2010). The International Groundwater Resources Centre

Table 4.9 The proportion of groundwater abstraction used for irrigation, industry and domestic supplies in 2010 for selected countries.

Country	Total extracted groundwater (km^3/year)	Groundwater used in different sectors (%)		
		Irrigation	Domestic use	Industry
India	251.00	89	9	2
China	111.95	54	20	26
USA	111.70	71	23	6
Pakistan	64.82	94	6	0
Bangladesh	30.21	86	13	1
Mexico	29.45	72	22	6
Saudi Arabia	24.24	92	5	3
Turkey	13.22	60	32	8
Russia	11.62	3	79	18
Italy	10.40	67	23	10

Source: Adapted from Margat and van der Gun (2013).

(IGRAC) identified 445 TBAs in 2013, but the number may well increase as international technology identifying the extent of aquifers improves. At present, many of the TBA boundaries are not properly delineated, especially in Africa and Asia. Wada and Heinrich (2013) report that many major TBAs are being overexploited and that their current use is unsustainable. With these aquifers, the only way to sustain their use into the future is to reduce abstraction. This would be difficult in many cases, such as in India and Pakistan, as the aquifers are in some regions of scarce surface water resources and water is needed for irrigation and food production. In these regions, a more efficient strategy for irrigation and surface water storage such as more reservoirs must be implemented. In other regions, increases in surface water storage capacity to meet demand are not feasible as the climate is too arid and there is not enough precipitation to make up the deficit, e.g. in the Arabian peninsular. Rijsbermann (2006) suggests that when the annual renewable water resources fall below 500 m^3/capita then water scarcity occurs. Some of the more arid regions have TBA recharges of less than 50 m^3, which also have low precipitation. The use of groundwater or surface water irrigation can be reduced if improvements in irrigation efficiency are introduced. Although some TBAs may be abstracted relatively evenly by the sharing countries, this is not always so. For example, China is by for the main user of the Tacheng basin aquifer that is shared with Kazakhstan. The United States of America is the main user of the Cuenco Baja del Rio Colorado aquifer shared with Mexico. Groundwater abstraction as increased greatly during the past half century, and to supplement this use, more fossil groundwater is, when available, being used.

The use of aquifers, both national and transboundary, has increased dramatically over the past 50 years. This has placed many aquifers in many parts of the world under great pressure as recharge rates do not match the abstraction rates. Much of the increase in use has been by humans and the need for irrigation water. This will only increase in the future as the global population grows and the demand for food increases. This is placing great demands and stress on many aquifers, including some of the worlds' largest aquifers. Two-thirds of the largest aquifers have had their reserves depleted in the past two decades, some to a point where water availability is jeopardised. Excessive groundwater pumping can also result in streams and even rivers to dry up. Proper management of all aquifers requires knowledge of the size of the resource, and unfortunately this is seldom accurately known. Equally often unknown are the flows into and out of the aquifers. There are both ecosystem and human pressures causing depletion. Examples include irrigation and growth of large cities in the northern Indian Ganges Basin, irrigated agriculture in the Arabian Aquifer underlying Saudi Arabia and abstraction for the mining industry in the Canning Basin of northwest Australia (Walton 2015). Early estimates (1969 and 1974) assumed that soil characteristics and water movement in aquifers were the same when making estimates of volume. In more recent estimates, taking into account local characteristics (Walton 2015), volumes were found to be much less by between 10 and 1000 times. It is important in aquifer management to have as accurate as possible estimates of their characteristics.

Surface Freshwater Resources

There is considerable variation in the amount of water available in different regions and within those regions how much there is available per person. Table 4.10 shows the estimated distribution based upon data from 2000 and 2003. These differences are

Table 4.10 Surface waters available internally in selected countries.

Country renewable (10^9 m^3/year)	Internal water produced (10^9 m^3/year)	Surface water entering country (10^9 m^3/year)	Total water
Africa			
Egypt	0.5	84.0	58.3
Ethiopia	1201.0	0	1221.0
Ghana	29.0	25.9	56.2
Libya	0.2	0	0.7
South Africa	43.0	6.6	51.4
South Sudan	26.0	50.0	49.5
Sudan	2.0	99.3	37.8
Americas			
Brazil	5661.0	2986.0	8647.0
Canada	29 021.0	2840.0	52.0
USA	2662.0	251.0	3069.0
Asia			
Afghanistan	37.5	10.0	65.3
Bangladesh	83.9	11 221.0	12 271.0
China	2712.0	17.2	28 401.0
India	14 041.0	635.2	19 111.0
Israel	0.25	0.3	1.8
Japan	420.0	0	430.0
Occup. Palestinian	0.15	0.8	0.072
Territory Pakistan	47.4	265.1	246.8
Saudi Arabia	2.2	0	2.4
Europe			
France	198.0	11.0	2111.0
Germany	106.3	47.0	1541.0
Hungary	6.0	98.0	1041.0
Russian Federation	4037.0	203.4	45 251.0
UK	144.2	2.0	1471.0
Oceania			
Australia	440.0	0	492.0

Source: Adapted from FAO (2014).

emphasised when considering individual countries within regions. Even adjacent countries can have vastly different amounts of freshwater produced within their borders. Table 4.10 presents examples of adjacent countries within regions with quite different surface water resources.

About one-third of the precipitation that falls onto land drains to the sea. This run-off forms the rivers and streams that have been exploited by both humans and wildlife. Many rivers have large seasonal variations in flow. In arid regions, the variations may be very large, for example in Namibia the main rivers are ephemeral and completely dry up at some times of the year.

Where countries traditionally relied upon surface waters for domestic supplies and irrigation, those in more arid and semi-arid regions are likely to become, if not already, water scarce. As noted above, the surface water supply has increasingly been supplemented by groundwater which, although it alleviates the scarcity in the short term, is not necessarily a long-term solution. The main exploitable surface freshwaters in many countries are rivers. The total flow in rivers is globally several times greater than the amount abstracted for human use. Rivers have been used for thousands of years for all aspects of human activity, including drinking, agriculture, navigation and power. Unfortunately, the distribution of rivers does not always correspond to the population density, and hence local requirements can often exceed available river flow. Another factor is that river flows are variable either with season or with extreme climatic events when floods or droughts may occur. Often most of the floodwater cannot be captured and considered as a useful future resource. A further problem of these variations can be catastrophic losses to habitations, industry and agriculture. Some of the major rivers are shown in Table 4.11.

The length of a river is not necessarily the most important feature from both its use for humans and for wildlife. It is not only the length of a river or the area of land that it drains which determines it flow and the amount of water carried. The average total river flows

Table 4.11 Examples of some of the major rivers by continent.

	Catchment area ($km^2 \times 10^3$)	Length (km)	Average discharge (m^3/s)
Africa			
Congo	3680	4370	41 200
Nile	2870	6670	3750
Niger	2090	4160	5589
Zambesi	1330	2660	7070
Orange	1020	1860	[a]
Asia			
Ob	2990	3650	12 475
Yenisey	2580	3490	19 600
Lena	2490	4410	16 871
Amur	1855	2820	11 400
Yangtze	1808	6300	30 166
Ganges[b]	1746	5425	38 129 (16 648)
Amu Darya	1100	1415	2525
Indus	960	3180	6600
Mekong	795	4500	14 800

(Continued)

Table 4.11 (Continued)

	Catchment area ($km^2 \times 10^3$)	Length (km)	Average discharge (m^3/s)
Shatt al Arab[c]	750	2900	1750
Syr Darya	440	2210	1180
Limpopo	440	1600	170
Salween	325	2820	4876
North America			
Mississippi	2980	3780	16782
Mackenzie	1787	5472	10300
St. Lawrence	1026	3057	16800
Columbia	668	1953	7500
Colorado	637	2333	637
Fraser	233	1370	3475
South America			
Amazon	6915	6280	209000
La Plata	3100	4700	22000
Orinoco	1000	2740	36000
Sao Francisco	600	2800	2943
Rio Negro	130	1000	28400
Europe			
Volga	1380	3700	8060
Danube	817	2850	7130
Dnieper	504	2200	1670
Don	422	1870	935
Neva	281	74	2445
Rhine	252	1320	2900
Vistula	198	1092	1080
Loire	120	1010	835
Rhone	98	812	1710
Australia			
Murray	1072	3490	
Cooper Creek	285	2000	73
Sepik	81	1120	3804
Fly	64	1040	6000

A more comprehensive list can be found in Shiklomanov and Rodda (2003). Other publications consulted were
Rivers of the Earth (http://home.comcast.net/-igpl/Rivers.html).
[a] No precise figure was available at the time of writing.
[b] This includes the first figure that is for the Ganges only, and the second includes the Brahmaputra and Meghna rivers.
[c] This includes the Tigris and Euphrates rivers.
Source: Gupta (2007), Kumar et al. (2005), Benke and Cushing (2005) and Nezhihovsky (1981).

globally greatly exceeds the amount of water currently being withdrawn for human use. Even allowing for flood flows, as long as some holding measures are in place, e.g. dams, there is still an ample amount. The distribution of run-off and large rivers is very uneven across the continents. The regions of highest run-off occur in the Amazon basin, parts of Indonesia and the southern part of the Himalayas down through Malaysia. Other similar regions are northern Siberia, parts of central Asia, central and eastern Africa (the Congo area), eastern and a narrow north-eastern coastal strip of North America. Areas of least run-off are in subtropical regions, parts of central Asia, most of Australia and western North America. There are seasonal variations in all of these areas. In marginal areas where there is seasonal scarcity, water management and conservation are essential. Although the size of rivers has frequently been measured by their length from the point of view of water resources, their average discharge is usually more relevant (see Table 4.11 for some discharge data). This gives an indication of the amount of water available, but note that seasonal variations must also be taken into account as these are often quite large. The Amazon is one of the longest rivers together with the Yangtze, but the former has approximately six times the discharge as the latter. The Ganges is a little over half the length of the Mekong but has nearly twice the discharge volume. Some relatively short rivers such as the Danube and the Columbia rivers have a relatively high discharge volumes and are very important sources of freshwater in their regions. Rivers have been used since ancient times for activities other than just for drinking and cooking. Apart from irrigation, rivers have been, and still are, used for navigation and as a source of power. Large amounts of water is used for cooling and some for feeding boilers. Water is harnessed in hydroelectricity production. It is also used for washing equipment and products such as coal and mineral ores.

Many major rivers cross national boundaries. Because of this mid to lower basin, countries do not have control of their headwaters. Where possible, treaties and agreements may be in place for the flow and volumes of the water to satisfy both upstream and downstream users. The Amazon, although largely within Brazil, has headwaters in Peru, Columbia and Ecuador. India has control of the Ganges that passes into Bangladesh. The latter country has concerns over developments in the former that would affect the cross boarder flow and thus agricultural and industrial development.

Lakes and Reservoirs

Lakes and reservoirs are also important reserves of freshwater for many countries. Examples of the principal lakes in different countries are provided in Table 4.12 together with the total freshwater lake resource volume by country.

By far, the largest volume of lake freshwater is in Lake Baikal in eastern Siberia (Figure 4.6). Although some of the North American Great Lakes, for example Lake Superior, have a far greater surface area (84 500 km^2 as opposed to Baikal that is 31 500 km^2), they are little under one-quarter its depth. Lakes of various sizes occur on all continents. Most lakes are relatively small, but some are quite large, some with a surface area exceeding 1000 km^2 and many of the largest are in the northern hemisphere. The largest lakes in Europe are mainly in the northern countries, those in Asia are mainly in the former Soviet Union,

Table 4.12 Examples of major lakes and their volume.

Lake	Depth (m)	Volume (km^3)	Country
Ladoga	230	908	Russia
Geneva	310	88.9	Switzerland, France
Victoria	84	2750	Tanzania, Kenya, Uganda
Tanganyika	1471	17 800	Tanzania, Zaire, Zambia, Rwanda, Burundi
Nyasa	706	7725	Malawi, Mozambique, Tanzania
Chad	10–11	72	Chad, Niger, Nigeria
Superior	406	11 600	Canada, USA
Huron	229	3580	Canada, USA
Michigan	281	4680	USA
Great Slave Lake	156	1070	Canada
Erie	64	545	Canada, USA
Titikaka	281	893	Peru, Bolivia
Eyre	20	—	Australia
Baikal	1741	23 000	Russia

Source: Adapted from Shiklomanov and Rodda (2003).

in Africa apart from Lake Victoria they occur in the Rift Valley, in North America they occur along the USA/Canada boarder and in Canada, in South America in Venezuela, Peru and Bolivia and in Australasia mainly in New Zealand. The figures shown in Table 4.12 do not take into account possible changes in water levels in the lakes. Several of the large lakes (and smaller ones) were formed by glacial dams and moraine deposits. Others were formed in tectonic depressions, and these are often very deep, e.g. Baikal.

Reservoirs, which are man-made standing bodies of water, have long been used as a method of storing water, often at a time of high river flows, to be used at a time of low precipitation and low river flows. Although reservoir construction has been taking place for millennia, the numbers constructed increased greatly during the twentieth century. These have been both large and small. Avakyan et al. (1987) estimated that there are about 30 000 reservoirs worldwide with a volume in excess of $1 \times 10^6 m^3$. More than half the world's rivers have one or more dams along them and in excess of 7000 km^3 of water is stored in them. As much as 40% of the water that would have flowed from continental lands directly to the sea is intercepted by dams, and they hold back in the reservoirs much more water than is present in natural river channels (Vörosmarty et al. 2003, 2005). Dams have become a major tool in the storage and management of water. As many as 800 000 small dams exist in rural communities as well as for industry and local agriculture. (Vörosmarty et al. 2005). This storage of water has dramatically changed both the volume and distribution of stored water on the continents. Most reservoirs have been constructed by damming a suitable river valley channel. The largest valley reservoir is that formed by the Three Gorges Dam, but others include the Bratskoye Reservoir on the Angara river

Table 4.13 Some of the principal reservoirs of the world.

Reservoir	Country	Basin	Volume (km^3)	Use
Bratskoye	Russia	Angara	169.3	HNTWFR
Nasser	Egypt	Nile	169	IHANF
Volta	Ghana	Volta	148	HNIF
La Grande 2	Canada	La Grande	61.7	H
Ataturk	Turkey	Euphrates	48.7	HI
Sanmenxia	China	Huang He	35.4	HISN
Itaipu	Brazil, Paraguay	Parana	29.0	H
Mica	Canada	Columbia	24.7	HA
Fort Peck	USA	Missouri	23.0	SHIN

Key to symbols: H, hydropower; N, navigation; W, water supply; I, irrigation; F, fishery; T, timber rating; R, recreation; A, accumulation; S, flood control.
Source: Adapted from Shiklomanov and Rodda (2003). For a more comprehensive list refer to this reference.

(Russia) and the Volta on the Volta River in Ghana but many other large reservoirs have been constructed (see Table 4.13).

Some lakes have had a dam constructed across their exits in order to raise the existing lake level. An example of this is Lake Victoria in Africa. Reservoirs, as Table 4.13 shows, have a number of uses apart from storage of water for use in drier times of the year. The most common use generally of the stored water is for irrigation, but energy production through hydropower is very common. They may also be used for flood prevention, navigation, fisheries and recreation. In more recent years, especially during the period of massive dam construction in China and India in the second half of the twentieth century, many objections to their construction have been raised. The downside of dam construction and the flooding of land to fill the subsequent reservoir that is produced is that many people will probably be displaced, useful land inundated and the ecology of the area both upstream and downstream of the dam changes, often adversely such as altering the silt load and affecting the stability of their estuaries. Reservoirs and lakes are subject to similar ecological problems affecting water quality. The main one is eutrophication. Indeed, the building of dams itself affects the distribution of nutrient compounds such as nitrates and phosphates both in the reservoir and the downstream river.

Sand Dams

All reservoirs, be they large or small, should be constructed where the river banks are high enough to hold the river flow, especially when the river in flooding, and prevent the water from spilling around the dam potentially causing erosion to its structure. The only flow downstream should be that which is allowed through monitored release or along constructed spillways. Before construction of a dam, its cost and sustainability, including the cost of repair and maintenance, must be considered and discussed with potential users of the water. The potential yield of water from the dam must be carefully estimated, and possible increases in use through population changes should be taken into account and explained to the local community. The life of the dam should be estimated of at least 20 years. This will allow users to decide whether it is worth the economic, environmental and socially disruptive costs. Not all dams are large-scale projects. Where there are marked seasonal differences in flows in smaller rivers in smaller communities, the scale of construction will be different. One important method for capturing and conserving seasonal water flows in developing communities is to use sand dams.

Sand dams (see Figure 4.7) are relatively simple and usually fairly small structures generally built above the surface across a seasonal sandy river where the bedrock is both

Figure 4.7 Profile of a typical sand dam.

accessible and impervious. The seasonable river must be one where the solids' load is mostly sand. A percentage of silt and clay in the sediment should not be greater than 5%. They provide stored surface water as well as help recharge the groundwater/aquifer. Sand dams are a very cost-effective method of collecting available water in dryland areas. Their construction is simple and consists of building a stone, rubble, reinforced concrete or any other impermeable material barrier across and bedded into the bottom of a seasonal sandy river. Their maintenance costs are low and, if made properly, can have a life of 60 years or more. When the seasonal rains occur, the stream/river fills with water that flows along its channel carrying not only water but also silt and sand. The sand and heavier silt is deposited at the upstream side of the dam, but the light silt may be carried downstream in any overflow water. The sand and silt accumulate until the dam is full, but this deposit will hold many litres of water (as much as 40 million – enough for the needs of 1000 people for a year). This water can be extracted into an infiltration chamber or protected shallow well. As the water is stored beneath the surface so there is no surface water, it is not only filtered through the sand but is also not open to the atmosphere and so the risk of contamination by mosquitoes, parasites, etc. is minimal. Sand dams can provide a valuable supply of water in semi-arid regions. The water provided can be used not only for human consumption but also for washing, cooking, growing vegetables and for animals. Because the water from a sand dam has already been filtered through sand, it is generally fairly clean so the health of the population using it is frequently quite good. Because crops can be watered, food security is less of a problem. Moreover, the water is more readily available close to inhabited areas, so less time is spent in collecting it. This allows more time for spending on economic activities and, ultimately, education for the children. Sand dams often benefit the environment and local ecosystems by raising the local water table (Brandsma et al. 2009). Raising the water table in turn helps fill other wells in the area as well as helps support natural vegetation.

Rainwater and Fog Harvesting

A further water resource that could be exploited in some regions is rainwater and/or fog harvesting. This can be used in regions with adequate seasonal precipitation or predictable fogs. In many parts of the world, there is a shortage of water for at least some months of the year for both human and agricultural needs. Such shortages not only affect human health but also food production and economic development. These regions would also greatly benefit from better water availability. In areas such as Sub-Saharan Africa, as much as 70–85% of rainwater that falls is lost due to evaporation. Much of this could be saved with appropriate technologies and management. This is especially true in rural areas.

Rainwater harvesting has been defined in various ways. Worm (2006) define it as follows: 'Water harvesting is the collection of run-off rainwater for domestic water supply, agriculture and environmental management'. Augmentation of water resources by rainwater harvesting is used in many parts of the world, especially in countries where rainfall is markedly seasonal and in more remote communities. It has been practiced in these communities, especially in semi-arid and arid regions, since ancient times. One of the goals of the International Drinking Water Supply and Sanitation Decade (1981–1990) was the provision of readily available clean drinking water for everyone. This goal impacts on several other goals where water availability can affect health and well-being and economies. This could be a challenge in arid and semi-arid areas and in some poorer rural communities. In such locations, traditional water resources are not always plentiful so to meet these goals additional and alternative supplies are needed. Rainfall harvesting offers a valuable alternative or supplement to water supplies. Although it was less popular during the pre and post Second World War decades when building of dams flourished, it has now been rediscovered and is becoming more popular in many countries, including China, Kenya, India, Germany and the United States. The main source of freshwater on land is from rainfall which, either directly or indirectly, is essential both for humans and ecosystems. The total global rainfall upon land is estimated at approximately 113 000 km^3 annually. About 36% of this goes in run-off, and the remaining 64% is evaporated from the land and water surfaces or by vegetation (UNEP 2009). Management of rainwater and the way in which it is partitioned must be carefully planned so that the proportion being diverted to one sector does not adversely deplete others, especially ecosystems. Ecosystems do provide benefits in the control of the hydrological cycle. Adequate water supply to the ecosystem will improve vegetation productivity which in turn will reduce soil erosion. It will also help flood control. These benefits must be balanced against other downstream needs. As water is essential for all living things and because the distribution of freshwater is uneven, rainwater can play an important role in meeting this diversity of needs. Water availability will affect human well-being in four main ways, which are as follows:

- A good clean supply for human consumption will improve health, hygiene and sanitation.
- Water is needed to produce food both for humans and livestock.
- Water supports a variety of economic activities and can thus improve the income of people.
- Water can also cause disasters and loss of life through floods and so must be properly managed.

Rainwater harvesting can make a significant contribution to all four of the above in many regions. Rainwater harvesting systems are orderly schemes in which organised components and techniques harness the precipitation and make it available for human and environmental use. Typically, a harvesting system consists of six basic components: a collecting area, a conveyance system, storage facility, a filtration mechanism, water treatment and a delivery system. Of these, the storage system can be the largest cost, especially in systems designed for more than one dwelling, and as such must be carefully planned and constructed (Khoury-Nolde 2016). These components are illustrated in Figure 4.8 with various examples in Figure 4.9, which also includes fog collecting.

The efficiency of any system depends upon the materials used as well as their design, construction and maintenance. For example, household roofs made of cement tiles can be 75% efficient, but plastic or metal roofs can be as high as 80–90% efficient (Gould and Nissen-Petersen 1999). Rainwater harvesting is often used in agriculture. Ex situ harvesting is the run-off water from outside the farming unit. This often involves storage units such as dams. Water of ponds or wells is frequently used for crop irrigation, commercial use and domestic supplies. In situ systems are those in which all rainwater harvesting is completed within the farming unit and frequently involves tank storage but also, in agriculture,

Figure 4.8 Typical domestic rainwater harvesting for a house with a corrugated roof and a collecting barrel.

aspects of tillage, terracing and other soil conservation techniques encouraging soil infiltration. Examples of in situ and ex situ harvesting technologies are provided in Figure 4.10.

Rainwater harvesting is being encouraged in countries other than those in arid and semi-arid zones in the developing world. Examples include the United Kingdom, Germany and the United States of America (Texas Water Development Board 2005; UK Environment Agency 2010). Although the perception in the United Kingdom is that it rains a lot, water resources are often strained in some parts of the country at certain times of the year. This is partly due greater demand from an increasing population, greater demand through life-style changes and partly through the effects of climate compared with less available water per capita than in some Mediterranean countries. Rainwater harvesting could ease the demand on public water supplies and is being practiced in many countries worldwide (Guerquin 2010). Many projects are being developed in India and Asia as well as Africa. The UK Environment Agency currently advocates rainwater harvesting for non-potable use although this is not so in the United States of America and Germany where only in some cases may the water be used for drinking. Where harvested rainwater is used, it has the additional benefit of saving the consumer money as they do not have to pay the water company for that water.

Rainwater has been used for human well-being for many years, and in many parts of the world rainfed agriculture is practiced. This latter use frequently represents the greatest proportion of land being cultivated. In sub-Saharan Africa, for example, rainfed agriculture is important to local economies accounting for 35% of the GDP and 70% of employment (World Bank 2010). As agricultural productivity is generally lower in rainfed areas, averaging 1.5 t/ha compared with 3.1 t/ha in irrigated areas, it is important to manage water including rain as effectively as possible (Rosegrant et al. 2002). In semi-arid and dry areas, it is not always the amount of rainfall that limits crop production, but the great variability of the rainfall often causing absolute water scarcity where the crop water demand exceeds the total rainfall (Rockström et al. 2010). Management of rainwater for agricultural purposes can be performed in two ways: (i) capturing water on site and in the root zone of the crop and (ii) capturing the water at a distant site and transporting it to the crop. In either

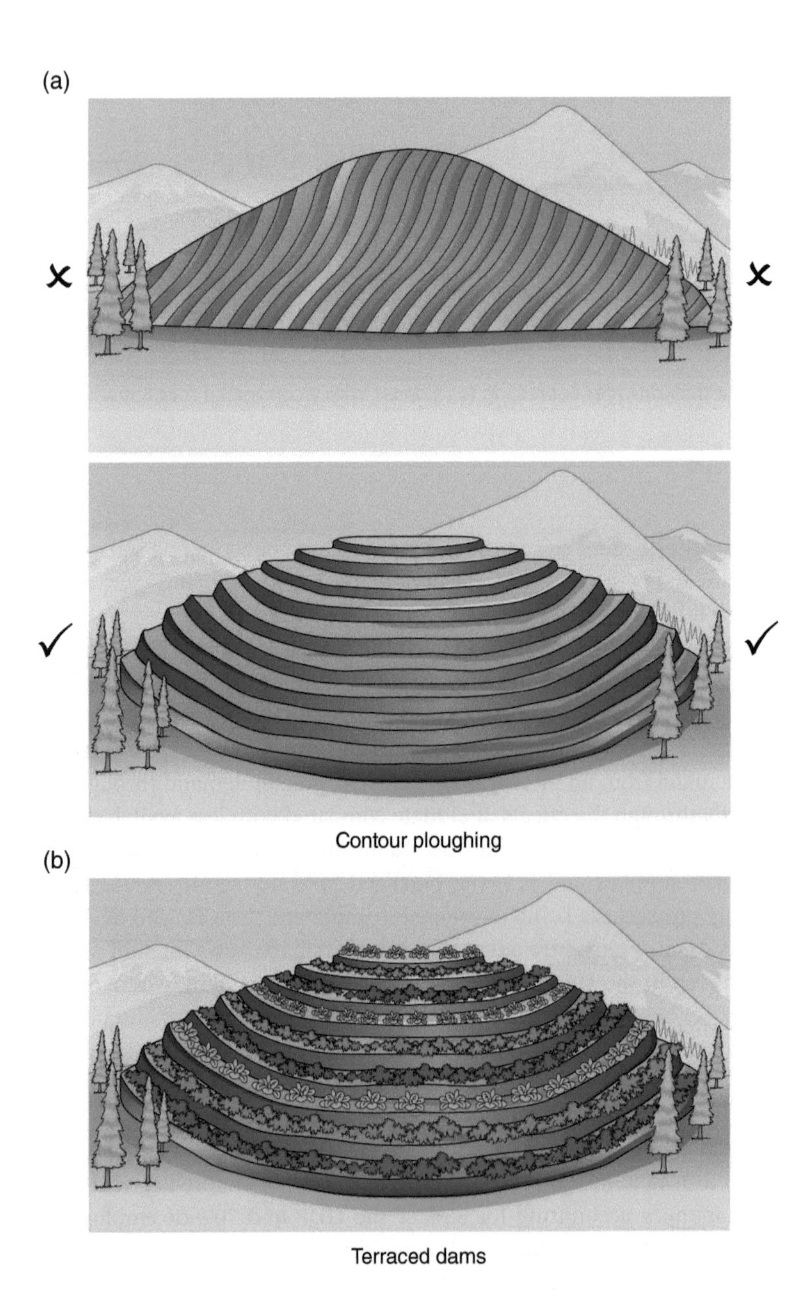

Figure 4.9 (a) Water conservation on land. Do not plough vertically up and down a slope. Plough along the land contours in many Asian countries as the hillsides have terraces of dams with crop-growing areas between each terrace. ✗, incorrect ploughing straight up and dowm the slope. ✓, correct non-erosive ploughing. (b) A typical fog-collecting screen.

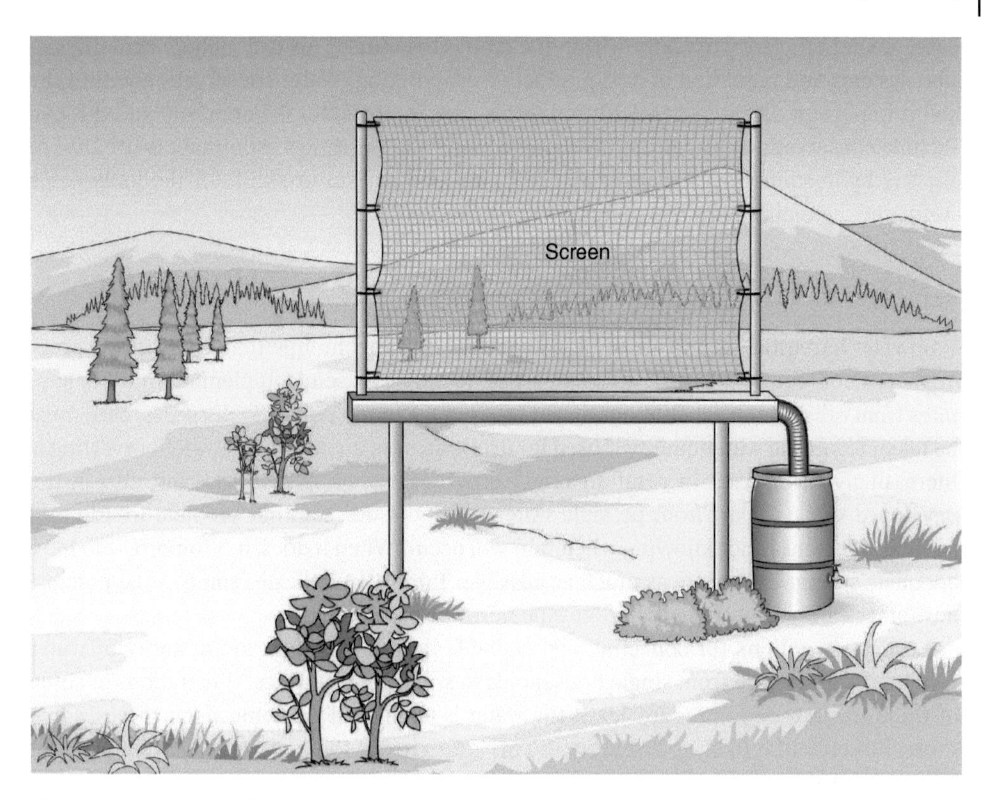

Figure 4.9 (Continued)

Figure 4.10 The typical range of collecting options for a house and surrounding area: (1) the house roof, (2) lawn or planted area, (3) gutter at property edge or edge of road and (4) sub-surface collecting tank with protected walled access pipe.

case, greater efficiency of water within the crop is essential. This can include better irrigation systems and reduction of non-productive evaporation. Water conservation within the soil is important. Improved agricultural techniques in dry zones is important, and this can include conservation agriculture involving, amongst other things, minimal disturbance of the soil by machinery (Landers et al. 2001). Rainfed agriculture will, for the foreseeable future, be the main method of agricultural production.

Rainwater Harvesting for Domestic Use

Rainwater harvesting is a relatively simple and low-cost technique that, if organised properly, does not require a large level of expertise. It can be a useful supplement to water supplies from other sources. Although it can be used for many domestic purposes, care must be taken regarding water quality if used for drinking or in cooking. Rainwater harvesting is increasingly being used in rural areas not only in developing countries but also in the developed world. Apart from possible water quality issues, another problem in arid and semi-arid regions is not knowing when rain will occur. When it does, it is important to have adequate storage to capture as much as possible. Table 4.14 indicates some of the possible advantages and disadvantages of domestic rainwater harvesting.

A range of options for collecting are available depending upon the quantity of water required. These range from single households to small communities. Illustrations of some of these are provided in Figure 4.9. If the water is to be used for domestic purposes, some form of treatment is likely to be required. This is also true if the water is used to supplement either sub-surface aquifers or otherwise clean surface waters such as rivers or lakes.

Rainwater Quality

Rainwater is usually of good quality but can contain unwanted pollutants, especially if the area is close to industrial zones. Except in the special circumstances of local industry, rainwater is considered adequate for most purposes although not necessarily for drinking.

Table 4.14 Potential advantages and disadvantages of domestic rainwater harvesting.

Advantages	Disadvantages
Simple construction but larger storage tanks can expensive to produce	Parts of the system can be expensive
Maintenance is generally quite easy	Proper regular maintenance is essential if good water quality is to be maintained
Rainwater is normally of good quality	Rainwater can be affected by pollutants, animal droppings, insects and dust
Renewable resource with low environmental impact	Supply depends upon rain and can be affected by long droughts
Provides supply at local level	Limited by rainfall amount and collection capacity
Systems flexible to suit local needs	Needs to be properly planned and built

Source: Adapted from Worm (2006).

Harvesting usually occurs in collectors open to the sky, and care needs to be taken concerning its bacteriological quality but that collected from properly maintained roof systems is usually as good or even better than traditional surface water supplies. This applies to many rural settlements in developing countries. There are a few examples of serious illness from drinking rainwater, but some form of purification and/or disinfection is recommended such as simple sand filtration and/or boiling before drinking. Cunliffe (1998) quotes the Australian government as saying 'Providing the rainwater is clear, has little taste or smell, and is from a well-maintained system, it is probably safe and unlikely to cause illness for most users'. If the user is in some way immunocompromised (i.e. the very young or the very old or with some other illness), boiling before consumption is recommended. Users could always collect a sample and send it away for microbiological analysis. Unfortunately, quality can vary with different intensities of rain and at different times of the year. Ward et al. (2019), sampling roof-collected rainwater from a UK-based office building, observed that physicochemical contaminants found posed little or no health risk as concentrations were below WHO standards. They did find that components in the collection system such as fittings and downpipes were susceptible to soft water corrosion and could result in elevated concentrations of copper, zinc and aluminium. This indicated that the choice of materials could be very important. No evidence of *Cryptosporidium*, *Salmonella* and *Legionella* was found; however, faecal coliforms were found throughout the study. Rainwater is usually slightly acidic and has a pH of about 5.6.

Floods and Droughts

Floods

Although water is essential for life it can, in the wrong quantity, be a hazard. Sidle et al. (2003) describe two groups of hazard. Chronic hazards include desertification, soil degradation and the melting of permafrost all occurring over a longer period of time, while periodic hazards are generally of a larger magnitude that occur over a short period of time, such as tsunami, volcanic eruptions, tornados and flash flooding. While chronic hazards are important to humans, some of the most devastating to human activities are the periodic type. Although these latter hazards gain headlines for being catastrophic, they can also be beneficial. For example, silt deposited on land by floodwater can add much needed nutrients for later plant growth. Such a case is the annual flooding of the River Nile in ancient times. In many countries, floodplains are managed to provide rich cultivated areas for agriculture. It should also be noted that floods can have a significant impact on public health. The response to both floods and droughts greatly varies from country to country. Bryant (2005) contrasts the response of Darwin, Australia, which was struck by the Cyclone Tracy in 1974, and Honduras, which was struck by hurricane Fifi in September 1974. Both storms were of similar intensity destroying around 80% of the buildings at the storm centres. 64 people died in Darwin but 8000 died in Honduras. This reflected the difference in infrastructure preparedness and emergency response organisation and ability in the two areas. Floods and droughts generally have a greater impact in third world countries compared with developed countries. Although it is difficult in many countries to estimate the economic impact of many floods, especially those in the past, the impact can be equated to the number of lives lost. Floods are defined in the European Union Directive 2007/60/EC, Art 2 (1) as 'the temporary covering by water of land not normally covered by water'. The degree to which floods may cause loss of life does depend upon the type of flood, the behaviour of the population in that area and the community infrastructure available. In recent years, floods have been responsible for about 12% of deaths caused by natural disasters. Many small floods have caused loss of life. The floods that have caused more than 100 deaths since 1960 are listed in Table 4.15.

Table 4.15 Floods causing more than 100 deaths.

Country	Year	Number of deaths
Africa		
Algeria	2001	921
Angola	2007 and 2011	218
Chad	2001	120
Congo	2014	154
Ethiopia	5 floods between 1997 and 2006	1434
Kenya	2006 and 2010	214
Malawi	1991 and 2015	750
Morocco	1995	730
Mozambique	3 floods between 1971 and 2000	1600
Namibia	2011	108
Nigeria	4 floods between 1988 and 2012	813
Sierra Leone	2009	103
Somalia	3 floods between 1961 and 1997	2611
South Africa	3 floods between 1981 and 1995	817
Sudan	3 floods between 1996 and 2007	353
Tanzania	1990	183
Tunisia	1969 and 1982	657
Americas		
Bolivia	1983	250
Brazil	18 floods between 1966 and 2011	5016
Chile	1993 and 2014	887
Columbia	5 floods between 1962 and 1988	812
Dominican Rep.	2004	688
Ecuador	1997	218
Guatemala	1982	620
Haiti	2004	2665
Honduras	3 floods between 1993 and 2010	641

Table 4.15 (Continued)

Country	Year	Number of deaths
Mexico	4 floods between 1973 and 2000	478
Peru	4 floods between 1971 and 2009	840
Puerto Rico	1985	500
USA	3 floods between 1969 and 1976	891[a]
Venezuela	1979	30 000
Asia		
Afghanistan	9 floods between 1963 and 2014	2751
Bangladesh	20 floods between 1960 and 2004	39 498
Bhutan	2000	200
Cambodia	5 floods between 1991 and 2013	1153
China	54 floods between 1980 and 2013	36 010
India	57 floods between 1960 and 2015	60 445
Indonesia	16 floods between 1966 and 2010	3526
Iran	6 floods between 1980 and 2001	2330
Japan	6 floods between 1961 and 1982	1799
Jordan	1966	259
Korea	6 floods between 1967 and 2007	1542
Laos	1966	300
Myanmar	2 floods; 2011 and 2015	254
Nepal	18 floods between 1968 and 2014	5459
Pakistan	23 floods between 1973 and 2015	11 774
Philippines	5 floods between 1972 and 2010	1662
Saudi Arabia	2009	161
Sri Lanka	2 floods between 1989 and 2003	560
Tajikistan	1992	1346
Thailand	8 floods between 1975 and 2011	2843
Turkey	1968	147
Viet Nam	9 floods between 1964 and 2001	2323
Yemen	1996	338

There are many other floods in these and other countries, but they caused less than 100 recorded deaths.
[a] This does not include deaths caused by events such as Hurricane Katrina, USA, in 2015.
Source: Data abstracted from EM-DAT, The International Disaster Database accessed February 2016.

The floods listed are of a number of types, including riverine, flash floods and floods causing mudslides. Sea surges are not included. Many catastrophic floods occurred over history in many countries although in earlier times reports were sometimes less detailed. In China, for example, a number of anthropogenic pressures can be identified. They have had to endure catastrophic floods on several occasions in the past. The Yellow River flooded in 1887 and 1938, on both occasions, killed a million or more people. In 1975, the Banqiao dam failed and 86 000 people died. The Netherlands, a country that is low lying with large areas below sea level, has frequently been subject to flooding from the eleventh century to more recent times. Many were caused by storm surges from the North Sea, but some were freshwater.

India, Pakistan and Bangladesh have major rivers arising in the Himalayas which, together with heavy monsoons, can flood causing extensive problems. In Venezuela, a flash flood in 1999 affected over 48 000 people killing some 3000 of them. Floods cause enormous problems globally. From 1990 to 2000, floods accounted for approximately 12% of all deaths caused by natural disasters. There are several types of flood that have been identified, which are as follows.

- Flash flood: These occur after very intense, sudden and local rainfall. This results in a rapid rise in water level causing rivers to overspill their banks and threaten lives. These floods often occur in hilly or mountainous areas and within a few hours of the intense rain. They can also occur if there has been a dam or embankment failure.
- Riverine flood: This is when a river carries so much water that it overflows its normal banks. The excessive flow can be caused by prolonged and high rainfall within the catchment, rapid melting of snow and blockage of the riverbed.
- Drainage problems: It is caused by heavy rainfall in the catchment where the run-off cannot be coped with by the normal drainage systems. This type of flood is not usually so destructive or danger to life but is more likely to cause economic damage.
- Coastal flood: These, sometimes called a storm surge, occur along the coasts of seas or large lakes. Strong winds in the same direction cause a build-up of water levels that overspill the shore and cause flooding.

Floods do occur naturally, and where they do usually cover a clearly delineated stretch of land known as the floodplain. This needs not be a problem except that land is often fertile and suitable for agriculture and human settlement. Problems can then arise when it floods. There is no complete definition of floodplains. Floodplains are generally low-lying areas along a river, which are sometimes covered by water. They are in a constant state of exchange of water with the river. They usually occur when the river overspills onto the surrounding land and can occupy any area from small to large. A floodplain can arise anywhere along a river. In remote regions, floods in floodplains are unlikely to cause significant problems but in densely populated areas such as in Europe the damage can be very large. In such areas, there is a need to manage the floods. It must also be recognised that floodplains often have considerable ecological significance and can have unique ecosystems associated with them. An example is floodplain forests.

One of the functions of a floodplain is that it is able to hold a large amount of excess flow from the river preventing damage lower down. However, there are considerable pressures imposed on floodplains through human activities. In Europe, because of the pressure on land availability, its use as arable land and pasture, permanent cropland, forests of various types and urban areas, including buildings and green spaces such as playing fields, hard artificial surfaces and inland waters, can all cause problems. Land reclamation has also had an effect. For example, the floodplains of the Danube and Tisza rivers in Europe have had their floodplains modified for navigation and the installation of hydropower schemes and for flood-retention measures although the result has sometimes been a disconnect with the original floodplain. The last major intervention was the construction of the Gabčikovo dam on the Hungarian–Slovakian border. As a result of these uses, changes to the hydrological regime are likely to occur. The flow dynamics of water entering the floodplain can be affected as can connections to groundwater. An important change can be a reduction in the ability to retain water, thus increasing the rate and volume of flow downstream. Compaction of the soil through use of heavy agricultural equipment can be the cause of compaction.

Floodplains have also functioned as repositories for silt and, in the past, a range of pollutants from mining and other industrial activities and sewage discharges. This contamination greatly increased during the nineteenth and the first half of the twentieth centuries. Water quality in most European rivers has improved significantly in recent years mainly through the urban wastewater and water framework directives of the EU. Unfortunately, floodplains can both act as a sink for current pollutants and a repository for past ones. Floodplains can, however, still carry out the important function of water purification as well as acting as a repository (Schulz-Zunkel et al. 2013; Schulz-Zunkel and Kreuger 2009).

Floodplains need to be protected not only to help manage floods, but they are highly diverse habitats often with many endangered species. As such they are listed in Annex 1 of the EU Habitats Directive from 1992. Floodplains need to be managed both for wildlife and flood protection.

Floods can cause considerable social disruption and economic loss. The financial costs not only apply to damage of individual livelihoods and belongings but also to property, business and agriculture. There is also the cost of flood control measures for future prevention and the additional cost to the population of increases in insurance premiums. If a property is deemed to be at risk from flooding, then the cost of insurance can be prohibitive.

Droughts

Although droughts are perceived as hazards, unfortunately areas where droughts occur are still often exploited by human developments. This is because they usually have less vegetation and often have areas of open land making them easy to clear for building and development. Droughts have been a problem for many agricultural societies throughout history. Although a great cause of harm to both ecosystems and humans, droughts, unlike floods, usually develop slowly and often persist for a longer period of time. A drought is a temporary thing, but aridity is a permanent climatic feature of a region. Some parts of the world, e.g. Africa, regularly have droughts while in others it is rare. Russia, during the twentieth century, was frequently affected by droughts impacting food production (Dronin and Bellinger 2005). Many definitions of drought exist, but simply it is an extended period when rainfall is absent, causing a severe reduction in plant and animal production. In heavily populated regions, a drought can have a severe effect on water supplies. The length of time proposed for lack of rain to cause a drought can vary depending on country, but in the United Kingdom and Canada the period is taken to be 30 days but generally four categories of drought have been recognised. These are meteorological drought, agricultural drought, hydrological drought and socio-economic drought (Institute of Food and Agricultural Sciences 1998).

Table 4.16 Major droughts, a summary of their number and the number of deaths caused.

Continent	Number	Deaths	People affected	Economic losses
				US$ 1000
Africa	139	1119007	243267579	5271332
Asia	113	9663043 1541783427	24026695	
Europe	33	1200002	19865575	17888600
America	90	73	61003351	18377739
Oceania	17	660	8027635	13303000

Source: Adapted from Sheffield and Wood (2012).

Meteorological drought is defined as a deficiency of precipitation from normal levels over a prolonged period of time. Typically, meteorological drought occurs when less than 25% of the long-term average precipitation occurs. If the deficit is between 26% and 50%, it is said to be of moderate intensity, and if more than 50% it is severe.

Agricultural drought is when soil moisture and precipitation are not enough in the growing season, causing stress to the crop, poor growth, wilting and eventually death.

Hydrological drought concerns both the surface and underground water supplies. It is measured by stream flow and lake/reservoir water levels. These measures are not instantaneous as there is a lag between a drop in the amount of precipitation and the flow of streams and levels of lakes.

Socio-economic drought occurs when the shortage of water has a significant effect on people.

A drought lasting a few months in some parts of the world may be devastating for agriculture, but has little effect on fish and some wildlife (Cooley 2006). Although it is difficult to recognise the start of a drought episode measurements of precipitation, soil moisture and temperature can be used. These, or any one of them, may be used in a drought index. An example of this is the Palmer Drought Index that uses soil moisture. There are many examples of droughts worldwide (see Table 4.16).

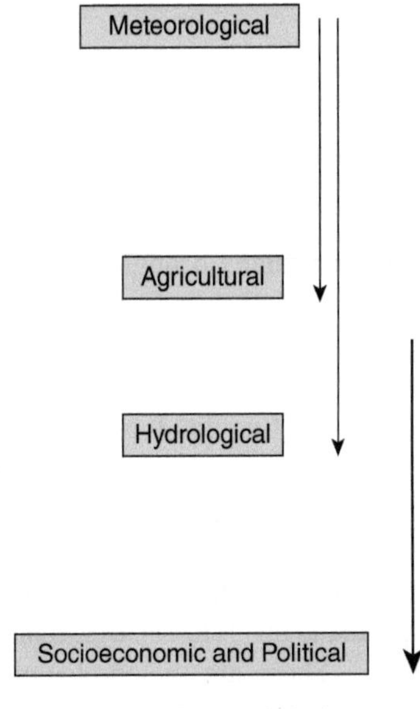

Figure 4.11 Causal chain between different types of drought.

Some droughts are widespread, for example the Dust Bowl of the 1930s in the United States of America. This lasted for nearly 10 years, whereas in Africa in 1991 there was a relatively shorter one. Figure 4.11 from the Drought Mitigation Center (2005) points to linkages between different types of drought.

In many developing countries, and in particular Africa, most of the rural population relies upon rainfed agriculture, so if the rains fail it will inevitably lead to food shortages and famine. As much as two-thirds of Africa is classified as drylands or desert, and thus are totally dependent upon the seasonal rains. These vulnerable regions are mainly in the Sahelian region and the Horn of Africa. The lack of rain and some local agricultural practices, such as felling trees for firewood, are partly the cause. The problem is also exacerbated in these areas by population growth. For example, the population of Nigeria has grown from 33 million in 1950 to 134 million by 2006. High population increases place greater demands upon already fragile resources. Rainwater catchment systems are used in many areas for the domestic supply of some 485 million people, which represents about 65% of the entire African population. Within Africa there is an increasing likelihood of drought that will certainly also lead to an increase in the area of degraded land. The current climate change predictions are that currently dry regions will become even drier and that freshwater resources will dwindle even further (UNESCO 2006.). In the drought-prone regions of Africa, this has led to mass migrations of the populations. This has also meant that in the concerned countries there has been socio-economic instability and the inability of the governments to build adequate infrastructures to mitigate the situation.

References

Avakyan, A.B., Saltankin, V.P., and Sharapov, V.A. (1987). *Water Reservoirs*. Moscow: Mysl Publication (in Russian).

Barnett T.P. and Pennell, W. (eds.). (2004). Impact of global warming on Western U.S. water supplies. *Climate Change*. 62.

Barnett, T.P., Adam, J.C., and Lettenmaier, D.P. (2005). Potential impacts of a warming climate on water availability in snow-dominated regions. *Nature* 438: 303–309.

Benke, A.C. and Cushing, C.E. (2005). *Rivers of North America*. Academic Press.

Berner, E.K. and Berner, R.A. (1987). *Global Water Cycle: Geochemistry and Environment*. Englewood Cliffs, NJ: Prentice Hall.

Brandsma, J., Hofstra, F., Masharubu, B., and Mallu, D. (2009). *Impact Evaluation on Sand Storage Dams*. OL Wageningen UR and Van Hall Larenstein UR.

Brunner, A.D. (1998). El Nino and World commodity prices: warm water or hot air? *Review of Economics and Statistics* 84 (1): 176–183.

Bryant, E. (2005). *Natural Hazards*. Cambridge University Press.

Cech, T.V. (2010). *Principles of Water Resources. History, Development, Management and Policy*. Wiley.

Cooley, H. (2006). Floods and droughts. In: *The Worlds Water 2006–2007* (ed. P. Gleick), 91–142. Washington, DC, USA: Island Press.

Cosgrove, W.J. and Rijsberman, F.R. (2000). *Making Water Everybody's Business*. London: World Water Council, Earthscan Ltd.

Cunliffe, D. (1998). Guidance on the use of Rainwater Tanks. National Environmental Health Forum Monographs. *Water Series 3, Public and Environmental Health Service, Dept. of Human Services*, P.O. Box 6, Rundle Mall SA 5000, Australia, 28 pp.

Devlin, J.F. and Sophocleous, M. (2005). The persistence of the water budget myth and its relationship to sustainability. *Hydrogeology Journal* 13: 549–554.

Dronin, N.M. and Bellinger, E.G. (2005). *Climate Dependence and Food Production in Russia 1900–1990*. New York: Central European University (CEU) Press.

Drought Mitigation Center (2005). Drought Types.
FAO (2003, 2014). AQUASTAT Global Information System on Water and Agriculture. https://www.fao.org/aquastat/en.
FAO (2015). World Water Resources by Country. http://www.fao.org/docrep/005/y4473e/y4473e08.htm (accessed October 2021).
Gleick, P. (2000). Water: the potential consequences of climate variability and change for the water resources of the United States. Pacific Institute for Studies in Development, Environment and Security, pp. 1–151. https://www.worldcat.org/oclc/45746538 (accessed 25 January 2024).
Gould, J. and Nissen-Petersen, E. (1999). *Rainwater Catchment Systems for Domestic Supply*. London: Intermediate Technology Publications.
Guerquin, F. (2010). *World Water Actions: Making Water Flow for All*. World Water Council Water Actions Unit, 1–162. Routledge: London.
Gupta, A. (2007). *Large Rivers: Geomorphology and Management*. Wiley.
Intergovernmental Panel on Climate Change (IPCC) (2007). *Climate Change, 2007: Impacts, Adaptation and Vulnerability. Contribution of Working Group II to the Fourth Assessment Report of the IPCC* (ed. M.L. Parry). Cambridge, U.K: Cambridge University Press.
Kääb, A., Reynolds, J.M., and Haeberli, W. (2005). Glaciers and permafrost hazards in high mountains. In: *Global Change in Mountain Regions (A State of Knowledge Review)* (ed. U.M. Huber, H.K. Bugmann, and M.A. Reasoner), 225–234. Dordrecht: Springer.
Kääb, A., Huggel, C., and Fischer, L. (2006). Remote sensing technologies for monitoring climate change impacts on glacier and permafrost related hazards. *Engineering Conferences International, Proc. On Geohazards* (18–21 June 2006).
Khoury-Nolde, N. (2016). *Rainwater Harvesting*. Fachvereingung Betriebs-und Regenwassernutzung e.V.
Korzun, V.I. (ed.) (1974). *World Water Balance and Water Resources of Earth*. Leningrad: Hydrometeoizdat (in Russian).
Kumar, R., Singh, R.D., and Sharma, K.D. (2005). Water resources of India (PDF). *Current Science (Bangalore)* 89 (5): 794–811.
Landers, J.N., Mattana Saturnio, H., de Freitas, P.L., and Trecenti, R. (2001). Experiences with farmer clubs in dissemination of zero tillage in tropical Brazil. In: *Conservation Agriculture. A Worldwide Challenge* (ed. L. Garcia-Torre's, J. Benites, and A. Martinez-Vileda). Rome: Food and Agricultural Organization.
Llamas, M.R. and Martinez-Santos, P. (2005). Intensive groundwater use: a silent revolution that cannot be ignored. *Water Science and Technology Series* 51 (8): 167–174.
Margat, J. and van der Gun, J. (2013). *Groundwater Around the World: A Geographic Synopsis*. Balkema: CRC Press.
McGuire, V.L. (2009). Water-level changes in the high plains aquifer, predevelopment to 2007, 2005-06 and 2006-07. *Publications of the US Geological Society* 17: 1–18. https://digitalcommons.uni.edu/usgspubs/17.
Nezhihovsky, R.A. (1981). *Neva River and Neva Bay*. Gidrometeroizdat.
Nijssen, B., O'Donnell, G.M., Hamlet, A.F., and Lettenmaier, D.P. (2001). Hydrologic sensitivity of global rivers to climate change. *Climate Change* 50: 143–175.
Peck, J.C. (2007). *Groundwater management in the High Plains Aquifer in the USA: Legal Problems and innovations: Chapter 14 in CAB International 2007* (ed. Giodano and Villholth). The Agricultural Groundwater Revolution: Opportunities and threats to development.

Rijsbermann, F.R. (2006). Water scarcity: fact or fiction? *Agricultural Water Management* 80: 5–22.

Rockström, J., Karlberg, L., Wani, S.P. et al. (2010). Managing water in rainfed agriculture – the need for a paradigm shift. *Agricultural Water Management* 97: 543–550.

Rosegrant, M.W., Cal, X., and Cline, S.A. (ed.) (2002). *Water and Food to 2005: Dealing with Scarcity*. International Food Policy Research Institute (IFPRI) and the International Water Management Institute (IWMI).

Schulz-Zunkel, C. and Kreuger, F. (2009). Trace metal dynamics in floodplain soils of the river Elbe: a review. *Journal of Environmental Quality* 38 (4): 1349.

Schulz-Zunkel, C., Kreuger, F., Rupp, H. et al. (2013). Spatial and seasonable distribution of trace metals in floodplain soils. A case study with the middle Elbe river, Germany. *Geoderma* 211–212: 128–137.

Seckler, D., Barker, R., and Amarasinghe, U.A. (1999). Water scarcity in the 21st century. *International Journal of Water Resources Development* 15 (1–2): 29–42.

Sheffield, J. and Wood, E.F. (2012). *Drought. Past Problems and Future Scenarios*, 1–192. Routledge.

Shiklomanov, I.A. (1990). Global water resources. *Nature and Resources* 26 (3): 34–43. (In Russian).

Shiklomanov, I.A. (2000). World water resources and water use; present assessment and outlook for 2025. In: *World Water Scenarios: Analysis* (ed. F.D.R. Rijsberman), 160–203. London: Earthscan.

Shiklomanov, I.A. and Rodda, J.C. (2003). *World Water Resources at the Beginning of the 21st Century.* Cambridge University Press.

Sidle, R.C., Taylor, D., Lu, X.X. et al. (2003). Interaction of natural hazards and society in Austral-Asia: evidence in past and recent records. *Quaternary International* 118–119: 181–203.

Sophocleous, M. (2000). From safe yield to sustainable development of water resources – the Kansas experience. *Journal of Hydrology* 235 (1–2): 27–43.

Sphhocleous, M. and Devlin, J.F. (2004). Is natural recharge relevant to groundwater sustainable development? *Groundwater* 42: 618.

Texas Water Development Board (2005). *The Texas Manual on Rainwater Harvesting*, 1–86. Austin, Texas: Texas Water Development Board.

Tol, R.S.J. (2009). The economic effects of climate change. *Journal of Economic Perspectives* 23 (2): 29–51.

UNESCO (2006). Water, a shared responsibility. *UN World Water Development Report 2.* World Water Assessment Programme, New York.

United Kingdom Environment Agency (2010). *Harvesting Rainwater for Domestic Uses: An Information Guide*. Bristol: Environment Agency.

United Nations Environmental Programme (UNEP) (2007). *Global Outlook for Ice and Snow.* Nairobi: UNEP.

United nations Environmental Programme (UNEP). (2009). Rainwater Harvesting: A Lifeline for Human well-being. *A report for UNEP by Stockholm Environmental Institute.*

United Nations Food and Agriculture Organisation. (2011). The state of the world's land and water resources for food and agriculture. Managing systems at risk.

United Nations World Water Development Report (WWAP) 4 (2) (2012). *Managing Water Under Uncertainty and Risk.* Paris: UNESCO.

Vörosmarty, C.J., Meybeck, M., Fekete, B. et al. (2003). Anthropogenic sediment retention: major global impact from registered river impoundments. *Global Planetary Change* 39 (1–2): 169–190.

Vörosmarty, C.J., Douglass, E.M., Green, P., and Revenga, C. (2005). Geospatial indicators of emerging water stress: an application to Africa. *Ambio* 14 (2): 230–236.

Vrba, J. and van der Gun, J. (2004). The World's Groundwater Resources. International Groundwater Resources Centre (IGRAC). *Report Nr 1P, 2004-1*, System 2, pp. 1–10.

Wada, Y. and Heinrich, L. (2013). Assessment of transboundary aquifers of the world- vulnerability arising from human use. *Environmental Research Letters* 8 (2): 1–13. 024003.

Walton, B. (2015). Groundwater depletion stresses majority of World's largest aquifers. Circle of Blue.

Ward, S., Memon, F.A., and Butler, D. (2019). Harvested rainwater quality: the importance of appropriate design. *Water Science and Technology* 61 (7): 1707–1714.

World Bank (2010). Spurring agriculture and rural development. In: *Can Africa Claim the 21st Century?* Washington, D.C.

World Water Assessment Programme (2003). *Water for People, Water for Life, A Joint Report by the Twenty Three UN Agencies Concerned with Freshwater*, vol. 1. United Nations Educational, Scientific and Cultural Organisation (UNESCO).

World Water Assessment Programme (WWAP) (2012). *UN World Water Development Report* 4 (2). (WWDR4). UNESCO: Paris.

Worm, J. (2006). AD43E Rainwater harvesting for domestic use. *Agrodoc 43 Agromisa Foundation and CTA*. Wageningen.

5

Water Scarcity

Water is essential for all life on this planet. We are fortunate in that there is a generous supply on this planet although most, nearly three quarters, is in the form of saltwater and only a relatively small amount is in the form of freshwater. Not only do we, as individuals, contain a large amount of water, about 70% of our body weight, it is also essential for the provision of our food, needed as a drink to prevent dehydration, required for our washing and hygiene, required by industry and many of our economic activities and it is also used to provide a large amount of energy that we use. Water has been described as 'the bloodstream of life and the biosphere' (Ripl 2003). It is a naturally occurring solvent that is capable of dissolving, to a greater or lesser extent, a very large range of chemicals. Like the bloodstream in our bodies, it transports these dissolved (and some in suspension) chemicals through the biosphere to both plants and animals, so it is essential for sustaining life and ecosystems.

For many thousands of years, water has been available in relatively plentiful supplies in most parts of the planet. During this time, human population numbers were relatively low compared with the present time, so water availability was not usually a general problem, and where it was scarce, people were usually able to adapt their lifestyles accordingly. Access to water was initially mostly free (Seckler et al. 1999), and this gave rise to the assumption in some communities that supplies were inexhaustible, so use was not always as careful as it should have been. In addition. the attitude arose that development costs should be kept to a minimum, and water contamination was not something that was considered to be an important problem. Little or no protection of water bodies was undertaken. Indeed, disposal of wastes into water was not a worry as many people thought that it would be infinitely diluted, which we now know is an entirely false premise. The situation is now changing, especially in semi-arid areas where water quantity and availability are the greatest threats to food production, human health and the survival of local and regional ecosystems. Attitudes have changed in recent years. Over the past century, anthropogenic influences on the hydrological cycle have affected lakes, rivers and aquifers, affecting both amounts available and water quality. Human population growth and their numerous associated economic activities now greatly influence access to water resources and availability to obtain good safe drinking water. This problem has been addressed by many international organisations, including UNESCO, FAO, UNEP, IWAR and others. All have initiated

Water: Our Sustainable and Unsustainable Use, First Edition. Edward G. Bellinger.

conferences and studies on water resources and access, but the problem is still growing. At the turn of this century, it was estimated that 1.4 billion people, approximately one-quarter of the global population or one-third of the population in developing countries, will suffer from severe water scarcity by 2025 (Seckler et al. 1998). More than one billion live in arid regions that have inadequate water resources and will face absolute water scarcity by 2025. In some countries, people will not be able to produce enough food for their survival, so they will have to rely upon importing food that costs money. In most countries where water scarcity is likely, population growth is high, so the problem can worsen.

Because of the ever-increasing world population, it will not be possible to allow everyone to have unlimited amounts of water. Hence, the term 'adequate amounts' when describing people's and societal needs is used. The question then arises as to what is adequate and what would be a minimum requirement. This will vary from country to country because of climate, diet, nature of work, culture and religion, which will change the individual requirements per person. These together with other factors, such as the economic status of a country, will all have a bearing on water availability.

What Is Water Scarcity?

If the requirements listed above cannot be met, a degree of water scarcity is said to exist. Rijsbermann (2006) and IWMMI-CA define water scarcity as 'when an individual does not have access to safe affordable water to satisfy her or his needs for drinking, washing or their livelihoods we can call that person water insecure. When a large number of people in an area are water insecure for a significant period of time, then we can call that area water scarce'. Thus, UN-Water and the FAO define it as 'Water Scarcity is the point at which the aggregate impact of all users impinges on the demand by all sectors, including the environment, cannot be satisfied fully'. This second definition importantly includes environmental requirements. While these definitions cover many things, they do not specify whether scarcity has arisen because of mismanagement, economics or just not enough water resources in that area. Developing these definitions gives rise to two other possible definitions, namely *economic water scarcity*, which is when investments needed to keep up with the growing demand are constrained by financial, human or institutional capacity. The second is *physical water scarcity*. This occurs when available water resources are insufficient to meet all demands. From these two definitions, it can be seen that an area can theoretically have plenty of water but will still be water scarce as water is not available to the population for practical reasons, such as poor infrastructure, bad institutional arrangements or other factors (Lautze 2014). This is an important distinction. It is reasonable to assume that mere survival is not enough for a person. That would require less than 3–5 l/day. The Mar del Plata conference stated that basic needs were that 'all peoples whatever their stage of development and their social and economic conditions have the right to have access to drinking water in quantities and of quality equal to their basic needs'. This was expanded at the Earth Summit in Rio de Janeiro to encompass ecological/ecosystem needs: 'In developing and using water resources priority has to be given to the safeguarding of ecosystems'. Four basic human needs were also defined. These were (i) drinking water for survival, (ii) water enough for human hygiene, (iii) water for sanitation services, and (iv) modest household needs for food preparation. Implicit in the last need was that water would also be needed for growing the food. Unfortunately, many nations still do not provide enough water for these four needs.

Our Right to Water

The Universal Declaration of Human Rights (UHDR) of the United Nations in 1948 emphasises the right to an adequate standard of living and health (UN 1948). Article 3 states that 'Everyone has a right to life, liberty and security of person'. Article 25 states that 'Everyone has the right to a standard of living adequate for the health and well-being of himself and of his family, including food, clothing, housing and medical care and necessary social services'. Although not explicitly stated, most of these rights, either directly or indirectly, require access to safe drinking water and sanitation. The UN General Assembly and the Human Rights Council recognised in 2010 resolved an individual's right to water (Winkler 2012). Although not unanimously accepted by all countries, none voted against it, and only a few were abstained. It was thus accepted by consensus. From this consensus, the right to safe drinking water also implies an adequate standard of living and a high standard of physical and mental health, the right to life and human dignity (Human Rights Council 2010). These rights are enshrined in international law. While the International Bill of Rights does not specifically mention water in order to achieve some of its provisions, access to water is required. Rights to water are guaranteed for states that have ratified the Social and Civil Covenants under the rights for housing, health and life. Other subsequent treaties explicitly mention the right to water, e.g. the African Charter on the Rights and Welfare of the Child and the Arab Charter on Human Rights regarding the right to good health and CEDAW (Convention on the Elimination of All Forms of Discrimination against Women 1979), requiring the right to an adequate standard of living. To provide this, states are obliged to respect the human right to water by protecting it from pollution and thus not preventing proper access or endangering human health. Water must be both accessible and affordable although it is not necessarily free from a reasonable charge. These rights should be achieved as quickly as possible, which has not always happened. No single use of water can be granted unconditional priority of use, and the basic requirements of humans have to come first. The right to food implies, however, that some allocation of water for food production is required and this, apart from human health, must be given some priority. As the provision of water is a basic need that is fundamental to human existence, access to it is a human right, and in order to provide adequate amounts to all peoples and users, it must be managed efficiently, equitably and most of all sustainably. Although climate change has come into prominence in recent times, the problems of providing drinking water and sanitation must be very close behind or, for some countries and international organisations, at the very top of their priorities.

Gleick (1996) attempted to define basic water requirements not just for individuals but also for society as a whole. He reported that water is used for removing and diluting wastes, growing food, producing and using energy and transport as well as other things, including recreation. The amount of water needed for each of these will vary from one country to another. Not all water for human use is available as a piped supply. The type of water access makes a difference in its rate of consumption by potential users.

Table 5.1 Domestic water use relative to distance to supply source.

Supply source	Water use (l/person/day)
Public standpipe more than 1 km away	<10
Public standpipe closer than 1 km	20
Connection to house with flush toilets	60–100
Connection to a house with toilets and gardens	150–400

Source: Adapted from Gleick (1996).

Table 5.2 A comparison of water uses by high- and low-income country groups and the global average.

	Agricultural use (%)	Industrial use (%)	Domestic use (%)
Low-income countries	82	10	8
High-income countries	30	59	11
World average	70	22	8

Source: Adapted from UNESCO (2003).

Table 5.1 clearly illustrates the impact of access to the water source on consumption and shows that the easier the access is, the more likely it is to be used. The very basic needs for a human will depend not only on their daily activities but also on the climate of the area. Dry warm conditions can double the amount needed as can those of a person undertaking vigorous activities. The excess water taken in is to replace that lost through the skin and lungs as well as excretion. Estimates of daily water requirements range from 2.0 l/day (USEPA 1976) to 5.0 l/day (Saunders and Warford 1976). It is important to remember that the demand for potable water is different from that for raw water. Table 5.2 compares the use by the three main sectors, agriculture, industry and domestic, for developed high-income countries, low-income developing countries and the world average.

The largest differences between low- and high-income countries are with industrial use and agricultural use.

Water Availability, Demand and Allocation

Although there is a large amount of water on this planet, indeed from space it appears to be a blue planet, and less than 3% is freshwater. Of this, a large amount is locked up in ice fields, glaciers and snowfields. Deep aquifers also hold a large quantity of freshwater, approximately 0.76% of the total water in the hydrosphere and only another 0.26% in lakes, reservoirs and river systems, and these are the most readily available to humans (Shiklomanov 1998). Soil moisture is also important for agriculture. Some resources that are being used are deep aquifers, which can be regarded as non-renewable. A further major water resource that may be more intensively used in the future is the desalination of saline water although this is not without potential difficulties. When estimating freshwater availability, consideration should only be given to renewable resources and not the total amount as non-renewable resources only have a finite life and are not, in the long run, sustainable. The estimated amount of renewable water is 42 750 km^3/year. What is very important, however, is its very uneven distribution in both time and space. A further problem is that the distribution of freshwater does not always coincide with the human population distribution. Asia, for example, has about 60% of the global human population but less than 40% of the available freshwater (Postel 1992). In order to assess the water distribution relative to the population, the total actual renewable water resources (TARWR) index for a region was developed. The aim was to take into account the population of a country, and knowing the amount of available water available, the resource per person can be calculated. Some countries, such as Canada and Brazil, have very large water resources (91 420 and 45 570 m^3/person/year). If the amount falls below 1700 m^3/person/year, there is said to be water stress, and demand for domestic, industrial, agricultural and environmental requirements may not be met. When the amount falls below 1000 m^3, water scarcity is said to exist. Extreme water scarcity exists when the amount falls below 500 m^3/person/year. There are a number of problems with the TARWR index in that it only uses average figures that do not take into account variations in time and space. For example, in countries subject to monsoon precipitation, such as India, nearly 90% of the annual rainfall can fall between June and September. It may be difficult to trap enough of this rainwater in reservoirs to allow people to manage during drier times of the year. Indeed, monsoon rainfall can swell rivers to levels that cause flooding. In addition, the TARWR is for entire countries and does not take into account local differences within that country, which can be quite large. In recent years, the situation has changed largely because of human interference with the hydrological cycle. This extends beyond altering rivers, exploiting aquifers and building reservoirs and includes dramatic global water withdrawals, land use changes, destruction of ecosystems and climate change. The annual water withdrawal has increased over fourfold in the last 50 years of the last century (Shiklomanov 1998). There are many areas of the world where human interference has not only depleted resources but also resulted in a deterioration in water quality. Because of its high profile, water is at the top of the agenda of many international organisations, e.g. UNESCO, UNEP, FAO, IWAR and others. All have sponsored conferences and symposia, and many studies have been published from many countries, but the problem of scarcity is not significantly decreasing.

There is an ever-widening problem of meeting the demands of an ever-growing global population, especially in arid and semi-arid regions where demand is growing progressively faster (Rodda 2001). The result is a growing need for policymakers to try to deal with this situation, but water scarcity is not just a question of there being insufficient amounts in a location at a certain time. Water scarcity occurs at different levels or to varying degrees. Different societies are able to cope and adapt differently, and distinctions must be made, for example, between domestic uses, industry needs and agricultural requirements. Table 5.3 shows examples of different levels of water use by type and hence demand in selected countries. The overall minimum water requirement must include sanitation services but not the amount of water required for food production. For an average human living in a temperate climate, the requirements for fluid replacement are about 2 l/day, but in hot

Table 5.3 Examples of water use in selected countries (l/day/person).

Use type	United States of America[a]	Sweden[b]	The Netherlands[b]	Ave. California[c]
Toilets	95	40	39	127
Kitchen	15	50	17	56
Bathing	75	70	27	99
Laundry	50	30	17	71
Garden/yard	11	25	4	178

Source: Data abstracted from Gleick (1996). Country references: [a]Postel (1992); [b]Kindler and Russell (1984); [c]California Department of Water Resources (1994).

climates, a 70 kg human will need much more because of water lost through sweat and breath. Their requirement is about 4–6 l/day.

As can be seen, the overall daily water consumption per person greatly varies between different developed countries, ranging from over 500 l to less than 20 l in developing countries. The average overall use, including all sectors listed in Table 5.3, is provided in Table 5.4. Because of this wide variation and estimating the impact of the wide variety of anthropogenic factors, it is very difficult to calculate a global or regional availability or demand. Many factors are local or regional, e.g. reservoirs and aspects of river flows, but the main overall global factor is climate change, which will impact water resources almost

Table 5.4 Examples of overall use per person in selected countries.

Country	Average water use per person per day (l)
United States of America	550+
Australia	490
Italy	390
Japan	375
Spain	320
Norway	300
France	285
Germany	190
United Kingdom	150
India	140
China	85
Nigeria	35
Uganda	20
Cambodia	<20

Source: Adapted from Watkins (2006).

everywhere. Because different economic and societal groups have different requirements, it is best, from the point of view of demands, to treat each separately.

Industry Demands

Different industries use water for different purposes. It can be used for power and cooling in the paper and pulp industry and metal industries and as a chemical constituent in other industries. Each may require water in different quantities and of different qualities. The amount consumed can vary from 0.3% to up to 40% of the water withdrawn for an area. Cooling water, for example, is usually discharged without too much loss but with a raised temperature, which can cause problems to the aquatic biota. For other industries, a large proportion of the abstracted water is discharged as wastewater due to contamination and is frequently only partially treated, which can rise pollution problems. Many of the most severe pollution problems arose during the industrial revolution and were not properly addressed at the time. The cost of cleaning those waters retrospectively can now be very high.

Municipal Water Demands

This is water withdrawn to meet the needs of the population. This may include private gardens, garden centres, allotments and market gardens. It can also include car washes and street washing. The amount of water required can depend upon the regional climate. Required volumes depend upon the size of the population being served and whether they live in a village or city. Urban populations tend to consume more water per capita than rural populations. In large cities, the amount of water withdrawn can amount to 300–600 l/person/day (Shiklomanov 1998). With greater use of washing machines, dishwashers, showers and domestic toilets in countries such as North America and Europe, this could increase to 500–1000 l/person/day. This is in contrast to water-scarce countries where, due to lack of water resources, withdrawal can be as low as 10–40 l/person/day. Much of the water withdrawn, especially in high demand areas, is returned to the environment either treated or not depending upon the presence and functioning of the local wastewater treatment systems. Gardens lose large quantities of water to the atmosphere through evapotranspiration. Street washing loses water by simple evaporation. The rates of these depend upon local climates and season of the year. In cities with a modern infrastructure, only about 5–10% of water withdrawn is consumed. A further problem, especially in older cities, is that they often have aged pipe distribution systems that can suffer from losses through leakage.

Water Demands for Agriculture

Water is needed not only to sustain people's individual needs but also to produce the food they eat. It is essential for the photosynthetic process. Plants take up their water mainly through their roots, so the availability of water in the soil root zone is of vital importance.

Different species of plants have different strategies for dealing with variations in root zone water concentrations. Solving the water availability problem is especially important in arid and semi-arid countries, such as parts of Africa, Asia and America. Because of the uneven distribution of water, it has long been a strategy of human societies to move water from distant sources to where people need it. Irrigation of land for agriculture has been practised for at least 5000 years. For humans, the problem is to provide enough water to meet both their personal needs and for agriculture and, especially since the industrial revolution, for industry. Food production is a continuing problem with the increasing global population. The amount of irrigated land continues to increase and may continue to do so with the changing climate. Indeed, with the existing population increasing at the current rate, the demand for food will also increase, as will the area of irrigated land, although the rate did level out somewhat in the 1980s. The water requirement for agriculture does, however, vary depending upon the species of plants and animals involved, type of cultivation being followed, irrigation methods, land management, geographical location and climate/seasonal variations. Estimated amounts for irrigation are between 70% and 80% of all water withdrawals. There is a trend to improve irrigation techniques to use less water in some countries. These use water more efficiently, so reducing the volumes extracted (FAO 2001). Unfortunately, the costs of these modern systems are generally higher. Water withdrawals for irrigation vary not only from crop to crop but also with the local climate. The smallest amounts used tend to be in northern countries, e.g. in northern Europe, where the amount used is about 3000–5000 m^3/ha, but in southern Europe, it can be between 7000 and 11 000 m^3/ha. In both cases, about 25% is returned to the environment. In rural areas, the amounts of water used by the human population are relatively small compared to what is used in irrigation.

Another factor that, although not a demand, is still a loss of potentially available water is evaporation from large surface area reservoirs. In recent decades, a large number of storage reservoirs have been constructed. Apart from changing the hydrological cycle in that area, the large surface areas created results in more evaporation. This loss should be taken into account when considering water resources and demand in an area. Currently, the total surface area of all large reservoirs (those with a total volume greater than 50 km^3) is approximately 500 000 km^2. This is a considerable area, especially for those in warm or hot countries (Shiklomanov 1998). Reservoirs are vital stores of water, providing most of the water consumed by industry, agriculture and hydropower.

Estimating Freshwater Availability and Scarcity Using Indices

Societies, especially in developed countries, use water in many areas of their lives. These reflect their multi-faceted livelihoods. To meet these societal expectations, the social aspects of modern life must be taken into account when estimating water demand as must their economic activities. Water managers must be aware of different pathways of activities of societies and where they intersect with water use (Tapela 2012). Although water scarcity as a concept has been broadly understood for many years and is of serious concern to both international agencies and national governments, there is no generally accepted definition of exactly how it can be measured (White 2014). For management purposes, there needs to be a measure of the degree of water scarcity in different regions so that allocation can be fairly managed. The purpose of any indicator is to inform the user, be it a policymaker, water manager or an interested member of the public. To achieve this the index must be simple as many features of the aquatic environment and water use are quite complex. If an index is intended for use by policymakers, it must cover four main objectives, namely (i) to provide information on the topic being investigated and to determine the seriousness or otherwise of the situation, (ii) to support the making of a policy and set priorities by targeting key factors causing pressures, (iii) to monitor the effectiveness of policy responses and (iv) to raise public awareness. Inform the users of the driving forces and their impacts, thus gaining public support for policy measures. Over the past two decades, a number of indices have been proposed. Examples of water scarcity indices are provided in Table 5.5. Some indicators of water scarcity try to embrace differing requirements, including how different societies have, and still are, adapting to different degrees of scarcity. In order to define scarcity, one first needs to establish how much water per capita is needed (Chenoweth 2008). Chenoweth states that 'an indicator of water scarcity is a specific sustainability indicator useful for directing policy information and resource allocation in the water sector within the overall context of sustainable development'. He also divides different indicators of water scarcity into a number of categories. His first group is classed as Neo-Malthusian indicators. These view all water resources as being fixed, but demand will increase as the population grows. The second are economic indicators of water scarcity (Feitelson and Chenoweth 2002), which are based upon the cost of supplying all segments of a nation's population with an adequate supply of clean sustainable water and sanitation services and comparing this cost with the national income. The third category is weighted indicators of water scarcity. This was proposed by Ohlsson (1999) and consists of a combination of a water quality index and the human development index (HDI; UNDP 2011). Another category is the standard measure of water scarcity (Seckler et al. 1998; Falkenmark 1986; FAO

Table 5.5 Some examples of water scarcity indices.

Index	Author/reference
Falkenmark indicator	Falkenmark (1989)
Basic human requirements	Gleick (1996)
Social water stress index	Ohlsson (2000)
Water resources vulnerability index	Alcamo et al. (2000)
Watershed sustainability index	Chaves and Alipaz (2007)
Water supply stress index	McNulty et al. (2010)
Physical and economical water scarcity	IWMI (2008), Seckler et al. (1998), and Molden (2007)
Environmental water requirements	Sullivan (2002) and Vorosmarty et al. (2005)
Population growth impacts on water availability	Asheesh (2003)
Water stress indicator	Smakhtin et al. (2005)
Water footprint	Hoekstra (2003) and Ridoutt et al. (2009)

2003). These show the increasing danger of a world water scarcity, particularly in countries with a low gross domestic product (GDP). Other indices are provided in Table 5.5.

This table is by no means exhaustive but does outline the authors and publications of the more common indicators.

Falkenmark Index

The most commonly used indicator of water scarcity is the Falkenmark index or water stress index (WSI) of 1986. This targeted the link between water availability and food security (Falkenmark 1989). This was based upon events in Sub-Saharan Africa and was intended as an early warning system for avoiding catastrophic drops in food production, especially in the light of predicted future droughts. It also recognised that any solution to water scarcity would have to be a combination of scientific and technological approaches in their widest sense as well as good informed management and education of the public. This index concentrates upon the provision of freshwater and the role of 'green water' and 'blue water'. In addition to water is needed to sustain our lives, it is needed for food production, including plants and animals. It is essential for the process of photosynthesis in plants. Plants take up water mainly through their roots, so water in the root zone is of great importance. Falkenmark defined water stress as occurring when the resource fell below 1667 m^3/capita. If the resource fell below 1000 m^3/capita, economic development, human well-being and human health could be threatened. If the resource fell below 500 m^3/capita, then absolute water scarcity would exist. Even if a country was currently in the 1000 m^3 range as the population grew, it would approach absolute water scarcity. The difficulty with this indicator is that it does not distinguish between different patterns of use in different countries. In addition, it does not take into account the ability of a country or community to adapt to a changing situation. It is important when calculating an index to properly assess how much water is needed not only by individuals but also for modern life. This should include, where appropriate, potential improvements in standards of living, lifestyle and the general economy. This then needs to be equated with the resource available to give the scarcity index. Although these factors are not included in the Falkenmark index, it has, because of its simplicity, been widely used in many situations. The WSI was originally based upon how many people do one unit of water supply. This was defined as 10^6 m^3/year and is called the hydraulic density of a population. Falkenmark (1989) suggested that the threshold for water scarcity was <600 people per flow unit. This water density does not include water for agriculture or for energy production. If these are included and they amount to 20 times the domestic demand of 40 m^3/year, the overall requirement increases to approximately 840 m^3/capita/year. The IPCC Third Assessment Report (2001) states that water stress may be a problem if a country or region has less than 1700 m^3/capita/year. Although this index was concerned with human and food needs, Allan (2011) showed that in the Middle East they have compensated for declining water resources per capita by importing certain foods such as grains. This is considered to be a 'virtual water' import. This trend is increasing even with lower development countries, especially now that the global market for food is so well developed. This, to a certain degree, breaks the link between water and food production for self-sufficiency.

Water Poverty Index (WPI)

One of the many problems of devising and using an index is that, especially in developing countries, the relevant data is not always available, so any index either has to be simplified or only calculated on a national scale, thus missing any effects at a local level. The WPI is designed to try to overcome this difficulty and take into account local geographical variations and target individual villages or even households. Salameh (2000) defined the WPI as ‘the ratio of the amount of available renewable water to the amount required to cover food production and household uses of one person in one year under the prevailing climate conditions’. This does not cover poverty, so Sullivan et al. (2003), Sullivan (2002), Sullivan et al. (2006) and Lawrence et al. (2003) devised a more detailed WPI in order to overcome some of these problems although they still did not include virtual water. The WPI index was designed after many consultations with different agencies to provide a more efficient and realistic method for better community water management. It combines information on the amount of local water resources, ease or otherwise of access, level of water use, socio-economic aspects, including the time spent and potential economic costs and lost income caused by having to collect water. While it is acknowledged that no index is perfect, this tries to give guidance on achieving the Millennium Declaration Goals 1, 7 and 10, which state that it aims to (1) eradicate extreme poverty and hunger, (7) ensure environmental stability and (10) halve by 2015 the proportion of people without sustainable access to safe drinking water. None of these can be obtained without proper access to adequate amounts of clean safe drinking water. An index can help administrators allocate capital resources to help the community, but many indices do not allow for the true value of water (Leach et al. 1999). Water is fundamental to economic development as well as health and well-being, especially in poor communities. In these communities, the burden of collecting water mainly falls upon women and children, where as much as one-quarter of their time is spent on it, as well as the children potentially missing school (Curtis 1986). The MGDs also recognise the importance of sufficient water being made available to ecosystems. Global access to adequate improved drinking water sources and sanitation is currently being monitored by the WHO and UNICEF (2000, 2006). In this context, ‘improved’ means that the water is delivered through a household connection, public standpipe, borehole, protected dug well and protected spring or rainwater collector. There must also be a minimum amount of 20 l/person/day, and the source must be no more than 1 km away from the persons dwelling. Although this seems reasonably comprehensive, there are some over simplifications and shortcomings. To overcome these, Sullivan et al. (2003) suggested a more comprehensive approach, including the following issues:

- Measures of access
- Water quality and availability
- Water for food and other productive purposes
- Capacity to manage the water
- Environmental aspects

Questions of spatial scale also arise. Each of these is considered below.

Access

This is a measure of the distance from a dwelling to the source of water. This is oversimplistic as it does not take into account the amount of time it takes to collect the water. If there is a single tap at a reasonable flow rate per household, this would be acceptable, but if there are only one or two taps for an entire village of 400 people, the time spent queuing would be large. This time could be usefully spent on other productive tasks. It is also unacceptable if at certain times of the day or year the flow rate is very slow as this would further increase the time taken to fill their containers. In both rural and some city communities, there may be no water at certain times of the day or on certain days in a week. Even if a person could obtain water, it may not always be free, so the ability to pay can also be a constraint. If the cost is too high, that source would have to be excluded (Kasrils 2001).

Water Quality and Availability

Although access to water should only be from 'improved' sources or facilities, this does not always guarantee that the quality is of a high enough standard. For example, many wells in Bangladesh have water with unacceptably high concentrations of arsenic. Water of a good safe quality is a vital factor for good health (UNESCO 2003). Seasonal variations in availability can also affect water quality as well as quantity, which in turn can lead to health problems. This can happen when summer droughts restrict the use of some sources, so water must be collected from more distant locations that are often of poorer quality. Variable supplies can also affect the stability of the supply system and perhaps lead to leakage in the pipework, wasting water, pollutant ingress and even a total breakdown. The supply infrastructure in many poorer regions is frequently underfunded, badly maintained and poorly managed.

Water for Agriculture and Other Productive Activities

About 70% of global water abstractions are used by agriculture and food production although this is not, however, always being carried out on a large scale, especially in developing countries. In those countries, many small agricultural units are managed by the family for personal use with any excess production being sold. There may also be 'cottage industries', such as brick making, textiles and brewing, all of which help local economies and lift people out of poverty. All of these activities require water in various amounts that may be larger than that for purely domestic uses. It is also true that all of the above activities can result in water pollution as wastes are often disposed of locally and can cause problems for other water users downstream. These activities also need proper management. When larger industries such as mining and commercial agriculture are involved, not only is there greater competition for the available water but also, unless properly regulated, there is the possibility of causing significant water pollution in the area.

Capacity to Manage Water

It is important that both managers and public users have knowledge and skills and, in the case of the public, have explained to them the importance of the reasons for careful use and management so that they are willing to engage with the process of keeping the whole system as efficient and clean as possible. Public participation is essential. As part of this, it is important to examine demand management. Globally, agriculture is by far the greatest user of water, with domestic use often being the smallest. National and local governments need to provide proper training for all systems managers. It is also advantageous to encourage the formation of local user groups or associations. All of these need to have policies and regulations explained to them in a constructive way so that they can disseminate the information to their whole community.

Environmental Aspects

Local residents and user groups need to be given proper and full explanations of why it is essential to maintain the integrity of their environment and how this connects with not only water resources but also a healthy environment for them all. This includes making sure that so-called improvements to the water system as a whole do not result in environmental damage as it is the local ecosystems that provide their life support systems. Critically, this includes both large and small wetlands, many of which are threatened globally as they have an important role in both the local ecology and the water cycle. Wetlands and surface waters can be very productive and can provide food, building materials, grazing, water purification and flood control.

Questions of Spatial Scale

The location of water resources of all types is quite variable geographically, even from one village to another. Even systems that are close to one another can vary considerably in their characteristics. This not only applies to the chemical composition of the water but also their physical properties, social use and economic value. They can also vary in their water availability either seasonally or have a sustainable rate of abstraction and address the ease of access. All of these variations need regular monitoring by competent authorities and, in consultation with local users, be considered when drawing up management plans.

Monitoring all of these issues is essential in both formulating a management plan and creating and using an index of water use. All are needed for creating a meaningful WPI and, looking at it from the opposite direction, maintaining and being confident in the WPI, which means that all parameters must be regularly monitored. Correct and continuous monitoring will allow not only modifications to water use as needed but also highlights areas in the local water cycle that might require remedial action. This information is also vital for policymakers and economists when costing any modifications. This sort of management at the local level is important. The data obtained from smaller locations can also be aggregated into meaningful national indices.

Table 5.6 Index of components used and data required for each component.

WPI issues	Data required
Resources	Freshwater flows within country Flows from outside country Population now and projected
Access	% population with access to clean water % population with access to improved sanitation % population with access to irrigation adjusted By per capita water resources
Capacity	Per capita income Mortality rates of under fives Education enrolment rates Coefficients of income distribution
Use	Domestic water demand in litres per day Industrial and agricultural share of water Adjusted by sector's share of GDP
Environment	Use the following indices of water quality: pollution as a measure of water stress; environmental regulation and management; informational capacity, biodiversity based upon threatened species

Source: Adapted from Lawrence et al. (2003).

Lawrence et al. (2003) compared the WPI score with the HDI and Falkenmark scores using the data given in Table 5.6. The results for the 147 countries examined showed that most of the countries with the best scores were either developed countries or richer developing countries. Table 5.7 presents examples of both developed and developing countries and shows that even with developed countries not every country scored highly and with developing countries not every category scored badly.

Lawrence et al. also calculated the WPI for 147 countries, which included a range of developed and developing countries. They then compared them with sub-indices, such as the HDI and Falkenmark index. Their results from selected countries are shown in Table 5.7.

Various indices use different measures and different weightings on the components, hence the quite large differences in some of the results. The Falkenmark water stress index, for example, measures per capita water availability and concludes that chronic water scarcity occurs between 500 and 1000 m^3 of water per person.

The aim of the WPI is to include a wider variety of factors. The difficulty is, however, that obtaining information on all of the issues included is often difficult in some countries and is also a more expensive process. It can be used at a national scale but equally importantly can also be used at a community level, thus highlighting local differences and problems. A final point that is often not included in the considerations is the affordability of the water. This is particularly important in developing countries where water may not be abundant and the cost of provision is high. Feitelson and Chenoweth (2002) stressed the importance of cost in assessing water availability as this could not only include the cost of supply for human needs but also for the environment. They also stressed the importance of

Table 5.7 The water poverty index (WPI) and the component sub-issues compared with the human development index (HDI) and the Falkenmatk index (F) for selected countries.

Country	Resources	Access	Capacity	Use	Environment	WPI	HDI	F
Algeria	3.4	11.2	14.5	12.2	7.8	49.7	0.693	0.4
Chad	8.3	3.1	7.8	8.4	10.9	38.5	0.359	3.8
Ethiopia	6.6	3.1	8.0	8.1	9.5	35.4	0.321	1.8
Egypt	3.4	18.3	13.3	12.5	10.5	58.0	0.635	0.4
South Africa	5.6	12.2	12.7	10.1	11.6	52.2	0.702	1.2
UAE	0.0	18.6	17.1	5.5	10.9	52.0	0.809	0.1
Australia	11.9	13.7	17.6	6.5	12.5	62.3	0.936	18.2
Brazil	13.5	14.6	12.5	9.7	11.0	61.2	0.750	36.4
United States of America	10.3	20.0	16.7	2.8	15.3	65.0	0.934	8.9
France	7.9	20.0	18.0	8.0	14.1	68.0	0.924	3.1
Denmark	5.5	15.9	17.6	7.6	14.7	61.3	0.921	1.1
Germany	6.5	20.0	18.0	6.2	13.7	64.5	0.921	1.7
United Kingdom	7.3	20.0	17.8	10.3	16.0	71.5	0.923	2.5
China	7.1	9.1	13.2	12.1	9.7	51.1	0.718	2.2
India	6.8	11.0	12.1	13.8	9.5	53.2	0.571	1.9

Source: Adapted from Lawrence et al. (2003).

not only quantifying the amount of water resources but also keeping the data and calculation of the index simple. It is also important to know the amount of the resource, and it is even more important to understand and quantify its availability. This can arise from the location of the water and from seasonal variations. In the neo-Malthusian view, the amount of global freshwater is constant, and demand will increase as the population increases, which is the basis of the Falkenmark index. Although numbers can be helpful, they do not necessarily reflect behavioural differences and water use patterns between countries, areas within countries and economic status of groups within a population. It is also important to note how much of a country's water comes from outside its national boundaries (transboundary water). This latter is subject to national and international agreements and can involve a financial cost. If water is treated as a commodity, then as demand rises so can the price. While the desire to take people out of poverty is admirable, the general effect of raising the standard of living in most countries does have a perhaps unforeseen effect on demand. The Malthusian scenario would say that with finite water resources, the increasing population and increasing standards of living will inevitably increase demand and will lead to scarcity. Examples of rich humid countries include Western and Northern Europe, Canada and parts of the United States of America. Rich arid countries include oil-rich Middle Eastern States. Poor humid countries include those with a low GDP so that they do not have the capacity to provide an appropriate infrastructure. Poor and semi-arid countries, although they may have adapted in the past, are now being overtaken by a rapidly growing population demand.

Ohlsson (1998) identified 'first-order' and 'second-order' situations that could lead to water scarcity, which are defined as follows. *First -order resources* are any natural resource (such as land, water and minerals) with which a country can either be well or poorly endowed. The degree of scarcity and/or abundance can be spatial, temporal or quality. One area could be stressed in one or more aspects, but its neighbouring area may have no problems. A *second-order* resource is social, not natural, and deals with the need as perceived by society, their administrative organisations and the management responsible for dealing with the natural resource, especially in the event of water shortage. It is also important for society to be able to (i) plan for short-term changes, including dealing with floods as well as droughts, and (ii) plan for potential future long-term changes. It is also important to educate all user groups to use water more efficiently (demand-side management). There are various combinations of first- and second-order situations (Turton and Warner 2002). The situations where scarcity is concerned are as follows:

1) Where water resources are good, but second-order social/institutional ability is poor. This could lead to poor public health, a lack of economic growth and poor infrastructural development. The situation would be worse if it was also combined with high population growth. This overall situation is termed structurally induced relative water scarcity. Examples include Angola and Zambia.
2) Where water resources are low, but the second order infrastructure/institutional ability is good. In this case, the water that is available is managed well, so within these limits, the population is being well served. There could still be economic growth, and public health is maintained. Infrastructural development is still good. This situation is termed structurally induced relative water abundance. Examples include Israel and South Africa.
3) Water poverty occurs when there are poor water resources and relatively poor institutional ability to deal with it. Countries that find themselves in that situation are unable to deal with it and result in under development, economic stagnation, poor public health and little infrastructural development, all leading to social instability. The situation is made worse by population growth.

Governments and political parties must have the support of the population so that decisions by the government on water management will be supported by the population. Public understanding and support is important in water management. Governmental support for selected groups, for example certain industries or more wealthy districts, can give rise to negative effects by the population as a whole. If the occasion arises, when all available resources have been tapped and abstraction rates cannot be increased, then the reduction of water to higher use activities will have to be considered.

The cost of providing good clean safe water will only increase with an increasing population because the provision of the additional infrastructure becomes more complex. In addition to this, an increasing population will inevitably result in the generation of more waste. This, and the need for more food and hence increasing intensive agricultural production, can also result in more water pollution and an increased need for more sophisticated water treatment to meet the required standards. If the economy of a country does not grow, then the ability to both pay for the required infrastructure, resulting in increasing costs for the water, would need properly trained people to operate it. The increased cost of water will affect the consumer's ability to pay. Superimposed upon this are the potential effects of climate change on water supplies and the economy of the country. Feitelson and Chenoweth (2002) reported several potential problems in gathering data for a structural WPI. They raised the issue of the cost of continuously providing a sustainable supply of potable water. The question of the standards for being safe and potable can be an issue in some regions. The supply should be monitored regularly to meet WHO or any other nationally agreed standards and be of sufficient quantity to meet the needs of the people. Sustainability will depend not only upon the source but also on the proper operation of the treatment works. Similar questions arise concerning the provision of sanitation and wastewater treatment so that environmental pollution is minimised. The overall cost of meeting these requirements cannot, in some communities, be passed on to consumers but must be largely financed by national governments. It is also true that the supply must be more than just for basic human needs, and the additional needs of agriculture and other parts of the economy must be taken into account. Assuming that some, if not all, of the cost of providing safe drinking water and sanitation would be met by governments. Feitelson and Chenoweth (2002) estimated that the amount of money per capita in US$ would be available if a certain percentage of GNP, e.g. 1% and 5%, were made available for water infrastructure. They looked at 39 countries ranging from those with high incomes to those with low incomes. Some examples are provided in Table 5.8.

The countries in heavy print represent examples of developing countries although China, because of its progressive industrial development, would now be regarded as one of the countries that would, in many measures, be included within developed countries. Saudi Arabia is an oil-rich country, and the rest are developed countries. The figures are for the turn of the millennium.

Table 5.8 Money available to finance water infrastructure projects assuming that 1% and 5% of GNP were made available.

Country	Population	Per capita internal renewable water	1% of GNP per capita (US$)	5% of GNP per capita US$
Congo	2894336	78668	6	28
Bangladesh	131269850	10940	4	18
Philippines	82841518	4476	9	47
China	1273111290	2231	8	38
Pakistan	144616639	1678	4	22
Saudi Arabia	22757092	119	61	306
Russia	145470197	29115	23	113
The Netherlands	15981472	635	228	1142
New Zealand	3864129	88859	138	690
UK	59647790	1219	235	1176
United States of America	278058881	8983	319	1597

Source: Adapted from Feitelson and Chenoweth (2002).

Well-established developed countries potentially have more money to invest in water infrastructure and resource development irrespective of the amounts of internal water resources available to them. For poorer developing countries, especially those with more rapidly growing populations, the only option available is to receive international aid and/or loans. Even in countries such as the United Kingdom (UK), increasing demands for water may not be able to be met in some regions within the next 25 years (CIWEM 2020; Anierobi et al. 2022), so further measures including plans for the future need to be considered. With oil-rich states, although they lack internal resources, they can afford technology such as desalination to cope with shortages. Addressing the problems of economic development and available monetary resources will always be an important component in water management, so planners and politicians need these to be taken into account in any useful index as do environmental requirements and the problems of the needs of rural populations. In addition, as Cook and Bakker (2012) reported that indices are a useful tool, but they need to be 'defined in a broad integrative manner and embedded within good governance practices necessary for achieving secure water for all'.

The resource component combines both the internal and external water flows expressed on a per capita basis. Because it was thought that external flows were less secure, their value was reduced by 50%. The availability of both resources is an important consideration especially if it is variable. The % of the population that has access not only just to water but to safe drinking water is as important as access to improved sanitation. The proportion of irrigated land must also be considered. Water quality is also important, and access should only be considered in the context of the water only being from an improved source; otherwise, the quality may not be of a high enough standard for domestic use. There are examples of where this is a problem. An obvious one is from Bangladesh, where many wells are from groundwaters that have unacceptably high concentrations of arsenic, potentially affecting between 35 and 77 million people and causing skin, lung, urinary and kidney cancers. Water of good safe quality is an important factor in the health and well-being of all people (UN World Water Development Report 2003). Seasonal variations in availability can also affect water quality. This can happen when summer droughts restrict the use of some sources, so water has to be obtained from inferior and sometimes more distant sources often of poorer quality.

Capacity is concerned with income and the ability to pay for water. Here, the under-five child mortality rate is also included as it is closely related to access to clean water. Use is a measure of the domestic use per capita, industrial use per capita and agricultural use per capita. In both of the latter two, the proportion of GDP derived from industry or agriculture is divided by the proportion of water used by each.

The Social Water Stress Index

Ohlsson (1998, 2000) developed the Falkenmark index to try to take into account the fact that human societies have the ability, to a degree, to adapt to changing situations, such as the arrival of new technologies, their economic situation and anything that may alter the water resource situation in their country. He considered that factors such as education, political participation and distribution of wealth within the country all have a role in

Table 5.9 Minimum water requirements for some Middle East countries.

Country	Population in 2025	Water resource (million m^3/year)	MWR in 2025 (million m^3/year)	Water shortage (million m^3/year)
Israel	95 649	1500	2 152 113	−6 521 127
Jordan	136 088	1100	1 117 621	−176 206
Palestine	58 629	300	481 494	−1 814 945
Turkey	949 321	105 000	7 796 279	+972 037 927
Egypt	1 357 695	60 000	11 150 072	+488 499 283

Source: Adapted from Asheesh (2003).

determining the ability of a society to adapt to change. Ohlsson used the HDI to assess the ability of a society to react to a situation. The HDI was combined with the Falkenmark index and called the social water stress index. In many aspects, this is also a bridge from the Falkenmark index to water poverty index of Sullivan (2002). The new index does not measure the ability of a country to develop and use new technologies and technological developments to deal with water scarcity and whether they have adequate infrastructural investment (Chenoweth 2008). Water stress can occur even though the water resources are theoretically adequate, but other factors, such as access or water quality, are a barrier. Stress will change both seasonally, inter-annually and regionally within a country. It is also likely to change over time because of climate change. For an index to be of value, these variations must be taken into account. Asheesh (2003) calculated a scarcity index for Middle Eastern countries based upon the minimum water requirement per capita (MWR) and population growth. Table 5.9 shows selected examples of minimum water requirements for some countries.

The scarcity index is based upon available water divided by the demand, including that by agriculture, industry, etc., which should also be projected for the future, allowing for population growth and any losses to the system. Where there are shortages either within a country or from one country to another, such as can occur in the Middle East, not only more efficient use should be encouraged but also more cooperation between countries, especially as many resources are transboundary. The table clearly shows that the minimum water requirements for the population at that time, 2025, could probably not be met with internal water resources in some of the countries, e.g. Israel and Palestine. Others, such as Turkey and Egypt, have more than enough internal water to meet their needs. For water-poor countries, Asheesh reported that one serious result of the situation is migration to better endowed countries, especially those in the developed world.

Water Resources Vulnerability Index (WRVI)

If the estimated water resource is divided by the population numbers, a simple index can be obtained, but this does not cover all aspects. It takes into account renewable water resources but no other factors. Raskin et al. (1997) used the amount of water withdrawn instead of water demand as demand can vary from one area to another depending upon culture and the nature of society (Rijsbermann 2006). The WRVI is designed to overcome these potential inaccuracies. On the basis of this index, a country could be considered to be water scarce if the annual withdrawals are between 20% and 40% of the annual supply, and severely water scarce if the withdrawals exceed 40% (Raskin et al. 1997).

Water Resources Availability and Food (Especially Cereals Imports)

Globally, the largest water user is agriculture. A number of indices have been created to examine the relationship between available water resources and agricultural food production. Many countries import certain foodstuffs, the main one being cereals. This, to an extent, compensates for any internal water scarcity as they do not then need to provide the water to produce that crop (Yang and Zehnder 2002). This relationship makes it important to ascertain the size of the total freshwater resource and the volume of all imported foods both currently and in the future, allowing for any future population growth. This would allow a water-deficit indicator to be developed beyond which a greater volume of food imports could be required to meet population needs. Yang and Zehnder (2002) used yearly average water data for a country and confined their calculations to Africa and Asia. The combined cereal imports of these two were 110 million tons in the 1990s. They stated that 'This grain would require all excess freshwater resources from all other countries'. Unfortunately, countries on those two continents are home to many of the people in the world living in poverty and food insecurity. Yang and Zehnder's analysis only considered countries with a population of more than one million people. There are some options for alleviating the water shortage problem in such countries (Yang et al. 2003). More food production within a country would be desirable, but with competition for water resources, there is a strong competition from other sectors. These authors came to the conclusion that food imports would have to continue and expand to meet population growth in the future, and this could ultimately lead to greater pressures on global food production. This pressure could well grow larger with the effects of both climate change and international conflicts. There is no internationally agreed price for water as there is for oil, so the export of water-consuming commodities from water-abundant countries to water-scarce countries is currently an enormous help to the poor (Oki et al. 2007).

Index of Local Relative Water Use and Reuse

The use of satellites for geospatial mapping with improved computer modelling, better weather prediction and GIS has enabled water use and reuse to be estimated on both large and small scales. Vorosmarty et al. (2005) used these tools to examine water stress in Africa. The study areas were divided into 8 km cells (n) producing a map that was independent of political boundaries. Local data was obtained on an index of water use and reuse for each cell for domestic water use (D), industrial use (I) and agriculture (A). Water discharges were calculated on a local grid cell basis as a product of run-off and the area of the cell. All corridor discharges within a cell were added together (O_c). Thus, the local water index of water use is calculated as follows:

$$\frac{DIA_n}{Q_{cn}} \tag{5.1}$$

To calculate the water reuse index, the total water use from all cells is divided by the corridor discharge:

$$\frac{\text{Sum of } DIA_n}{Q_{cn}} \tag{5.2}$$

Vorosmarty et al. considered that a high degree of stress occurred when the local water use exceeded 40%. For any region, the degree of water stress is important when considering the viability of rain-fed agriculture and local water supplies. It is possible to transfer water from some areas identified as having no water stress to those with high stress. Such redistribution of water, if possible, could be important as approximately 75% of Africans live in arid areas and about 30% are also subject to interannual variability in discharge, while others have low variability.

Vorosmarty et al. (2005) carried out their assessment for Africa as this is a continent with, in parts, high population densities as well as high aridity. Correct and appropriate engineering and management could help many areas in mitigating water stress and improving food production, health and the general economy.

Inter-basin water transfers do not occur very often. It is not possible in many countries, and for those where costs could be too high as well as there being potential political problems. Desalination is being used in richer countries, especially oil-rich countries, such as Saudi Arabia, but the energy costs are often too high for most other countries in a water stress situation. The richer countries often use their oil to generate the energy for the desalination plant and the water obtained for non-agricultural purposes. Another option receiving more attention generally but particularly in water-poor countries is treated wastewater effluent reuse. The reuse and recycling of wastewater is already being used, particularly in water-poor countries, but it is essential that proper treatment of sewage is given in the first place. It is unwise to use untreated wastewater as this can not only cause soil pollution but also be a serious health risk to humans. The WHO guidelines for the Use of Wastewater in Agriculture and Aquaculture (WHO 1989) provide an excellent basis for the reuse of this water, which can make a valuable supplement to the resources of a country. Following these guidelines will make the wastewater safe to use. It must always be remembered that wastewaters can nearly always contain polluting contaminants, including metals, pesticides and harmful microorganisms. It can be unwise to apply them too intensively to soils. It is also important to test both water and soil and crops to ensure that they are perfectly safe. There are also concerns for the future as populations in water-poor countries will increase, forcing them to increase agricultural production, which could push them closer to the calculated threshold of 1500 m^3/capita/year (Yang et al. 2003). Countries in danger of this occurrence include Algeria, Ethiopia, Libya and Tunisia. In some of these countries, irrigation efficiency could be improved, which would provide more crops per drop.

The problem of water pollution could be more significant in some developing countries as in many countries the environmental regulations and pollution are either not strictly enforced or do not even exist (Yang et al. 2003). Future climate change could also have a detrimental effect on food production, causing food insecurity to increase as well as water availability.

Watershed Sustainability Index and Water Resources Vulnerability Index

Most of the indices described so far concentrate on the water requirement of humans and its availability, usually on a national scale, whereas most countries have considerable local variations. They have not often considered renewable water and national annual demand for water (Rijsbermann 2006). Often, the numbers are based upon 'blue water' only and neglect the contribution that 'green water' makes to both global food production (Savenije 2000) and ecosystems. Many indices also reduce data to annual averages, thus missing important temporal as well as spatial variations. Chaves and Alipaz (2007) suggested that a watershed sustainability index (WSuI) that includes not only hydrology but also human and environmental needs as well as policy management would be more useful. This index would be created for a specific river basin and is not averaged over a whole country. Each of the parameters included is given a score as indicated in Table 5.10.

The WSI has four defined indicator areas. These are hydrology, environment, life and policy. The hydrology section has two sets of variables: water availability per capita (quantity) and biological oxygen demand (BOD) as a simple measure of quality. Vanle et al. (2015) considered the NHUE-Day River Basin in Vietnam and showed that the upper reaches of the basin were more sustainable for all four parameters compared with the lower reaches. They also concluded that the lower reaches must take precedence in efforts to remedy any problems. They reported that some of the pollution originated upstream, so needed to be addressed to solve the problems in the lower reaches. The water quality has four levels based upon Falkenmark and Widstrand (1992). When availability falls below 1700 m^3/person/year, a value of 0.25 (poor) is given. When it is 1700–3400, the value was 0.5 (medium); when it is 3400–5100, the value was 0.75 (good); when it is >5100, the value was 1.00 (excellent). BOD is chosen as a water quality measure as it is probably the easiest available measure. It also correlates well with dissolved oxygen concentrations, turbidity

Table 5.10 Water sustainability index parameters and scores.

Indicators	Pressure parameter	State	Required response	Level	Score
Hydrology	Δ1-variation in the basin	Basin per capita water availability (average) Basin	Improvement in water use efficiency	Δ1 < 20%	0
	Availability of water	BOD (long-term average variation)	Improvements in WWT		
Environment	Basin's EPI	% of basin area	Progress in basin		

Table 5.11 Human development index calculation.

Index	Maximum value	Minimum value
Life expectancy at birth	85	25
Adult literacy rate	100	0
Aggregate enrolment ratio	100	0
Per capita GDP (US$ purchasing power)	40 000	100

Source: Adapted from UNO (1997) and Sonekar (2019).

and organic matter. It also provides an indication of the efficiency of wastewater treatment or the level of improved sanitation. It can also be associated with nutrient and pesticide concentrations (Hunsaker and Levine 1995; Hunsaker et al. 1992). This study showed that land use had a large impact on river water quality and, in particular, the proportion of natural vegetation to land taken over for agriculture. This relationship is often more easily calculated than trying to measure non-point pollution sources. It is also important in all environmental studies that account for future possible changes that might occur through climate change. The life pressure indicator is based upon the UNDP Chhattisgarh Human Development Indicators (2011). These were outlined in Sonekar (2019), the main points of which are outlined in Table 5.11.

Each of the four dimensions can be calculated as follows:

$$\mathrm{WSI} = \frac{\mathrm{Withdrawals}}{\mathrm{MAR} - \mathrm{EWR}}$$

Smakhtin et al. (2005) suggested four categories of environmental water scarcity and the proportion of WSI with resulting degrees of environmental water scarcity of river basins, which are as follows:

a) $\mathrm{WSI} > 1$: Overexploited with current water tapping into EWR_water basins.
b) $0.6 \leq$: Heavily exploited (0–40% of utilisable water is still available in basin before EWR is in conflict with other users); environmentally water-stressed water basins.
c) $0.3 \leq \mathrm{WSI} < 0.6$: Moderately exploited (40–70% of the utilisable water still available in a basin before EWR is in conflict with other users).
d) $\mathrm{WSI} < 0.3$: Slightly exploited.

Applying these criteria, they found that more basins had a greater degree of water stress when ecosystem water requirements were taken into account, in particular, parts of the United States of America around the Great Lakes, the Murray Darling basin in Australia and the Yellow River Basin in China.

Environmental Sustainability Index (ESI)

Consideration of potential environmental impacts is important and is based upon the ESI (World Economic Forum 2011). Water is essential for sustaining ecosystems. Human existence on earth depends upon ecosystems to provide an environment suitable for our life.

The Dublin Conference of 1991 stated that 'since water sustains all life, effective management of water resources demands a holistic approach, linking social and economic development with protection of natural ecosystems' (ICWE 1992). As natural water resources become depleted, they will be less available to ecosystems, which are then likely to degrade and not contribute to the planet's life support systems. The term 'environmental sustainability' was first coined by Goodland in Goodland (1995). He stated that this index is designed to measure and improve human welfare by protecting the sources of raw materials used for human needs and ensuring that the sinks for human wastes are not exceeded in order to prevent harm to humans. This was his formula for environmental sustainability but equally applies to the scarcity of water. Molden et al. (2012) indicate two key sectors in measuring environmental sustainability. These are (i) biological diversity and (ii) the biogeochemical integrity of the biosphere, which involves the conservation and proper use of air, water and land resources. Whatever indicator is being used, it is not the number that results as a measure but, because the indicator must be measured over time, it is the trend that is shown that is important. The first measurement can be considered as a baseline from which the trends are compared and management decisions are made in order to reach a target environmental situation. The European Environment Agency in 1999 concluded that an indicator needs to tell us (i) what is happening to the environment and humans? (ii) does it matter? (iii) are we improving? and (iv) are we on the whole better off? A further consideration with environmental issues is that many countries do not have adequate databases on ecosystems and species in their areas and so are not able to construct a proper database for future measurements and comparisons.

The environment comprises a wide spectrum of features, and to include all of them in the index is difficult. It will require a combination of the following five separate indices.

- *Water quality index*: This is relatively straightforward and includes the main physical and chemical concentrations. These are dissolved oxygen concentrations, phosphorus concentrations, suspended solids and electrical conductivity. These cover the general chemical composition and, in the case of phosphorus, one of the main causes of eutrophication.
- *Water stress index*: This includes the amount of fertilizer used per hectare of arable land, pesticide use per hectare of arable land, industrial organic pollutant concentrations in the available water and the percent of a country's area under severe water stress.
- *Regulation and management capacity index*: This includes environmental regulatory stringency, environmental regulatory innovation and the percent of land area under protected status and the number of sectoral EIA guidelines.
- *Informational capacity index*: This is a measure of the availability of information on sustainable development at the national level, the existence of action plans and an environmental strategy and the percent ESI variables missing from public global data sets.
- *Biodiversity index*: A biodiversity index is based on the numbers of threatened mammal and bird species.

Molden et al. (2012) stressed the importance for humans of all ecosystems in providing good quality water.

Most of the countries with the best scores are either developed or rich developing countries. The WPI as presented does cover a broad spectrum of parameters and, as a first attempt, is by no means comprehensive, but it is still a useful start. Data for other factors that could have been added was not because not every country could provide them. Lawrence et al. and Sullivan et al. regarded this as a work in progress and recognised that more needs to be done. They acknowledge that no index covers every possible factor and is therefore not perfect, but this is a step in the right direction. It does give pointers to meeting the Millennium Declaration Goals numbers 1, to 'eradicate extreme poverty and hunger', 7 'to ensure environmental sustainability' and 10 'to halve by 2015 the proportion of people without sustainable access to safe drinking water'.

Water moves over the earth's surface as well as through its upper layers and the atmosphere in a continuous cycle that is partly driven by gravity and partly by energy from the sun. This movement is termed the hydrological cycle (see Chapter 2). This cycle involves water as a liquid, solid and gas/vapour. It is needed to sustain ecosystems that provide ecosystem services that are used by all human societies (Jimenez-Cisneros 2015). The water cycle is foremost in stabilising and moving heat energy across the earth. It is also important in shaping our landscapes through erosion and weathering of rocks and soils as well as transporting mineral materials. Its interaction with the entire biosphere is a self-organising and self-sustaining system (Ripl 2003). Another key feature of freshwater is its uneven distribution. The amount and availability of freshwater varies in both time and space as does the climate. The most widely used classification of the various climate zones on the planet was derived by Köppen and dates from the early 1900s. This has now been developed into the Köppen–Geiger classification (Peel et al. 2007). A summary of these climate zones is provided in Table 5.12, which clearly indicates the differences between countries.

These maps indicate that in Africa three main climate zones are present with type B at just over 30% in area dominating. Human societies have traditionally grown up close to available supplies of water, such as rivers, lakes and springs. Water has provided food in the form of fish and has been harnessed to increase agricultural production. It is now also used for recreational activities as well as being significant in a number of religious activities. Freshwater systems are basically confined to continents, and as such they were exclusively controlled by factors driving fundamental Earth Systems, such as climate, lithology, topography and ecology. This natural variability is now affected by anthropogenic factors, such

Table 5.12 The earth's climate zones according to the Köppen–Geiger classification.

Climate zone	Main characteristics of zone	Area covered (%)
A – Tropical	High temperature never below 18 °C High precipitation	19
B – Arid and semi-arid	Evaporation greater than precipitation	30.2
C – Temperate	Average temperature in coldest month <18 °C but >−3 °C. Warmest month >10 °C	13.4
D – Cold	Average temperature in warmest month >10 °C Average temperature of coldest month <−3 °C	24.6
E – Polar	Temperature below 0 °C for most of year	12.8

Source: Adapted from Sheffield and Wood (2011) and Peel et al. (2007).

as land use changes, population change, changing food demands and improvements in the economic status of many people, which can lead to destabilisation in national systems. In many regions, these anthropogenic drivers have become more important than the earth drivers. With population growth and the increased demand for resources, building land and agricultural land, there is an increased demand for controls to be introduced on these adverse effects to the hydrological cycle, such as floods and droughts. However, any modification to the water cycle must take into account wider implications of change for other stages of the cycle. The combined effect of all of these drivers is most likely to lead to an increase in natural hydrological disasters, such as the melting of glaciers, causing landslides. These factors together with the increased demand placed upon freshwater resources by an increasing human population result in a potential scarcity in supply. The provision of basic needs is fundamental for human existence. Access to water is a human right. In order to provide adequate amounts of water to all peoples, it must be managed efficiently, equitably and most of all sustainably. Although climate change has come to prominence in the past decades, the problems of water and sanitation should be very close behind or, with some international organisations and countries, at the very top. There have been a number of international conferences during the past 50 years paving the way to more sustainable water management. The main conferences are listed in Table 5.13 together with some of the water topics discussed.

There have also been a number of covenants agreed over the years that either directly or indirectly imply the right to an adequate supply of safe freshwater for everyone. These include the 1948 UHDR, the 1966 International Covenant on Economic, Social and Cultural Rights (ICESCR), the Inter-American Convention on Human Rights and the European Convention on Human Rights, but there are also many others (see Gleick 2007). Although these generally agree that water availability is central to life, health, food and economic development, many people still do not have proper access to water and sanitation. Water should no longer be regarded as a cheap commodity. Its availability and supply are likely to become more difficult and expensive in the future. Either more of the resource has to be tapped to meet demands or each person and user must adapt to having a smaller allocation. Although the 1948 UHDR was not binding upon member states, in subsequent years the United Nations produced a covenant that was binding and stated that all member states should recognise the right of individuals to have an adequate standard of living that includes food, clothing and housing. They should also strive to continuously improve living conditions and control disease. The right to life was explicitly stated in the International Covenant for Civil and Political Rights (Tomuschat 2008). It has also been stated that the term ‘Inherent right to life’ must not be too narrowly interpreted and that states should adopt positive measures to ensure this right. In more recent years, many international meetings have further reinforced different aspects of water, including the need for Integrated Water Resource Management (see also Mitchell 2005; Thomas and Durham 2003).

It is clear that there is international agreement that drinking water and sanitation problems need to be addressed. Water is a thread that draws all human activities together. This thread runs through food production, health, energy, industry and all aspects of the economy and recreational activities. The water requirements of ecosystems must also be taken into account and not ignored. Climate change effects on the hydrological cycle are already affecting human requirements for water supply and availability. Although some progress

Table 5.13 Examples of major international conferences on freshwater from 1972 to 2015.

Date	Conference	Water-related topics
1972	UN Conference on Human Development	Human actions and their environmental impacts
1977	UN Conference on Water. Mar del Plata	Water resources and their systematic measurement
1981–1990	International Drinking Water and Sanitation Decade	All people should have adequate drinking water and satisfactory disposal of excreta by 1990
1990	Global Consultation on Safe Water and Sanitation, New Delhi	Safe water and proper means of waste disposal
1992	International Conference on Water and the Environment, Dublin	Freshwater is a finite resource, essential to maintain; management should be participatory and the role of women should be recognised; water has an economic value
1994	Ministerial Conference on Drinking Water Supply and Environmental Sanitation, Noordwijk	High priority to provide basic sanitation and excreta disposal to urban rural areas
1997	First World Water Forum, Marrakech	Emphasising the basic human need for clean water and sanitation; management of shared waters
1998	International Conference on Water and Sustainable Development	Improve coordination and encourage political commitment for sustainable development
2003	UN General Assembly proclaimed The International Decade for Action 'Water for Life'	Focus on achieving water-related goals in the UN Millennium Declaration
2007	IPCC Working Group II	Climate change and its impacts on water
2014	IPCC Working Group II	Climate change 2014; impacts and vulnerability on water
2015	Paris Declaration	Emphasising climate change impacts on water

This table is not a comprehensive list. There are several other meetings that could be included. In spite of the international interest, much work concerning water management still needs to be done.

has been made over the past few decades, there is still a large gap in the universal provision of safe drinking water and sanitation in many parts of the world and affecting millions of people.

Because of the ever-increasing population, it is not possible to continue to allow everyone to have unlimited amounts of water. Hence, the term often used 'adequate amounts'. The question then arises as to what is adequate and what is the minimum requirement? The Mar del Plata conference stated that basic needs were that 'all peoples whatever their stage of development and their social and economic conditions have the right to have access to drinking water in quantities and of quality equal to their basic needs'. This was expanded at the Earth Summit in Rio de Janeiro to encompass ecological/ecosystem needs: 'In developing and using water resources, priority has to be given to the satisfaction of basic needs and the safeguarding of ecosystems'. Four basic human needs were also defined. These were (i) drinking water for survival, (ii) water enough for human hygiene, (iii) water for sanitation services and (iv) modest household needs for food preparation. Implicit in this last need was that water would also be needed for growing food. Unfortunately, many nations still do not provide enough water for these four needs.

A number of indicators of water scarcity have been proposed to try and embrace differing requirements, including how different societies have, and still are, adapting to different degrees of scarcity in order to define scarcity. One first needs to establish how much water per capita is needed. Chenoweth (2008) divides various indicators of water scarcity into a number of categories. His first group is classed as Neo-Malthusian indicators. These are based upon the ideas proposed by Thomas Malthus, an economist who in 1798 published his essay 'On Principles of Population' in which he argued that the population was growing at a faster rate than the resources available. To survive, there must be a degree of self-restraint in consumption. Chenoweth grouped a number of scarcity indicators that were based upon the premise that scarcity was a function of demand and availability, i.e. resources are finite, but demand is increasing as the population grows. These indicators are calculated on the basis of renewable water resources per capita. For an average human living in a temperate climate, the requirement for fluid replacement is about 2 l/day, but in a hot climate, a 70 kg human will need much more as more loss occurs through sweat and breath, so they would need 4–6 l/day. The overall minimum water requirement must include sanitation services but not the water required for producing the food consumed. Estimates of water requirements per person for various uses taken from a range of authors include food preparation and dishwashing 10–50 l/day, sanitation 2.5–26 l/day and bathing 5–150 l/day. Use will vary in different countries as shown in Table 5.14.

The overall daily water consumption per person ranges from over 550 l to less than 20 l. Examples are provided in Table 5.15.

Table 5.14 Examples of water use in different countries.

Use type	United States of America[a]	Sweden[b]	The Netherlands[b]	Ave. California[c]
Toilets	95	40	39	127
Kitchen	15	50	17	56
Bathing	75	70	27	99
Laundry	50	30	17	71
Garden/yard	11	25	4	178

Source: Data abstracted from Gleick (1996). Country references: [a]Postel (1992); [b]Kindler and Russell (1984); [c]California Department of Water Resources (1994).

Table 5.15 Examples of water demands in selected countries.

Country	Average water use per person per day (l)
United States of America	550+
Australia	490
Italy	390
Japan	375
Spain	320
Norway	300
France	285
Germany	190
Philippines	160
United Kingdom	150
India	140
China	85
Nigeria	35
Uganda	20
Cambodia	<20

Source: Adapted from Watkins (2006).

Water use is also greatly affected by the price paid. Where prices are low, as in Canada, the United States of America, Italy and Spain, use is high. Where price is relatively high, as in Sweden, Denmark, France and The Netherlands, use is low (McDonald and Mitchell 2014).

It is important that societies adapt to actual or potential water scarcity. Educating the public firstly to the problem and secondly to what they can do to help. This means that societies must be encouraged to modify their approach to water use without adversely affecting their needs but understanding that there may be unforeseen knock-on effects (Ohlsson 1999). Three stages of adapting to water scarcity were identified. The first is to use engineering hydrology to try to obtain more water. This could include building more dams and water transfer from water-rich to water-poor regions, usually by building transfer pipes. It could also include drilling deeper boreholes or increasing the depth of existing boreholes and trapping more rainwater and floodwater. These can involve either small- or large-scale projects. The second stage is introduced when the first approaches are unable to meet the demand. This involves instigating greater end-use efficiency (more per drop). These factors can eventually lead to increased water pricing. For the third level, a society would be forced not to require food self-sufficiency. Obviously, enough food is needed to meet the needs of the population, so an alternative way of meeting the food demand is needed that does not require as much water, especially from internal resources. One possibility is to develop an economy with enough income to pay for imported products, i.e. develop a virtual water trade. These alternatives are not without problems. The first depends upon more internal water resources being available to develop and having the funds to build the infrastructure required. Further exploitation of resources must be done in a sustainable way and include the needs for the environment and the development of an efficient and knowledgeable institutional management; otherwise, it will just delay the problem and not solve it. Educating the public is essential and should involve recognising different demands from different sectors. This could be a problem when encouraging economic development to raise the GDP to pay for imported food. In addition, as standards of living increase, often so does the demand for water. Importing foods also needs to be managed and perhaps diets changed. This could be a problem in some societies. A further problem is that developing the economy would quite likely mean that there would be most growth in cities requiring more water to be diverted to them. It would be very important for city development to be as water efficient as possible, meaning more efficient design of dwellings, offices and factories as well as transport systems. All options for each stage are a great challenge that must be planned for well in advance.

References

Alcamo, J., Henrichs, T., and Rosch, T. (2000). World Water in 2025: Global Modelling and Scenario Analysis for the World Commission on Water for the 21st Century. Kassel World Water Series Report No 2, *Center for Environmental Systems Research*, Germany; pp. 1–49.

Allan, T. (2011). *Virtual Water, Tackling the Threat to Our Planet's Most Precious Resource*, 368pp. London: I.B. Tauris & Co.

Anierobi, C.M., Okeke,F.O., and Efobi, K.O. et al. (2022). Waste to Wealth Activities of Scavengers: A Panacea to the Environmental Impact of Global Climate Change in Enugu Metropolis, Nigeria, pp. 1–12.

Asheesh, M. (2003). Allocating the gaps of shared water resources (The Scarcity Index). Case study Palestine Israel. *IGME*; pp 797–805.

California Department of Water Resources. (1994). The California water plan update. *Bulletin 60-93*, Sacramento, CA, USA.

Chaves, H.M.L. and Alipaz, S. (2007). An integrated indicator based in basin hydrology, environment, life, and policy. The watershed sustainability index. *Water Resource Management* 21: 883–895.

Chenoweth, J. (2008). A re-assessment of indicators of national water scarcity. *Water International* 33 (1): 5–18.

CIWEM (Chartered Institution of water and Environmental Management) (2020). The Environment. What goes around: Inside the circular economy, April 2020.

Cook, C. and Bakker, K. (2012). Water security: debating an emerging paradigm. *Global Environmental Change* 22: 94–102.

Curtis, V. (1986). *Women and the Transport of Water*. London: Intermediate Technology Publications.

Falkenmark, M. (1986). Fresh water – time for a modified approach. *Ambio* 15: 192–2000.

Falkenmark, M. (1989). The massive water scarcity threatening Africa – why isn't it being addressed? *Ambio* 18 (2): 112–118.

Falkenmark, M. and Widstrand, C. (1992). Population and water resources: a delicate balance. *Population Bulletin* 47 (3): 1–36.

Feitelson, E. and Chenoweth, J. (2002). Water poverty: towards a meaningful indicator. *Water Policy* 4: 263–281.

Food and Agricultural Organisation (FAO) (2001). *Crops and Drops. Making the Best Use of Water for Agriculture*. Rome: FAO.

Gleick, P. (1996). Basic water requirements for human activities: meeting basic needs. *Water International* 21: 83–92.

Gleick, P. (2007). The human right to water. *Water Policy* 1 (5): 487–503.

Goodland, R. (1995). The concept of environmental sustainability. *Annual Review of Ecology and Systematics* 26: 1–24.

Hoekstra, A.Y. (2003). Virtual Water Trade. *Proceedings of the International Expert Meeting on Virtual Water Trade*. Value of Water Research Report Series No 12. Delft, The Netherlands: UNESCO-IHE.

Human Rights Council (2010). Human rights and access to safe drinking water and sanitation, 6 October 2010, A/HRC/Res/159; para 3.

Hunsaker, C.T. and Levine, D.A. (1995). Hierarchical approaches to the study of water quality in rivers. *Bioscience* 45: 193–203.

Hunsaker, C.T., Levine, D.A., Timmins, B.L. et al. (1992). Landscape characterisation for assessing regional water quality. In: *Ecological Indicators* (ed. D. McKenzie, E. Hyatt, and J. McDonald), 997–1006. New York: Elsevier Applied Science.

International Conference on Water and the Environment (ICWE) (1992). *The Dublin Statement and Record of the Conference*. Geneva: WMO.

IPCC (Intergovernmental Panel on Climate Change) (2001). *Third Assessment Report-Climate Change*. Cambridge University Press.

IWMI (2008). *Areas of Physical and Economic Water Scarcity*. UNEP/GRID-Arendal Maps and Geography Library.

Jimenez-Cisneros, B. (2015). Responding to the challenges of water security: The eight phase of the International Hydrological Programme, 2014–2021. Hydrological Sciences and Water Security: Past, Present and Future. *Proceedings of the 11th Kovacs Colloquium,* Paris, France (June 2014).

Kasrils, R. (2001). *Keynote Speech to International Conference on Freshwater*. Bonn 3–7 December 2001.

Kindler, J. and Russell, C.S. (ed.) (1984). *Modelling Water Demands*. Toronto, ON: Academic Press.

Lautze, J. (2014). *Key Concepts in Water Resource Management. A Review and Critical Evaluation*. Earthscan: Routledge.

Lawrence, P., Meigh, J. & Sullivan, C. (2003). The Water Poverty Index; International Comparisons. Wallingford: Centre for Ecology and Hydrology.

Leach, M., Mearns, R., and Scoones, J. (1999). Environmental entitlements: dynamics and institutions in community based natural resource management. *World Development* 27 (2): 225–247.

McDonald, A.T. and Mitchell, G. (2014). Water demand, planning and management. In: *Water Resources. An Integrated Approach* (ed. J. Holden). Oxford: Routledge.

McNulty, S., Ge Sun, Moore Myers, J., Cohen, E., and Caldwell, P. (2010). Robbing Peter to pay Paul: Trade-offs between ecosystem carbon sequestration and water yield. *Proceedings of the Environmental Water Resources Institute Meeting*, Madison, WI, 12.

Mitchell, B. (2005). Integrated water resource management, institutional arrangements and land-use planning. *Environment and Planning A* 37: 1335–1352.

Molden, D. (2007). *A Comprehensive Assessment of Water Management in Agriculture.* Colombo: International Water Management Institute (IWMI).

Molden, B., Janouskova, S., and Hak, T. (2012). How to understand and measure environmental sustainability: indicators and targets. *Environmental Indicators* 17: 4–13.

Ohlsson, L. (1998). Water and social resource scarcity. An issue paper commissioned by FAO/AGLW. Presented as a Discussion Paper for the 2nd FAO Email Conference on Managing Water Scarcity. WATSCAR 2.

Ohlsson, L. (1999). *Environment, Scarcity and Conflict: A Study of Malthusian Concerns.* Goteborg: Department of Peace and Development Research, Goteborg University.

Ohlsson, L. (2000). Water conflicts and social resources scarcity. *Physics and Chemistry of the Earth* 25 (3): 213–220.

Oki, T., Yano, S., and Hanasaki, N.A. (2007). Economic aspects of virtual water trade. *Environmental Research Letters* 12: 044002.

Peel, M.C., Finlayson, B.L., and McMahon, T.A. (2007). Updated world map of the Koppen–Geiger climate classification. *Hydrology and Earth System Sciences* 11: 1633–1644.

Postel, S. (1992). *Last Oasis: Facing Water Scarcity*, 239pp. Worldwatch Institute Series. London: W.W Norton.

Raskin, P., Gleick, P., Kirshen, P. et al. (1997). *Water Futures: Assessment of Long-Range Patterns and Prospects.* Stockholm: Stockholm Environmental Institute.

Ridoutt, B.G., Srellahewa, J., Simons, L., and Bektash, R. (2009). Product Water Footprinting: How Transferable are the concepts from Carbon Footprinting? *Australian Conference on the Life Cycle Assessment*, Melborne, Australia, pp. 16–19.

Rijsbermann, F.R. (2006). Water scarcity: fact or fiction? *Agricultural Water Management* 80: 5–22.

Ripl, W. (2003). Water the bloodstream of the biosphere. *Philosophical Transactions of the Royal Society of London. Series B: Biological Science* 358: 1921–1934.

Rodda, J.C. (2001). Water under pressure. *Hydrological Sciences* 46 (6): 841–854.

Salameh, C. (2000). Redefining the water poverty index. *Water International* 25 (3): 469–473.

Saunders, R.J. and Warford, J.J. (1976). *The Goal of Improved Health. Village Water Supply. Economics and Policy in the Developing World*, 31–55. Baltimore: World Bank/John Hopkins University Press.

Savenije, H.H.G. (2000). Water scarcity indicators; the deception of numbers. *Physics and Chemistry of Earth* 25: 199–204.

Seckler, D., Amarasinghe, U., Molden, D., de Silva, R., and Barker, R. (1998). World Water Demand and Supply, 1990–2025; Scenarios and Issues. *Research Report 19.* Colombo, Sri Lanka (IWMI).

Seckler, D., Molden, D., and Barker, R. (1999). *Water Scarcity in the Twenty First Century.* Colombo: International Water Management Institute (IWMI).

Sheffield, J. and Wood, E.F. (2011). *Drought. Past Problems and Future Scarios.* Earthscan Publications.

Shiklomanov, I.A. (1998). World Water Resources. A new appraisal and assessment for the 21st century. *Monograph prepared for the International Hydrological Programme of UNESCO*, Paris.

Smakhtin, V., Revenga, C., and Doll, P. (2005). *Taking into Account Environmental Water Requirements in Global-Scale Water Resource Assessments.* IWMI The Global Podium.

Sonekar, B.L. (2019). Human development index in Chhattisgarh state: an analytical study. *Journal of Economic & Social Development* 15 (1): 104–110.

Sullivan, C. (2002). Calculating a water poverty index. *World Development* 30 (7): 1195–1210.

Sullivan, C., Meigh, J., Giacomello, A.M. et al. (2003). The water poverty index: development and application at the community scale. *Natural Resources Forum* 27: 189–199.

Sullivan, C., Meigh, J., and Lawrence, P. (2006). Application of the water poverty index at different scales: a cautionary tale. *Water International* 31 (3): 412–426.

Tapela, B.N. (2012). Social Water Scarcity and water. *WRC Report No. 1940/1/11.*

Thomas, J.S. and Durham, B. (2003). Integrated water resource management; looking at the whole picture. *Desalination* 156: 21–28.

Tomuschat, C. (2008). International Covenant on Civil and Political Rights. United Nations Audiovisual Library of International Law. UN.

Turton, A.R. and Warner, J.F. (2002). *Exploring the Population/Water Resource Nexus in the Developing World.* Change and Security Project. ECSP Publications.

UNDP, Chhattisgarh Economic and Human Development Indicators. (2011).

UNESCO (2003). *Water for People, Water for Life.* Paris: UNESCO.

UNO (1997). Report to UNDP on Human Development. https://www.undp.org/publications/human-development-report-1997.

United Nations (1948). *Universal Declaration of Human Rights.* UN General Assembly.

United Nations World Water Development Report (2003). Water for people water for life.

USEPA (1976). EPA-570/9-76-003.

Vanle, T., Duongbui, D. and Buurman, J. (2015). Development of a watershed Sustainability Index to assess water resources in the NHUE-DDAY River Basin, Vietnam.

Vorosmarty, C.J., Douglas, E.M., Green, P.A., and Revenga, C. (2005). Geopolitical indicators of emerging water stress: an application to Africa. *Ambio* 34 (3): 230–236.

Watkins, K. (2006). Human Development Report 2006. Beyond Scarcity, Power, Poverty and the Global Water Crisis. *UNDP Human Development Report.*

White, C. (2014). Understanding water scarcity: definitions and measurements. In: *Global Water: Issues and Insights* (ed. R. Q. Grafton et al.). Canberra: ANU Press.

WHO and UNICEF (2000). Water Supply and Sanitation Assessment. Geneva.

WHO and UNICEF (2006). Meeting the MGD Drinking Water and Sanitation Target. The Urban and Rural Challenge of the Decade. Geneva.

Winkler, I.T. (2012). *The Human Right to Water. Significance, Legal Status and Implications for Water Allocation.* Oxford: Hart Publishing Ltd.

World Economic Forum (2011). *Water Security: The Water-Food-Energy-Climate Nexus.* World Economic Forum, Island Press.

World Health Organisation (WHO) (1989). Health Guidelines for the Use of Wastewater in Agriculture and Aquaculture. *Technical Report Series 778.* WHO Geneva, p. 74.

Yang, H. and Zehnder, J.B. (2002). Water scarcity and food import: a case study for Southern Mediterranean countries. *World Development* 30 (8): 1423–1430.

Yang, H., Reichert, P., Abbaspour, K.C., and Zehnder, A. (2003). A water resources threshold and its implications for food security. *Environmental Science & Technology* 37: 3048–3054.

6

Water Quality: Some Management and Use Issues

It is rare to find pure water, just H_2O, in the environment with no other chemicals present. As rain falls, it will dissolve certain gases in the atmosphere and perhaps collect any particulate materials so that when it reaches the earth's surface it already contains chemicals and minerals other than just H_2O. Even if no other pollutants are present, rainwater will dissolve a certain amount of CO_2 from the atmosphere. This will make the water slightly acidic, frequently between pH 5 and 6. When rainwater reaches the ground, it will dissolve other substances it comes into contact with because water is a good general solvent. In natural conditions, the chemical composition and quality of surface and groundwaters will depend upon factors such as the geology of the area, topography, land use, climate and biology. When precipitation occurs, the water will either evaporate directly or return to the atmosphere by evapotranspiration. If it does not first percolate into the soil or be taken up by the vegetation, it will run downhill under the influence of gravity or percolate through the soil, forming groundwater. All of these processes are linked as part of the hydrological cycle (see Figure 2.1). The combination of these naturally occurring substances in the water is not necessarily consistent throughout the year, and seasonal differences commonly change the water quality. Water quality criteria will also be viewed differently depending upon their intended use, from an international, national or local perspective, but wherever and whatever the water is used for, quality is important. Because the world's population continues to grow, so both the demand for water and the amount of waste products being produced, including human, agricultural and industrial wastes, will increase. Much of this waste is disposed of directly or indirectly into water courses. The result is that more people die from drinking unsafe water than from all other forms of violence, including wars, each year (Gleick 2012). A deterioration in water quality severely reduces the quantity that is safely available for the above uses. An adequate supply of clean water is essential for human life and ecosystem survival. Poor water quality will threaten not only human health and life but also economic and social development. Over 3.5 million people die each year from drinking either chemically or biologically contaminated water. Almost 450 km^3 of wastewater is discharged into the environment each year. The majority of this will be discharged into rivers and lakes. If this contamination is not curtailed, many world 9+ rivers would be unsuitable for human use in the future. Fortunately, the situation is being controlled, but vigilance is still needed. The situation is worse in Africa and Southern Asia and developing countries in general. In India, many of the major rivers that are used for drinking by the adjacent towns and cities are

Water: Our Sustainable and Unsustainable Use, First Edition. Edward G. Bellinger.

severely contaminated with untreated sewage and rubbish. The World Health Organization (WHO) produces guidelines for the quality of drinking water (World Health Organization 2011), which recommends minimal standards for safe drinking water. These, or variations in them, have been adopted by all countries. Depending upon local circumstances, the standards for certain substances may, however, be stricter in some countries. Other countries, such as those in the European Union, have their version of the WHO standards (EU Drinking Water Directive, 98/83/EC 1998; Revision of the Drinking Water Directive, February 2018). Poor water quality will threaten not only human health and life but also the economic and social development of society and ecosystems. A large number of pollutants find their way into surface waters either by direct discharge (point sources), indirect seepage (diffuse sources) or even atmospheric deposition. Metals and compounds, such as polychlorinated biphenyls (PCBs), can enter by any or all of these routes. Mercury contamination, for example, can come from point sources and by short- and long-range atmospheric transport through the burning of fossil fuels and industrial discharges. Mercury and other elements and compounds, if present in water, can be biomagnified through the food chain, increasing their concentrations with each consumer organism from micro-organisms through invertebrates to fish, which, when consumed by humans and animal predators, can be poisonous (Meybeck et al. 1989). Managing lake water and river water quality requires consultation between all relevant actors with stakes in water usage in the region. It is inevitable that there will be a degree of disagreement between different parties that can be resolved with the additional requirement of conforming with national and international standards.

Whatever the perspective, it is essential to have a good comprehensive database in order to produce and execute an effective management system for the provision of an adequate supply of good safe drinking water for all users as well as ecosystems. Water quality is as important as water quantity. Poor quality renders much water unusable and thus reduces the quantity available. The effects will be on ecosystems with many native species of fish in some regions being threatened with extinction. It can also threaten the general functioning of ecosystems and thus ecosystem services. There will also be economic implications to health, tourism, agriculture and water treatment costs. In addition to geology affecting the chemicals that will dissolve in a particular water land use, temperature, organic waste, persistent organic materials and sediment loads will all have an effect. The range of contaminants in water is very large. Examples of pollutants include nutrients N and P, sediments, pesticides, persistent organic compounds, heavy metals and acids. Many of these contaminants can act synergistically and may have a multiplying impact on the recipient. Some degrade naturally and become less harmful. Much poor water quality is caused by human activity. For example, about 70% of industrial wastes in developing countries are disposed of untreated into surface waters, causing severe contamination to existing water supplies and water usage (UN-Water 2009). The estimated cost of this for countries in the Middle East and North Africa ranges between 0.5% and 2.5% of Gross Domestic Product and up to 5% for the rest of Africa (UN WWAP 2009). Although ecosystems can tolerate certain levels of contamination, the continued inputs of pollutants overcome their ability to resist, often resulting in irreversible change. Because most plants are at the bottom of the food chain and are the main food source for most groups of animals, the impacts of pollution on them will be discussed first. Many species of plants are able to accumulate metals in their tissues. When consumed, this contamination is passed on to the consuming animal. Many experiments have been carried out on the rates and amounts of contamination. The easiest group to work on is the relatively simple group of plants, algae. Algae have been used as biological indicators by many authors as they can provide information on the surrounding physical and chemical environment, i.e. the water quality (Bellinger and Sigee 2015). The presence or absence of different species, cell sizes and physiological states all reflect environmental conditions in lakes, rivers and wetlands. Changes in the dominance of certain species can possibly indicate biochemical and physical changes within the water body. The changes in the algal community are often called a 'biomarker' of changing conditions. These changes can come from the increase in a number of pollutants, including heavy metals. Cattaneo et al. (2004), examining pollution from mining activities in Lac Dufault (Quebec, Canada), found that the diatoms *Tabellaria flocculosa* and *Asterionella formosa* dominated in pre-mining conditions but significantly reduced in abundance when pollutants from mining containing Zn, Cd and Fe entered the lake. Then, another species, *Fragilaria*, became important. As the contamination reduced and partial recovery occurred *A, Formosa* reappeared in relative abundance (see Figure 6.1 illustrating these diatoms; see also Bellinger and Sigee 2015). Cattaneo et al. also noted changes in cell size and some deformations during the pollution period. Similarly, Cattaneo et al. (1998) reported significant species changes in Lago d'Orta in northern Italy in response to pollution from discharges of metals from an electroplating industry and ammonium sulphate from a rayon factory. They not only examined changes in the diatom populations but also animal populations of cladocerans and thecamoeba. All three groups of organisms showed changes coincident with the increases in pollutants. They also concluded that smaller organisms with faster reproductive cycles could adapt and tolerate pollutant levels to a greater extent than larger and slower growing species. The response of algae to these environmental changes is illustrated in Figure 6.2. Of commercial as well as environmental concern is the effect of acidic and metal-containing pollutant discharges. Quite often where such discharges occur, fish are completely absent. Examples include part of the Obey River in Tennessee, USA (Carruthers and Barlow 1973), and below coal mining and gold mining areas in South Africa (Warner 1971), although this is not always the case. Greenfield and

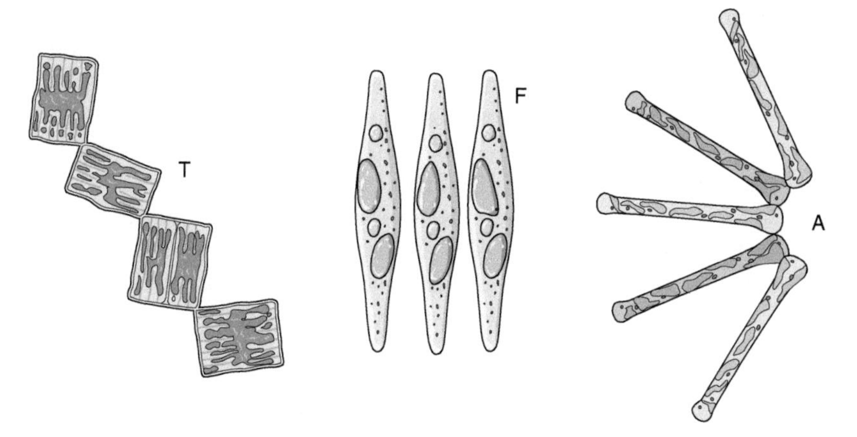

Figure 6.1 Examples of some of the diatom species affected by water quality. T, Tabellaria; A, Asterionella; F, Fragilaria.

Lake type	Spring	Summer	Autumn	Winter
Oligotrophic	Diatoms, e.g.	Blue green species, e.g. Anabaena, Osciilatoria. Some Blue greens and Diatoms		
Mesotrophic	Diatoms, e.g. Asterionella, Tabellaria.	Dinoflagellates, e.g. Ceratium	Diatoms, e.g. Asterionella, Stephanodiscus, Fragilaria Melosira.	
Eutrophic	Diatoms, e.g. Asterionella	Ceratium, Anabaena, Microcystis some green algae e.g. Scendesmus, Volvox	Stephanodiscus, Melosira Eudorina	
Hypertrophic	Stephanodiscus	Sceneedesmus, Pediastrum, Chlamydomonas	Blue green Aphanocapsa	

Figure 6.2 The changes in algal populations from oligotrophic to eutrophic lakes or reservoirs. It must be realised that these species are examples of the groups of algae that may occur. Depending upon which part of the world the lake is can, and probably will, alter the species that are found. Adapted from Sigee (2005) and Reynolds (1990).

Ireland (1978) carried out experiments with caged fish in 1978 in the River Irwell, Lancashire, United Kingdom, and showed that there were significant impacts on the health of the fish, including the production of excess mucus and the formation of deposits on their gills. It has been generally found that where the pH of a discharge is <3.0, both the flora and fauna become very restricted. Some highly specialised organisms can live in these conditions, for example the algae *Chlamydomonas acidophila* and *Euglena mutabilis*. These do not seem to occur elsewhere. Generally, it seems that specialised plants are more likely to survive than animals. What is clear is that firstly there must be frequent and accurate monitoring of both discharges and the receiving waters to assess any effects, and there must be controlled to neutralise or remove any substances that could harm the biota from the effluent at source. Water temperature can be changed by many human activities, including discharges from reservoirs, power station cooling systems and some industrial discharges. This can affect the spawning of fish, oxygen concentrations in the water and fish preferences to amounts required. It will also affect the metabolic activities of many aquatic organisms. Freshwater plants and animals cannot tolerate raised salinity in the water that can arise from drainage from certain groundwater systems, oil and gas drillings and some industrial processes. These raised levels can also arise from some municipal water treatment processes. Trace metals such as zinc, copper and selenium are found naturally but can be increased significantly through mining as well as agriculture when large concentrations can be mobilised, resulting in higher concentrations that can adversely affect fish and birds. The human chemical industry produces many hundreds of new compounds each year, some of which are released into the environment without knowing their possible adverse effects. Some contaminants react with each other, resulting in compounds that are even more toxic. Some of these can enter groundwaters where they are extremely difficult to remove. These include pesticides, dioxins, PCBs and furans. Table 6.1 indicates the main industrial activities that produce organic pollutants found in freshwaters.

From this table, food production in both high- and low-income countries can be seen to be the largest contributor of organic water pollutants. Pulp and paper industries feature high on the list for high-income countries and textile industries for low-income countries. Contaminants can be either inorganic or organic chemicals, either microbial or non-living, in suspension or in solution. In developed countries, effluent from wastewater treatment plants is not only from human wastes but also from a multitude of household cleaning

Table 6.1 Proportions of organic pollutants produced by different human activities.

Source	% contaminants produced	Economic status of country
Food industry	39.6	High
	53.9	Low
Pulp and paper	23.0	High
	10.1	Low
Metal	10.2	High
	6.7	Low
Chemical	8.8	High
	7.2	Low
Clay and glass	0.2	High
	0.3	Low
Textile	6.6	High
	14.6	Low
Others	11.5	High
	7.3	Low

High is high OECD income countries; Low is low-income OECD countries.
Source: Adapted from UN WWAP (2003).

products, personal care products and pharmaceuticals, many of which are not removed in normal wastewater treatment plants. One of the largest users of water is the power generation industry, where water is used for cooling. The discharge will often be a different temperature from the receiving water and may also contain some chemical treatment used in keeping their cooling systems free from contamination. The contaminants in industrial wastes can, potentially, be of any of the above categories. Globally, industry dumps about 300–400 million tons of heavy metals, solvents and toxic sludge into surface waters each year (Palaniappan et al. 2012). Environmental problems caused by mining can arise at several stages in operation. These are the exploration and development of the site for production, extraction of the ore, concentration of the ore and processing of the ore. All but the final stage are usually carried out at the site of the mine, but the last stage can be carried out remotely from the mine (Ripley et al. 1979). These stages are similar for mineral mining and coal mining. Mine drainage waters, which are often very contaminated with metals and sulphur compounds, can cause severe pollution problems if they either drain or are pumped into the environment where they destroy the fish and plant biota. Some mine drainage water is very acidic and can contaminate streams for many kilometres downstream. In Colorado State (USA), there are an estimated 23 000 abandoned mines polluting 2300 km of streams and rivers (Banks et al. 1997). Palaniappan et al. (2012) reported that, through a breach in the impounding wall of a containment dam from the Church Rock uranium mine, over 1000 tons of radioactive wastes and 93 million gallons of radioactive effluent were released into the Puerco River. This was one of the worst radioactive pollution incidents in the United States of America.

Added Nutrients and Eutrophication

The world population is predicted to increase to about 10+ billion people within the next 50 years, a 30% increase over the 2000 population numbers. If current trends continue, about one-half of these will live in cities. More food will be needed, and that will, if current farming practices are used, probably require more land, fertilizers and water. The disposal of human wastes can frequently lead not only to the addition of nutrients to the water but also to the spread of microbial pathogens in the water. Untreated disposal of human wastes is not infrequent in countries such as China and India (Carr and Neary 2008). Over 2.5 billion people live without improved sanitation, and over 1.8 billion people lack sanitation at all. The result is a greater spread of disease, health effects from raised concentrations of certain nutrients and possible infections from other microbial and metal pollutants. Certain nutrient additions have become the world's most widespread water quality problem (UNWWAP 2009), and in lakes and reservoirs, this most likely results in eutrophication and a deterioration in water quality. The problem is widespread, possibly universal, as a survey reported in Chorus and Bartram (1999). They reported that in the Asia Pacific Region, 54% of the lakes were eutrophic. Estimates for other regions were 53% for Europe, 28% for Africa, 48% for North America and 41% for South America. Modern-day usage of water has changed. In the past, many people, especially in monasteries, had fish ponds for food production, which were often fertilised with animal manure to make them more productive and provide fish for eating. This still happens in some parts of Asia but is no longer generally practiced in developed countries. Certain industries give rise to effluents that are rich in nutrients. For example, currently, the brewing industry releases about 11 000 m^3/day of effluent rich in N and P (156 mg/l and 20 mg/l, respectively), usually into a sewer, but in developing countries some is discharged into rivers. Although eutrophication can be a natural process, it can occur relatively slowly over decades or even centuries. With human intervention, it can accelerate rapidly, and as a result of the increased nutrients, biological productivity increases to a greater degree. Before the fourteenth century, the rate of eutrophication was relatively slow, but with the introduction of improved agricultural techniques and the use of artificial fertilizers, the inputs of N and P into waters accelerated. Clearance of forests to provide more cultivated land and increased building of towns and cities both increased nutrient inputs and sediment transport from land to water. Currently, the main source of nutrients is point discharge and diffuse pollution run-off from agriculture. Other human activities, including increases in domestic wastewater effluents and industrial wastes, have led to more than a doubling of the cycling of nitrogen and phosphorus on land (Sutton et al. 2013). Both nitrogen and, in particular, phosphorus often limit biological productivity in natural unpolluted systems. The changes in water quality by these two key nutrients that are induced by human activities are called cultural eutrophication to distinguish it from the natural process. The rapid change that then occurs frequently harms water uses, such as recreation, agriculture, drinking water and industry. These changes are not exclusive to eutrophication and can be caused by other factors, such as mineral and inorganic pollution, colouration of the water through humic substances, low dissolved minerals and thermal discharges (Meybeck et al. 1989). The rate of change with cultural eutrophication as well as subsequent biological and chemical problems can impose considerable financial costs on users. In the nineteenth century, the major sources of these nutrients were farmyard manures, treated and untreated sewage and irrigation water, especially that using treated sewage water. From the early twentieth century, the use of manufactured mineral fertilizers surpassed the use of animal and human waste fertilizers. Land use changes such as clearing woodlands can cause erosion and sediment removal, which can introduce both organic and inorganic nutrients into the water. This can also affect other chemical and physical processes in water. It can also affect fish populations and benthic organisms. Nitrogen gas makes up 78% of the atmosphere but, apart from leguminous plants and certain species of algae that can fix the nitrogen, cannot be generally used by organisms directly as a fertilizer. The invention of the Haber–Bosch process allowed this atmospheric nitrogen to be economically harvested and has become predominant. This process allowed the large-scale production of ammonia (NH_3) by combining N_2 gas and H_2 in an energy efficient and relatively cheap operation. This not only results in an increase in the amount of nitrogen nutrients circulating in the environment but can also lead to some acidification of both soils and waters. Phosphorus, which is typically present in small quantities in freshwaters, often limits to plant growth in unpolluted waters and is typically obtained from mineral deposits, the most commonly mined being guano deposits. It also enters aquatic systems from sewage treatment plant effluents, especially where phosphorus-containing detergents are used in the community, animal wastes and fertilizers. Mined phosphorus is not very abundant, and the question has arisen as to whether there might be a shortage in the future and whether there may not be enough to meet the needs for the increased crop production required to feed the growing global population. Countries such as Germany and Japan have no significant reserves of phosphorus. Others have large reserves that, assuming their current consumption does not increase, should last many years into the future. Examples of such countries include Algeria and Morocco, which have reserves of perhaps a thousand years. Most countries only have enough reserves to last a few decades (Sutton et al. 2013). It should be noted, however, that some geographical regions, for example Sub-Saharan Africa, lack enough fertilizer of any sort to boost crop production to the level required to feed the predicted population increases. Mobilisation of both phosphorus and nitrogen has an impact on the natural nitrogen, phosphorus and carbon cycles. Some countries suffer because of the overuse of fertilizers, while others have insufficient amounts. The advent of the 'green revolution' with its requirement for both more fertilizers and irrigation water allowed the price of cereal crops to be reduced, which made it economical to use greater amounts of cereals for animal feed. This has resulted in the 'livestock revolution', which has accelerated since the 1960s (Sutton et al. 2013). This was a change from traditional practices to using intensively fertilised grass and grain production to feed an increasing number of livestock often in enclosed, confined feeding operations. This more intensive method of rearing livestock produces animal waste in a concentrated area and can create a problem of disposal. Some farmers installed anaerobic digesters to treat and break down waste as well as produce methane gas, which can be used as a fuel. This initially occurred in developed countries but is also spreading in developing countries. This increase in livestock production is also associated with a rapid increase in meat consumption and results in lower prices. Mineral nitrogen fertilizer use has increased by approximately 10-fold from 1960 to 2019 and is projected to increase a further 25% by 2050. The largest consumers of animal meat protein are North America and Europe, and the lowest consumers are Africa and Asia, excluding China (Westhoek et al. 2011). World agriculture, in general, is the major contributor from diffuse sources of N and P to

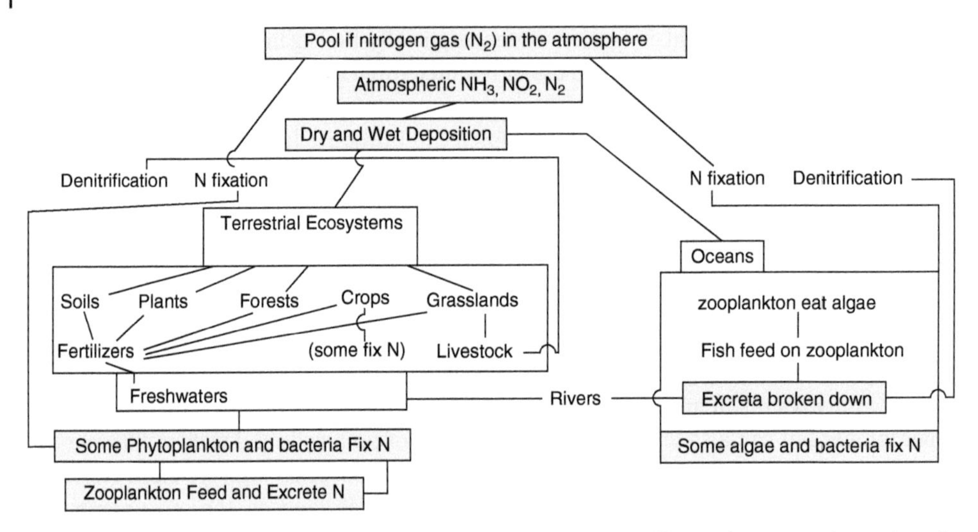

Figure 6.3 The global nitrogen cycle. This is a simplified diagram illustrating a complex series of interactions.

surface waters. Sewage and industrial inputs are generally contributors from point sources. In addition, there is a continuing amount of nitrogen being fixed by natural biological means (see global nitrogen cycle Figure 6.3), including the cultivation of leguminous crops that have a symbiotic relationship with the nitrogen-fixing bacterium *Rhizobium* that fix more nitrogen than would be expected from other plant systems. Nitrogen-based fertilizers are now, on average, adding four times as much nitrogen to the environment as burning fossil fuels (Galloway et al. 1995). There are a range of nitrogen inputs into crop production, including atmospheric deposition, irrigation water, animal's manure, seeds for crops, recycled crop residues, biofixation and fertilizers. There are also losses from the system, including harvested biomass, emissions of NO, N_2 and N_2O, volatilisation of NH_3, leaching from soils, soil erosion and losses from plants (Smil 2001) (see Tables 6.2 and 6.3). Human-generated nutrients can arise through a number of pathways. These are listed in Table 6.2. Apart from fossil fuel combustion, all of the pathways listed can directly lead to surface waters. It should also be assumed that airborne pathways can directly or indirectly enter surface waters.

Nitrates are relatively soluble in water, so when, for example, a fertilizer is applied to land, if there is either precipitation or irrigation, some will dissolve and enter surface waters through run-off and contaminate sub-surface aquifers through infiltration. Phosphorus, on the other hand, is less soluble in water, and as it is not available in the atmosphere as a gas, its availability in surface waters is more limited than that of nitrates. Quite often, phosphorus availability (or a lack of it) is a limiting factor to plant growth and thus to the productivity of the whole aquatic ecosystem. Phosphorus is transported in run-off from fields but is often associated with soil particles rather than in solution. Phosphorus loading, i.e. the amount of phosphorus entering a lake each year per m^2 of its surface area, can be used to predict the amount of growth of algae in that water (Burgis and Morris 1987). Phosphor molecules, because they are associated with sediments, are not always available

Table 6.2 The potential sources and pathways of nitrogen and phosphorus to surface waters.

	Pathway		
Source	**Surface water**	**Groundwater**	**Atmosphere**
Industry	+		+
Sewage treatment plants	+		
Septic tanks	+	+	
Agriculture	+	+	+
Livestock husbandry	+	+	+
Urban rainwater run-off	+		
Aquaculture	+		
Fossil fuel combustion			+
Industry	+	+	+

Source: Adapted from Selman and Greenhalgh (2000). Policy Note No. 3 World Resources Institute (2000).

Table 6.3 Global yearly amounts of nitrogen fluxes from various selected sources.

Origin of nitrogen flux	**$TgNyr^{-1}$**	**References**
Total fertilizer consumption	120	Fowler et al. (2013)
Crop fixation of nitrogen	50–70	Fowler et al. (2013)
Crops and grass for animal production	100	Billen et al. (2013)
NH_3 emissions from crops and grass	15	Sutton et al. (2013)
NH_3 emissions from livestock	22	Sutton et al. (2013)
Crops for human and animal Consumption	28	Billen et al. (2013)
Fish landing	3.7	Voss et al. (2013)
Food waste	13	Billen et al. (2013)
Human sewage	19	Billen et al. (2013)
Wastewater treatment	13	Billen et al. (2013)
Rivers input to estuaries and sea	40–66	Voss et al. (2013)
Surplus in agricultural soils	120	Billen et al. (2013)
Denitrification in soils and aquatic Systems	77	Billen et al.(2013)
NO_x emissions from soils	23	Fowler et al. (2013)
NH_3 emissions, biomass burning, Soils and natural ecosystems	20.8	Fowler et al. (2013)
Combustion	30–40	Fowler et al. (2013)
Wet and dry deposition on soils	70	Fowler et al. (2013)
Wet and dry deposition of NH_3 Wet and NO_2 on agricultural soils	50	Dentener et al. (2006) and Duce et al. (2008).

in the water column and so can be a limiting factor to plant growth. However, because many lakes exhibit thermal stratification, some phosphorus is released in the anoxic lower layers (the hypolimnion) and, at the time of autumn mixing, is cycled throughout the lake's depth.

The global fluxes of nitrogen estimated by different authors are provided in Table 6.3.

As a result, the quality of freshwater is declining, and the waters are becoming eutrophic.

Very few lakes and other surface waters worldwide remain unaltered by human activity either directly or indirectly. The United Stated Environmental Protection Agency (USEPA) in its report to congress in 1996 stated that 40% of streams and rivers were impaired because of the nutrients N and P (Dodds and Welch 2000). Lakes and surface waters can be classified according to their nutrient status. This is frequently measured in terms of the two essential compounds for living organisms, nitrate and phosphate concentrations. Those that have low enough concentrations to limit biological activity are termed oligotrophic (oligo = few), and those with ample concentrations are termed eutrophic (eu = well). There is no clear cut-off between the two but a gradation. The enrichment status also depends upon the situation. As nutrient concentrations increase, so does eutrophication. Although N and P are the main targeted nutrients, increases in others such as silicates that are required by diatoms may also be involved. Both nitrogen and phosphorus are normally present in quite low concentrations in uncontaminated waters. The biological impacts of this process are outlined in Figure 6.2 and Table 6.4. Biological productivity, especially of algae and macrophytes, is much greater in eutrophic conditions than in oligotrophic conditions.

There is no sharp dividing line between oligotrophic and eutrophic states but a gradual change. Similarly, it is also important to realise that there are different degrees of eutrophication. This was recognised, and the terms oligotrophic, mesotrophic and eutrophic were introduced although this still did not overcome the problem of multiple gradations between them. They do not recognise the fact that there may be differences from one part of a lake to another and in many cases involve measuring many parameters. There is a relationship between phosphorus concentrations and algal biomass. In general terms, measurements of algal biomass can be regarded as an integration of the range of factors affecting the trophic state of a lake. Carlson (1977) concluded that Secchi disk transparency could be regarded as an integrated measure of a range of factors integrating algal productivity. It is also one of

Table 6.4 Main differences between oligotrophic and eutrophic lakes and reservoirs.

Characteristic	Oligotrophic	Eutrophic
Algal blooms	Rare	Common
Cyanobacterial frequency	Low	High
Main algal groups	Chrysophytes, diatoms	Cyanobacteria
Characteristic zooplankton	Larger Daphnids and Cyclops	Other Daphnids, Cyclops and protozoans
Plankton density	Low	High
Types of fish	Game fish	Coarse fish
Lake depth	Often deeper	Often shallower

the simplest and easiest available measures in limnology. It is also more easily understood by the authorities. The resulting Carson index scale is from 0 to 100 with major divisions of 10, 20, 30, etc. It also points out that the trophic index is not the same as a water quality index as this has a largely different set of parameters and is concerned with the ultimate use of that water, including consumer safety. More recently, Landsat data has been used to predict the trophic status of lakes although there can be problems such as cloud cover and small lake size (Lillesand et al. 1983).

Various indicators for estimating the trophic status of lakes have been suggested, but probably the most widely used indicator is the Carlson index (Carlson 1977).

Although many articles have been published about eutrophication, the problem still exists and is widespread as are the problems caused and the remedial costs of overcoming the problem. As with plants and animals on land, those in aquatic systems are affected both by their abundance and species present by these changes to their environment. We discuss climate and soil nutrients as important factors in terrestrial ecosystems, but the aquatic 'climate' is equally important to organisms in water. These aquatic climatic factors include not only nutrients and temperature but also water movement, light availability and the cell position in the water column. It also includes competition between species and predation (Imboden 1992). Algae are photosynthetic plants, so their main requirements are for nutrients and light. Many waters are affected by toxic industrial pollutants; most cases are local point sources and occur in restricted areas. Probably, the most widespread problem that occurs in most, if not all countries, is nutrient enrichment of freshwaters and coastal waters. This is the problem of eutrophication, which can impact all ecosystems, especially in the case of surface waters where the planktonic algal populations that change both in species and numbers often form dense scum or 'blooms'. The eutrophic populations are often dominated by species of cyanobacteria or blue–green algae (see Box 6.1 and Figure 6.4).

The species mentioned above represent typical species that may be found at those times of the year in a temperate climate. Species vary from one location and climate zone to another. A key to these and other freshwater algae can be found in Bellinger and Sigee (2015).

While nutrients, e.g. phosphorus and nitrogen, are important, different species such as diatoms also have specific requirements for silica, and their proliferation depends upon these requirements as well. These other factors include water temperature, carbon dioxide

Box 6.1 Typical Sequence of Algal Species in a Eutrophic Lake

Spring	Early summer	Mid-summer	Autumn
Asterionella	*Colonial green algae*	*Ceratium; Microcystis*	*Anabaena*
Fragilaria; *Steph. Ankyra*	*E.g. Pandorina, Eudorina*	*Gomphonema, Fragilaria* *Melosira, Closterium*	*Oscillatoria* *Ankistrodesmus* *Asterionella*
	Scenedesmus, Oocystis	*Anabaena, Oscillatoria*	*Stephanodiscus*

Abbreviations: Steph., Stephanodiscus; Eudor., Eudorina.

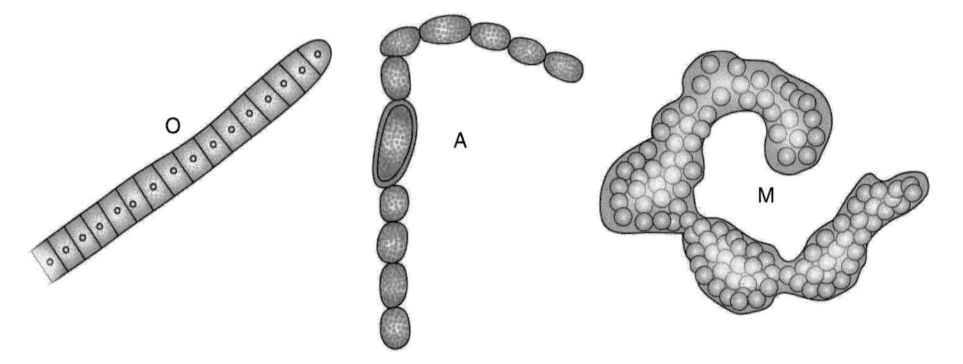

Figure 6.4 Examples of cyanobacterial species (blue–green algae) that can cause problems in water quality. O, Oscillatoria; A, Anabaena; M, Microcystis.

and for some a generally stable water column. When present in large numbers, they can reduce water quality by causing turbidity of the water, production of tastes and odour and, with certain species of algae, toxins harmful to humans and other animals. These include both domestic animals and wildlife, such as fish and birds. These toxins and their effects can render water unfit for use as a drinking water source. These large accumulations, called blooms, may be present at such high densities that they overwhelm drinking water treatment facilities. They have to be removed from the water during drinking water treatment, which can be expensive. If any animal deaths are reported quickly, and if possible, analysed for cyanotoxins, they can act as sentinels for human use of that water. When these blooms of algae die, they decompose and cause reductions in the oxygen concentrations in the water, which in turn have adverse effects on animal life. Eutrophication is now recognised as being the most serious water quality issue facing freshwater use globally. About 30–40% of all lakes are affected by it to some degree (Meybeck et al. 1989). In the United States of America, 40% of lakes and 37% of rivers are impacted by increased nutrient inputs. The problem of controlling eutrophication stems from the large number of possible sources of these nutrients although often the most affected lakes are those near human settlements, their economies and their agricultural activities. The cultural eutrophication increase is also linked to the growth in the human population, their wastes and food demands.

Examples of some of the inputs of nitrogen from agricultural sources are provided in Table 6.5. Apart from the problems of a large biomass of microalgae, a dense bloom at the surface of a lake, reservoir, coastal water or a slow moving river will prevent the photosynthesis of organisms below the surface, which can lower the oxygen content of the water, slow down oxygen transfer from the atmosphere and actively deoxygenate water when the bloom dies and decomposes. This adversely affects fish and other plants and animal lives (Ghosh and Mondal 2012).

There are approximately 1.477 Gha of cultivated land globally, but this figure is now expected to be an underestimate. Although in some countries, mainly in developed countries, crop residues are ploughed back into the land adding both organic matter and nutrients to the soil, but in developing countries, much of these residues is used as straw or burned for heating, liberating more nitrogen into the atmosphere. Not only gases but also dust, soot and several trace gases are liberated by burning. This is particularly true of many

Table 6.5 Nitrogen balance: inputs and outflows.

Source	Inputs (mean)	Outflows (mean)
Seeds	2	
Atmospheric deposition	20	
Irrigation water	4	
Crop residues	14	
Animal manures	18	
Biological fixation	33	
Inorganic fertilizers	78	
Harvested crops		85
NO_2, N_2, N_2 emissions		22
NH_3 volatilisation		11
Nitrate leaching		17
Soil erosion		20
Losses from tops of plants		10
Total	**169**	**165**

All figures are approximate and in $TgNyr^{-1}$.
Source: Data adapted from Smil (1999).

subsistence farmers who traditionally practice 'slash and burn' agriculture, e.g. in the Amazonian forest. This technique is often used for forest clearance to increase the area of croplands. It is also used in other regions for the disposal of rice crop residues. In recent years, there have been very hot weather periods in summer and drought conditions, which have given rise to wildfires on grasslands and woods in Australia, North America and Europe. It must also be remembered that the air pollution impacts may be felt many kilometres away and may even have global impacts. Ways of preventing this include reducing deforestation and, as rangeland burning is often used to produce younger grass growth to be more nutritious for extensively managed cattle, move to a partly more intensive method for livestock rearing although this has difficulties of its own. It is also necessary to help prevent global warming as much as possible. Just under 20% of the world's cropland amounting to 250 Mha was irrigated at the start of this century, with nearly 2/3 of this in Asia alone. Cultivated slope lands are prone to erosion, especially where the gradient is greater than 10–30% (Bruinsma 2003). Greenhouse gas emissions from agriculture can be important sources in some countries but, at the present time, the dominant source in many countries. Agriculture contributes about 30% of total anthropogenic greenhouse gas emissions.

Phosphorus

Phosphorus is an essential element for all living organisms. Its abundance in freshwaters is much less than that of nitrogen. The main phosphorus compounds likely to be found in these waters are phosphates, and they may be present either as an inorganic or an organic compound. Phosphorus generally enters water courses through run-off from land, not in solution as it is less soluble than nitrates but more frequently attached to sediment particles. In this form, it is not easily available to aquatic plants. Once in solution, however, it is easily taken up by plants, including algae and larger plants (macrophytes). There are sources other than agriculture. Many modern detergents contain phosphates that are disposed of by households into sewers and are not effectively removed in the wastewater treatment process. They will then be present in the treatment plant effluent and can contaminate surface waters. Detergents, together with agricultural fertilizers, have been a major source of phosphates in most surface waters. Pickering (2001) reported that the total annual input into Lake Windermere (English Lake District) is almost 20 tons/year. The majority of this comes from a number of sewage work effluent discharges. Efforts are now being made to control these inputs and slow further biological deterioration of the lake.

Controlling Nutrient Inputs

Nutrient inputs and the resulting eutrophication have a number of adverse effects on water use. These include (i) effects on drinking water quality, (ii) impacts on fisheries, (iii) reductions in biodiversity, (iv) reductions in reservoir holding capacity through dense algal growths (blooms), (v) unsuitable for recreation and (vi) increased costs of water treatment. As there are a wide range of nutrient inputs, it is necessary to consider a range of options to reduce unwanted impacts of these nutrient inputs into the environment. The main user of fertilizers is agriculture, so it is important that this and all other users of nitrogen- and phosphorus-based compounds improve the efficiency of their use. This would include not only agriculture but also transport, industry, waste recycling and human lifestyles, all of which emit both nitrogen and/or phosphorus into the environment. In the case of agriculture, it is important to only use the amount of fertilizer required to maximise the production of an individual crop so as to obtain the maximum efficiency per unit of food produced at every stage of the production cycle. This must include both plant and animal products. The ultimate aim of agriculture should be to produce as much food as is needed to efficiently (including the correct balance of nutrients) support the current and predicted future human populations. This, as there is a finite amount of agricultural land, will possibly need to involve hydroponics, which can be used, together with suitable lighting, in multi-story blocks of older no longer used factory buildings. While reducing the emissions from agriculture is important, there could be a general improvement if the whole production and supply chain efficiency were improved. It is important to remember that fertilizers are a vital part of the productivity of crops and animals at the present time if we need to feed the world. It is not disputed that the green revolution together with the increase in productivity brought about by using fertilizers and more irrigation produced enormous advances, but more improvements need to be made without damaging other aspects of the environment. The question is whether the fertilizers are both produced and used in an environmentally acceptable way. Not all fertilizers need to be produced by an industrial chemical plant or by mining finite reserves. There must be controls on the flows of all nutrients. This obviously covers a large proportion of all human activities, hence the need for a range of control strategies based upon nutrient use efficiency (NUE). This must encompass, where necessary, whole production chains. For example, in the case of agriculture, it is not just the production of a particular item but also its harvesting, including losses, transport to selling locations, use by the consumer and waste and disposal by the consumer (Sutton et al. 2013). Climate change will not only affect productivity directly but also increase the spread of pests and diseases requiring changes in our approach to control and use pesticides. Another major input of nutrients is sewage effluent. Most wastewater treatment works do not have specific phosphorus and nitrogen removal strategies. Their removal often requires alterations to the treatment process and additional stages in the treatment sequence. This would involve extra costs but would certainly be required if the final effluent was to be discharged into a sensitive lake or river. There are problems at the present time (mid-2030s) of sewage dischargers into rivers contaminating the water with nutrients and other chemicals.

Phytoplankton blooms have occurred over many years and on all continents at one time or another, for example Australia, China, Europe and the United States of America, in both freshwater and marine systems. These algal blooms frequently cause harm to all organisms in contact with them. Hence, the term (harmful algal blooms, HABs). These are predominantly cyanobacteria or blue–green algae in freshwaters (see Figure 6.4). It is now recognised that these may not only be harmful to animal species, including humans, but they also have serious economic implications. HABs can cause harm either through the large numbers of microalgae present in a bloom affecting treatment costs or the ability of certain species to produce toxins and release them into the water, which are harmful to humans. Because of their significant impact on human health, HABs cannot be ignored. Fifty years ago, HAB occurrences were occasional and widespread. Currently, they are quite common, their growths can be quite dense and they occur in most if not all countries. This is partly due to better monitoring of waters and recording relevant data and more detailed identification of the species involved as well as a better understanding of both the biochemistry and action of the toxins produced.

Water will naturally dissolve chemicals, including nitrogen and phosphorus, from the geological deposits with which it is in contact. Increases in the natural input of these nutrients can take decades or millennia and occur both in developed and developing countries, but the rapid increases in recent decades are due to anthropogenic activities.

Nitrogen

Nitrogen compounds are the most frequently found chemical contaminants found in groundwaters and aquifers as they are very soluble in water and can percolate through the soil to aquifers. The global nitrogen cycle, although of prime importance to all living things, has changed markedly since the invention of the Haber–Bosch process (see above), which initiated the widespread availability of useful nitrogen compounds for agriculture (Smil 2001). With an ever-increasing population, the need for more food also increases, meaning that food production must rise and traditional sources of nitrogen were not adequate to meet the demand, so more artificial sources, i.e. chemical manufacturing of fertilizers, have to be used. About 2% of the world's energy is used in the Haber–Bosch process, so fertilizer prices are directly linked to energy prices. The affordability of these fertilizers means that their use globally is not uniform. Some regions can afford them and have a lot, while others cannot and have very little. The green revolution, which arguably started when mechanisation using first coal powered and then petrol-driven machinery became available, has been heralded as a global breakthrough in crop productivity, but it is dependent upon there being more fertilizer and water to increase crop production. Fertilizer costs rose significantly in the early twenty-first century due to international conflicts, resulting in several farmers resorting to traditional farmhouse manure as a soil enrichment. Productivity has increased in many regions but not all. There has also been a requirement for more grains for food through increased livestock production, which in some regions has become more intensive. The millennium development goals aimed not only to raise more people out of poverty but also to ensure that they are adequately fed. One consequence of raising people's income is that, as their economic status improves, their diets tend to change, and they eat more meat. By 2050, these developing countries will account for about 75% of global nutrient consumption (Sutton et al. 2013). Aspiring to a better and greater nutrient intake is very praiseworthy to a point but ceases to be so when the amounts of obesity rise and more agricultural production damages ecosystems, impacts climate changes and adversely affects human health. Turning more land to agriculture is part of the answer, but greater crop productivity is also vital. One of the limiting factors to productivity is crop nutrient limitation, mainly nitrogen and phosphorus (the latter will be dealt with later). Although nitrogen is abundant in the atmosphere, it forms about 78% of the gaseous component; it is in the diatomic form N_2, which most organisms cannot use. Only a few species, such as certain bacteria and archaea, can convert this into a useable form, reactive nitrogen N_R. If this form is not available, ecosystems including agriculture become limited by a lack of available nitrogen, leading to greatly reduced primary production. It has been known since the nineteenth century that nitrogen was essential for increased crop production and that legumes could replenish nitrogen in soils, but it was not until later that it was recognised that the conversion of gaseous nitrogen (N_2) into useful nitrogen (N_R) was carried out by certain micro-organisms and by lightening (Smil 2001). Reactive nitrogen (N_R) is continuously moved through an ecosystem from soil to rivers or groundwater and to lakes or the sea. The use of nitrogen-based fertilizers in the EU increased greatly after the Second World War. Consumption in the 1930s was about 1 Mt/year and peaked at about 11 Mt/year in the 1990s. Although natural fixation of nitrogen occurs in anthropogenic inputs from motor vehicles, electricity production and the use of synthetic artificial fertilizers represent the main inputs in many regions (Billen et al. 2013). The average flux of nitrogen into watersheds in $TgNyr^{-1}$ is provided in Table 6.6.

Table 6.6 Average fluxes of nitrogen from different sources comparing 1970 with 2000 amounts.

	1970	2000
Natural and crop fixation	129	129
Fertilizer application	27	80
Atmospheric deposition	28	34
Total anthropogenic inputs	80	144
River flow to sea	37	43

Source: Adapted from Billen et al. (2013).

If one looks even further back to pre-industrial times, the differences in nitrogen fluxes are even more striking. In pre-industrial times, the estimated loading of nitrogen to the land masses was 111 $Tgyr^{-1}$.This doubled to 223 $Tgyr^{-1}$ by 2000 (Green et al. 2003). This increase is mainly due to anthropogenic activities. Pre-industrial nitrogen was mainly from biological fixation (89% of total loads). Currently, a mix of fertilizer use (15%), livestock increases (24%) and atmospheric deposition (15%) dominate in many parts of the industrialised world. Non-livestock agriculture still plays an important part. Recent estimates of Billen et al. (2013) were that 85% of the net anthropogenic input of reactive nitrogen occurs on only 43% of the land area. A large amount of this escapes from the land into rivers and the sea. NO_3N concentrations widely range in different parts of the world from below 20 in some arctic rivers to above 200 µg/l (the Mekong River) (Meybeck 1982). Nitrogen entering rivers can also be in other forms such as particulate organic nitrogen. This can be transformed either chemically or biologically into useable forms for the biota contributing to eutrophication. Worall et al. (2009) found that in recent years the main flux of nitrogen in British rivers has doubled when compared with pre-industrial levels. They also concluded that the United Kingdom rivers represent a hotspot of nitrogen flux to the surrounding marine environment. They calculated that the total fluvial flux from Great Britain varies between 230 and 631 kt N/ year. Allowing for correction for flow, the calculated flux of nitrate at average flow varied from 275 to 758 kt N/a with sharp rises in some years, such as 1977, 1996 and 2003. These peaks followed dry periods from the previous year. From their study of the R. Thames data from 1868 to 2008, Howden et al. (2010) also concluded that the loading of biologically available nitrogen had doubled to the terrestrial environment compared with pre-industrial levels. Nitrate levels rose during the Second World War and then stabilised at average concentrations of almost 8 mg/l and remained high for the remaining years in spite of large-scale interventions to limit nitrogen inputs. They attributed this to land use changes, some of which could be changed in the future, but if not carried out, the concentrations and fluxes will remain high. This has resulted in contamination of both surface and groundwaters that are used for drinking water and industrial supplies. Drinking water in particular has to be treated to the WHO standard, which would be quite expensive. Normal conventional treatment may not be effective enough, so additional techniques such as ion exchange resins are needed. While this is feasible in the long term, it would be better to reduce the nitrogen inputs into the source waters by controlling their flow into the water. In response to this problem, which is general throughout Europe, the European Commission brought in both the Water Framework Directive (60/2000/EC) in addition to the already existing Nitrates Directive (91/676/EEC), which was designed to protect waters against pollution caused by nitrates from agricultural sources. This was designed to tackle the high nitrate concentrations largely because, in some member states, the application of nitrogen fertilizers to crops was larger than the crop uptake, leaving excess amounts to be leached into water courses or through the soil (Monteny 2001). The directive should also encourage good farming practices. It recognises that nitrates are important for plant growth and improved productivity, but concentrations that are too high can cause both harm to humans and adversely impact ecosystems. An outline of the EU Nitrates Directive is given below.

This directive is designed to protect water quality in Europe from nitrates arising from agricultural sources entering and polluting both ground and surface waters. To reduce this type of pollution, all surface and underground waters must be regularly monitored.

This especially applies to those intended for drinking water supplies. In particular, surface waters should not contain more than 50 mg/l of nitrates. This also applies to lakes, estuaries and inshore marine waters, which should also be monitored. Land areas that are recognised as being vulnerable to nitrate pollution can be declared as nitrate vulnerable zones (NVZs). Farmers should be encouraged to follow an established good code of practice on a voluntary basis before stricter measures are taken. These should include (i) a limitation on the periods when nitrogen fertilizers can be applied to land. These should target those times when plants are actively growing and so will more rapidly take up the nitrates preventing them from being lost to water. (ii) Restrict or take adequate precautions when using fertilizers on steeply sloping land, in freezing conditions or snow covered land and near watercourses. These measures will prevent losses from run-off and leaching. (iii) Provide adequate storage capacity for livestock manure. (iv) Plant winter or cover crops to avoid bare soil at that time of the year. In addition, catch crops should be grown to minimise nitrate leaching and run-off. (v) Countries are required to produce action programmes to implement the actions to be taken by farmers within NVZs. This will include all codes of good agricultural practice and all other actions regarding fertilizer use and application, including manure and artificial compounds. (vi) Monitoring and reporting factors such as nitrate concentrations and the degree of eutrophication in lakes, rivers and estuaries. Update NVZ areas as needed and assess possible future trends. In addition, four yearly reports must be produced to the European Commission. The key part of the approach was to designate the NVZs. Each member state was required to designate any areas of their land vulnerable to nitrate pollution where nitrate concentrations in groundwater should not exceed 50 mg/l (equivalent to 10 mg/l nitrate-N). This approach was reinforced by the European Council Directive 98/83/EC of 1998. This required member states to develop and implement action plans to meet these standards and to report to the council every four years on progress to achieve this. Table 6.7 shows an indication of the area of designated NVZs in selected European countries.

Table 6.7 Examples of the areas of land designated as an NVZ expressed as a % of the total area of land per selected EU country.

EU state	Total area ($km^2 \times 10^3$)	% NVZ
Belgium	31	51
Denmark	43	100
Portugal	91	13
Germany	356	100
Spain	504	14
France	539	7
Sweden	448	10
UK	244	8
Netherlands	37	100
Finland	334	100

Source: Adapted from Monteny (2001).

From the above table, it can be seen that some countries declared their entire cultivated area as being an NVZ. For the whole of the EU, at the time of this report when the EU consisted of 15 countries, the total amount of land classed as NVZ was 37%. Partly because of its mobility in the environment, there are many pathways for losses in agricultural practice. Inputs include fertilizers, cattle and other animals, manure, roughage, deposition and N_2 fixation. Losses include crops, cattle and other animals, milk, manure, NH_3 emissions from manure, animals, soil, surface run-off, leaching and erosion. Leaching and oxidation of ammonia is the main cause of contamination in groundwater (Monteny 2001). Nitrogen passage through the soil can take many years, so even though mitigation measures were carried out, it could take several years for reductions to be seen in the groundwater. The run-off contribution is mainly from applied fertilizers on land close to watercourses and depends upon the slope of land, whether there are crops growing and weather conditions. Examples of nitrate concentrations in selected rivers in the United Kingdom are provided in Table 6.8.

In the River Loir (France), there has been a small increase in nitrate concentrations during the past 30 years, but during the same period phosphorus concentrations have decreased significantly (Minaudo et al. 2015). The Department for Environment Statistics service reported the annual average concentrations of nitrate nitrogen for the years 1980–2011 for different regions of the United Kingdom (enviro.statistics@defra.gsi.gov.uk). The results for mean concentrations for 1980 and 2011 are provided in Table 6.9.

The regions listed in Table 6.9 represent different topographies, different amounts of industry and farming and different population densities and numbers of cities. There are also different types of agricultural practices, ranging from hill farming with sheep predominating to pasture grazed cattle and croplands. The problem is to manage the amount of environmental contamination and effects on the biosphere but still boost the amount of food being produced to meet the needs of the growing population. Our knowledge of the cycling of nitrogen in the environment has greatly developed in the past century (Galloway et al. 2004, 1995), and the understanding of nitrogen availability controls the productivity of many ecosystems (Vitousek et al. 2002). The need for increased agricultural productivity meant that nitrogen limitation had to be overcome. More nitrogen-based fertilizers were needed. Initially, the only abundant source was mining guano deposits found on certain islands heavily colonised by sea birds off the coast of South America. These provided an extra 0.2 $TgNyr^{-1}$ (Smil 1999). This supply quickly became exhausted, and other ways became necessary to meet the needs. In 1913, the Haber–Bosch process was invented to

Table 6.8 River nitrate concentrations for selected rivers in the United Kingdom.

River	**Average N concentration**	
UK rivers	1970	2000
R. Stour (Essex)	2.0 mg NO_3 – N/l	7.5 mg NO_3 – N/l
R. Frome	3.0 mg NO_3 – N/l	7.0 mg NO_3 – N/l
R. Tees.	1.3 mg NO_3 – N/l	1.0 mg NO_3 – N/l

Source: Adapted from Burt et al. (2011).

Table 6.9 Mean nitrate concentrations reported for selected regions of the United Kingdom for 1980 and 2011 (units in mg/l).

Region	1980	2011
Anglian	38.44	27.52
EA Wales	6.20	5.27
Midlands	3.60	37.42
North East	14.28	15.80
North West	11.64	16.54
South West	15.04	16.87
Southern	20.25	23.58
Thames	37.62	36.18
SEPA East	11.39	9.66
SEPA North	5.67	5.81
SEPA West	6.54	4.31

SEPA, Scottish Environmental Protection Agency data; EA Wales, Environment Agency Wales data.
Source: Adapted from UK Department for Environment, Food & Rural Affairs (www.gov.uk; accessed November 2019).

produce NH_3 from N_2 and H_2. Although mainly used for fertilizer production, it was used in the First World War by Germany to produce ammonia for use in the explosives industry. The process initially produced as much as 20 tons/day, much of which was converted to nitrates for explosives. The processes are still used but now produce about 450 million tons of nitrogen fertilizer per year. It is estimated that about half the human population at the beginning of the twenty-first century depends upon fertilizer N for their food. It is estimated that the productivity of agricultural land and food production increased from being able to support 1.9 persons to 4.3 persons/ha between 1908 and 2008. Reactive fertilizer nitrogen is used for producing crops for human consumption and for animal feedstuffs (Sutton et al. 2013). Nitrogen can leave the farm as a marketable product (plant and animal) or it is released to the environment in an active form, e.g. nitrate (NO_3) ammonia (NH_4) and as oxides (NO_X and N_2O). This nitrogen is released to waters, both surface and groundwaters, soils and the air (Hartmann et al. 2008). Nitrogen losses to surface waters from atmospheric deposition in Western Europe amount to approximately 10–25 kg/ha/year (Addiscott et al. 1991). Much (about 50%) of this is in the form of NH_3. Agriculture is probably the main contributor of NH_3 to the atmosphere, with the largest portion coming from livestock waste Bussink and Oenema (1997). Many fields have subsurface drainage, usually by tile drains. As nitrogen compounds are readily soluble in water, drainage water can have nitrogen concentrations between 1 and 20 mg/l. This water generally flows into the nearest stream and lake. In addition, planned drainage by tile drain heavy rain can lead to surface run-off and soil erosion. This is particularly important on sloping land and hilly areas. With a changing climate, storms are likely to become more frequent, and during such events, up to 90% of the total nitrogen loss in a field could come from this run-off and erosion (Oenema and Roest 1997). Animals tend to concentrate around watering sites where they may urinate and defecate, contaminating both the water and soil. Because of the high use of nitrogen fertilizers in the past, many aquifers were heavily contaminated with nitrates over a number of years. This not only means that water has to have the nitrate reduced by treatment to an acceptable level before it can be used as a drinking water source but also that any surface water streams fed by that aquifer will also be contaminated.

The increased environmental concentrations invariably lead to problems, including eutrophication, acidification of soils and deterioration of air quality, and possibly contribute to climate change. There are national, regional and international policies aimed at reducing nitrogen emissions from industry, energy production and agriculture. One example of attempts to combat the adverse effects of these increases is by the European Community, which has set reduction targets for reducing NO_2 and NH_3 emissions by 49% and 15%, respectively, below the levels of 1990 by 2010. In addition to protect the north-east Atlantic marine environment, the OSPAR Convention committed to reduce nitrogen inputs into the North Sea by 50% below 1985 levels. Unfortunately, only Sweden and the Netherlands expect to meet this target by 2020 (Hartmann et al. 2008; Peter et al. 2006). The main problem is with reducing inputs from diffuse sources, particularly because of the high solubility of many nitrogen compounds and the fact that many come from diffuse sources. It is certainly necessary to introduce regulatory and economic incentives, if they do not already exist, to meet the targets. Such economic incentives could include a nitrogen tax on fertilizers or a manure tax (Hartmann et al. 2008). In addition, taxing agricultural losses from farmland changes to land use must also be addressed.

If the amount of dissolved chemicals in the water, in particular nitrogen and phosphorus, is low, biological activity will be reduced. Such freshwaters are termed *oligotrophic*. This term is derived from the Greek and means 'few feeding'. If the supply of these chemicals, especially nitrogen and phosphorus, is ample for the needs of the organisms present (in particular the phytoplankton), the water is described as *eutrophic*. The term eutrophic is also derived from the Greek eutrophic meaning 'well nourished'. This change in nutrient availability causes a shift in the biology of the water (see Reynolds 1984). What must be understood, however, is that there is seldom only one factor that determines an ecological shift in the dominant species in a population. It is widely acknowledged that the primary nutrient involved in this change is phosphorus which in most waters can be the limiting factor to primary production. It is also certain that nitrogen, an essential element, needs to be present in adequate quantities and, in certain cases, may itself be the limiting factor. Certain groups of algae (the main primary producers in aquatic systems) may have specific requirements for other chemicals, e.g. silica. In recent years, interest has grown concerning the natural background concentrations of nitrogen and phosphorus in freshwaters (Smith et al. 2003). Some knowledge of the natural background concentrations would enable us to determine the impact of humans on water quality. Nitrogen and phosphorus concentrations in particular have increased dramatically in both concentration and bio-availability in many aquatic systems both locally and globally (Vitousek et al. 1997). Nitrogen (and phosphorus) increases in the aquatic environment are caused not only by increases in fertilizer use but also by fossil fuel burning, increases in livestock and legume crops, and an increasing human population. As a result, the rate of nitrogen input into the terrestrial nitrogen cycle has approximately doubled; increased losses of soil nutrients, e.g. calcium and potassium; partly responsible for soil, stream and lake acidification in many regions; increased the quantity of organic carbon stored in terrestrial systems; and increased biodiversity loss (Vitousek et al. 1997). In the United States of America, pristine sites that might be used for background reference are non-existent. This is true of most, if not all, industrialised regions of the planet not only through land use changes, industry, agriculture and population growth but also through long-distance atmospheric transport and deposition of nitrogen compounds. Often, the best sites that can be used for reference are in small watersheds as there is less chance of them being developed. Even so, these are not truly background natural sites. However, from these reference sites, it has been found that mean annual run-off strongly correlates with background nutrient levels. Nutrient yields will also vary quite considerably from region to region and hence so will concentrations in water. Sprague and Lorenz (2009) estimated regional nutrient trends in streams and rivers in the eastern, central and western United States from 1993 to 2003 and found that the only significant trend was for total phosphorus in the central region, which corresponded to increases in phosphorus inputs from fertilizer (Sprague and Lorenz 2009). It is useful not only to chemically check the microbiology of streams, rivers, lakes and estuaries but also to relate these measurements to the nature and state of the ecosystem they contain. Indices of trophic status are often used in aquatic sciences, and their use dates back many decades (e.g. Kelly and Whitton 1993). Increases in nutrients, particularly nitrogen and phosphorus, result in increases in algal biomass and changes in the structure of both the plant and animal communities present. Algae and other plants and animals in aquatic systems exhibit many different responses to both nutrient enrichment and chemical and physical pollution. Lake classification based upon trophic systems has been in use for many years, but these were not applied directly to rivers and streams (lotic systems) as flowing waters have, in many ways, a different environment than lakes. As streams and rivers are the primary recipients of nutrient run-off from land and often the discharge point for wastewater and industrial wastes, it is important to both frequently monitor their condition and develop criteria and, if possible, an index of their condition to help with the regulation of inputs. It is important to include all criteria in the assessment. Originally, because phosphorus is often a limiting nutrient in freshwaters and increased inputs remove this limitation, lakes and then flowing waters were classified by their phosphorus concentrations (Dobson and Frid 2009; Ravera 1981). In the United States in the 1990s, the USEPA in their Water Quality Inventory found that 40% of streams and rivers surveyed had impaired chemical concentrations. Before the relevant authorities can arrange or impose limits for these nutrients in rivers, streams or lakes, the appropriate criteria for those particular locations must be determined.

Although eutrophication may take place naturally, it can be accelerated by anthropogenic activities such as land use changes, including agriculture, sewage and industrial waste discharges. This is termed 'cultural eutrophication'. Rivers, and lakes via rivers, eventually flow to the sea and carry with them these nutrients. This potentially results in estuaries and coastal waters also becoming eutrophic. Eutrophication is now widely recognised as one, if not the main, water quality problem for the future. There are a range of changes to biodiversity and water quality that could occur as a result of eutrophication, including increased biomass of phytoplankton, proliferation of potentially toxic species, decreases in water transparency, production of odours and smells in the water, fish kills and loss of

Box 6.2 Cyanobacteria or Blue–Green Algae

Cyanobacteria are photosynthetic micro-organisms generally known as blue–green algae. They can be found in a wide range of freshwater habitats, including streams, rivers, lakes and wetlands. Some species also occur in estuaries and oceans. Some, but not all, species are known to produce toxins that affect a wide range of other organisms, both animals and plants, as well as humans. HABs have been recorded and caused problems in all continents. The US Geological survey, for example, detected cyanobacterial toxins in nearly 40% of streams in the SE USA (US Geological Survey 2017). When large numbers occur in the plankton, they often tend to float near the surface, colouring the water and forming a scum. Under these conditions, they are said to form a bloom. Such large densities of algae can cause problems for water users and drinking water treatment works as well as potentially producing toxins.

some fish species, impairment of recreational activities and harm to humans. From the point of view of drinking water, a small amount of eutrophication per se is not harmful to humans, but it can cause problems of taste and odour persist in the water. Large growths or blooms, especially cyanobacteria (blue–green algae), give rise to problems (Box 6.2).

Eutrophication and Harmful Algal Blooms

In most countries, as mentioned above, waters are becoming more contaminated with nutrients, nitrogen and phosphorus. This makes them more biologically productive, and many lakes that are oligotrophic are gradually becoming eutrophic. One of the effects of eutrophication is that large growths (blooms) of blue–green algae (cyanobacteria) develop, colouring the water green and often forming dense scums on the water surface. They can occur as single cells, filaments or colonies that may be regular or irregular (Figure 6.4). They can occur in salt waters, but here we are only concerned with freshwater blooms. Apart from the reservoir and lake management problems that they cause, such as reducing the useful stored volume of a reservoir and adversely affecting fisheries and recreation, they may impact human health. The potential problems caused by different groups of algae, including cyanobacteria, were highlighted by Palmer in his 'Algae in Water Supplies' (Palmer 1962). Some species of these algae produce toxins that can affect livestock, birds and humans, but although many deaths of wildlife and cattle have been recorded, no deaths of humans by ingestion have been recorded. While effects on humans were relatively rare in 1931, for example, approximately 8000 people fell ill after drinking water originating from tributaries of the Ohio River, which was contaminated by a massive cyanobacterial bloom (Lopez et al. 2008). Human death through cyanotoxins has only been recorded in a patient with renal dialysis. Although the occurrence of these blooms has increased, there is evidence that they also occurred thousands of years ago (Züllig 1989), so the phenomenon is not new. What is new is that it is much more common, with almost every country having this problem. It is estimated that 53% of Europe, 28% of Africa, 48% of North America, 41% of South America and 54% of the Asia Pacific region of the lakes are eutrophic (ILEC/Lake Biwa Research Institute 1988–1993). Algae are not the only aquatic plant growths that are stimulated by increases in nutrients. The littoral zones of many lakes and reservoirs can be clogged up with macrophyte growth as can the water surface, e.g. with *Eichornia* in warmer climates. This latter is a problem in parts of Lake Victoria and Lake Nasser.

Some species of cyanobacteria or blue–green algae, apart from forming unsightly scums on the water surface, secrete a diverse group of toxins into the water. These toxins can affect various organs, including liver, nerves, gastro-intestinal tract and skin. A list of toxins, the organs they effect and the genera of cyanobacteria that produce them are provided in Table 6.10. There is growing evidence of an increase in HABs in lakes in the United States. An example of this is the large record-breaking growth in Lake Erie reported in Michalak et al. (2013).

As noted above, the increased occurrence of HABs is increasing worldwide. In the United States, there is evidence, as yet unquantified, that HAB toxins have caused significant human morbidity and mortality in recreational, commercial and potable waters (Hudnell 2010). HABs have a significant impact on the sustainability of aquatic ecosystems. He estimated that eutrophication and HABs are probably costing the US economy between 2.2 and 4.6 billion $US each year. He also noted that there was an urgent need to form a National Research Plan for freshwater HABs and to form a policy with regulations to deal with the problem. This should not only highlight the risks but also develop cost-effective strategies for dealing with the HAB problem. This would not only help the US economies of all other countries but also protect human and animal health and ensure that freshwaters were used more sustainably. Other estimates put the annual economic impacts at nearly 50 million $US for the period 1987–1992 for marine systems (Anderson et al. 2000). Studies

Table 6.10 The toxin groups, organs they affect and the cyanobacterial genera producing them.

Toxin group	Target organ	Cyanobacterial genera
Microcystins	Liver	*Microcystis, Anabaena, Planktothrix, Nostock, Anabaenopsis, Hapalosiphon*
Nodularin	Liver	*Nodularia*
Anatoxin-a	Nerves	*Anabaena*
Anatoxin-a (S)	Nerves	*Anabaena*
Aplysiatoxins	Skin	*Lyngbya, Schizothrix, Planktothrix*
Cylindrospermopsis	Liver	*Cylindrospermopsis, Umezakia, Aphanizomenon*
Lyngbiatoxin-a	Skin, gastro-intestinal tract	*Lyngbia*
Saxitoxins	Nerves	*Anabaena, Aphanizomenon, Lyngbia, Cylindrospermopsis*
Lipopolysaccharides	Potential irritant Affects any exposed Tissue	All

Source: Adapted from Chorus and Bartram (1999). Spon Press, London and New York on behalf of the World Health Organization.

have been conducted in other countries, including the United Kingdom (Filatova et al. 2021). They found that in addition to the listed toxins, many secondary metabolites were also present in the water that could potentially affect human health. The WHO set a guideline for a threshold value of 2×10^4 cyanobacterial cells/ml for recreational waters, which may correspond to a concentration of 20 µg/l of microcystins for a *Microcystis* bloom. The WHO also set a guideline value of 1 µg/l for total microcystins in drinking water (WHO 2011). Because cyanotoxins are soluble in water, the treatment of the water in drinking water treatment works will usually not effectively remove them unless they are bound to a cell or other particles. Reservoirs with blooms of these algae are often treated with copper sulphate to control them. This has often resulted in significant numbers of people drinking water developing symptoms of gastro-enteritis as has occurred in Australia and the United States of America (Chorus and Bartram 1999). This treatment was thought to kill the cells, and lysing them released the toxins into the water. The growths of cyanobacteria are not confined to standing water bodies. Graham et al. (2017), reporting for the US Geological Survey (USGS), sampled 11 large rivers ranging from oligotrophic to eutrophic in the United States of America during 2016 and found cyanobacteria present in all algal communities, but they were rarely dominant in the phytoplankton. In spite of the widespread occurrence of HABs, many parts of the United States still do not have sufficient monitoring programmes for HABs, and when problems arise, HAB impacts are not consistently examined.

It should be noted that not all species in the genera mentioned produce every one of the toxins recorded for that genus. It should also be noted that toxins generally remain inside intact cells and are only released when the cells die and decompose. This will happen to some cells in any population all of the time.

Humans can be exposed to these toxins through ingestion, inhalation and dermal contact. Surface blooms of these algae may be blown in one direction across a lake by the wind and form even larger accumulations where the concentrations of toxins are 1000 times greater than in the open water.

Mitigation of Nitrogen and Phosphorus Inputs to the Environment

The degree to which nutrient enrichment and subsequent eutrophication impact water quality depends, to an extent, on local conditions, such as the geology of the area, lake depth, land use in the catchment and population density. Controls must take all of these factors into consideration. Agriculture can generally be regarded as a diffuse source of pollution. Approved action programmes are a requirement for all NVZs in the EU. Although individual programmes are devised by individual countries, the EU has provided overall guidelines that are expected to be included by each country. Dodds and Welch (2000) outlined the different criteria by which streams, rivers and lakes could be classified. Although their conclusions were aimed at the United States of America, they do apply generally. They suggested five criteria that are as follows:

1) Adverse effects on humans and domestic animals
2) Aesthetic impairment
3) Interference with human use

4) Negative impacts on aquatic life
5) Excessive nutrient input into downstream systems

These may not be applicable to every stream/river or every site on each stream. There may also be reasons to have standards set at different levels for different situations.

Certain nutrients, for example Nitrates, at elevated concentrations have a harmful effect on humans, especially young babies. This effect is called 'blue baby' disease or methaemoglobinaemia. There have also been suggestions that high nitrates can be correlated with stomach cancer (Hartman 1983). Freshwaters are frequently used for recreational activities and sports. If water becomes highly eutrophic, it can result in large growths of algae, especially cyanobacteria. Not only are these blooms of algae unsightly but they may also be toxic to humans, domestic animals and wildlife. Eutrophication also causes a widespread cyanobacterial bloom in the Australian Murray-Darling river system, resulting in many cattle deaths (Bowling and Baker 1996). The best way to prevent HABs is not to provide the growth factors that they need, i.e. nutrients and light. Control and management of nutrient inputs is a key measure within the river basin. Control of light can be achieved by artificial mixing. Biological controls are also possible, such as biomanipulation.

There are a number of ways in which nutrient inputs into freshwaters can be controlled. Sutton et al. (2013) advocate a 'Five-element strategy' to improve NUE.

These are as follows. (i) Use the right fertilizer, the right amount, the right time of application, the right placement and low-emission precision application. (ii) Select the right crop cultivar with the correct spacing of plants and use the right crop rotation. (iii) Irrigate efficiently, use precise irrigation with drip feed, and where possible use soil water harvesting and prevent erosion. (iv) Weed and pest control management, minimise yield losses and protect the environment by targeting the pests. (v) Use site-specific mitigation measures to reduce nutrient loss, reduce erosion, manage the amount of tillage and use best practice for fertilizer and manure applications. Not only is better practice needed in the use of agricultural fertilizers but also better control of industrial wastes and domestic waste is needed. Secondary treatment alone, although it removes organic solids and improves the bacterial content of the effluent, does not efficiently remove phosphates, and as much as 70–100% can pass through the treatment works (Kallqvist and Berge 1990). It is of use to managers to have some idea what the carrying capacity, i.e. how much algal growth can they expect, of the water is, and this depends upon the amounts of N, P and light. Any one of these can be a limiting factor, and this can change with season. For example, the day length in summer is considerably greater than that in autumn. In addition, if a large growth of floating macrophytes is allowed to develop, then this will cut out most of the subsurface light. In order to manage water, decisions need to be made as to which factors to control to be the most effective. An example of a decision tree is given in Figure 6.5, which has been modified from Reynolds (1997). It is useful to have an estimate of how much biomass can be produced. This is often based upon the Redfield ratio (see Round (1965)). These ratios indicate the average amount of key elements that are required by organisms, which are as follows:

C42 H8.5 O57 N7 P1

Hydrogen and oxygen are never limiting in normal aquatic environments, and carbon can come from CO_2 and is used in photosynthesis. CO_2 can be an important factor in soft

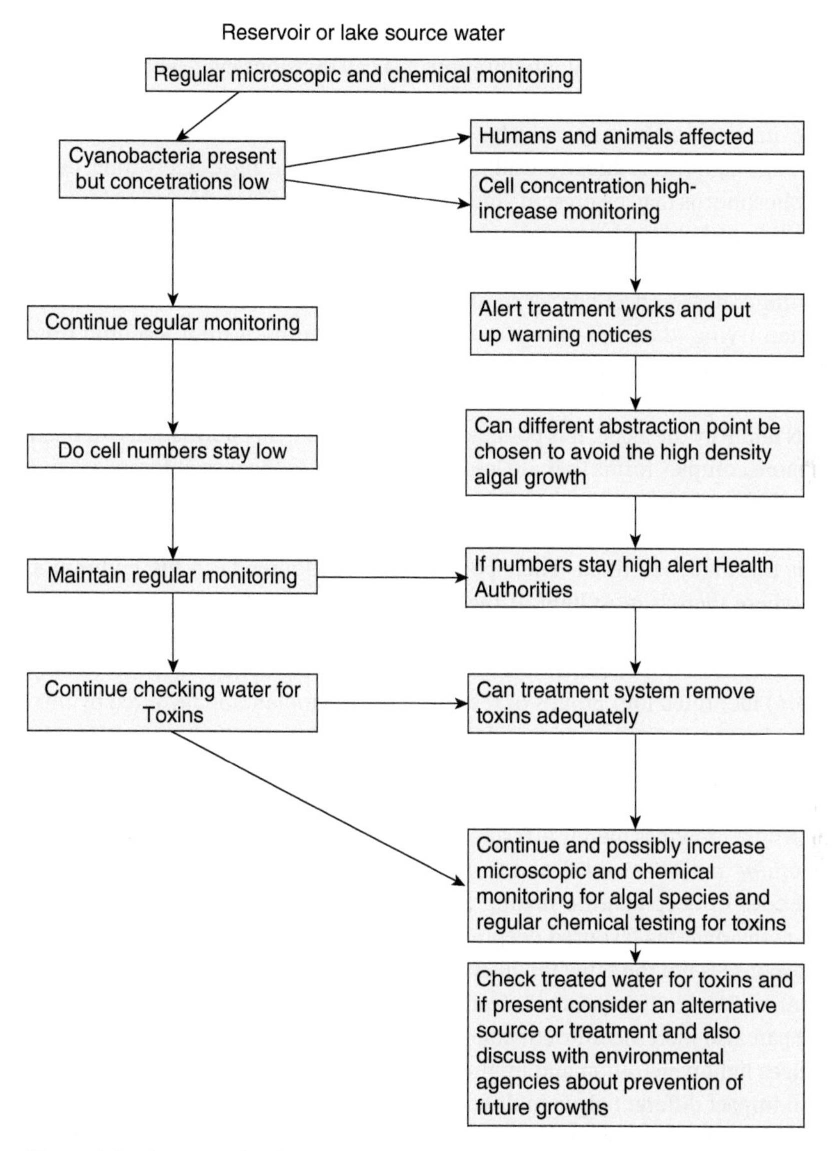

Figure 6.5 An example of a decision tree for action if cyanobacteria are growing in source water for drinking water supplies.

water lakes. Nitrogen enters the water mainly through leaching from soils, but it can also come from sewage treatment work outflows where tertiary treatment is not given to remove it. In some arid regions, nitrogen may be limiting. It must also be noted that some species of cyanobacteria, e.g. *Anabaena* and *Aphanizomenon*, have the ability to fix atmospheric nitrogen. Phosphorus can come from inadequately treated sewage, i.e. that which is not given tertiary treatment. As phosphorus readily binds to soil particles, it can enter a lake/reservoir through soil erosion. Although organisms need less phosphorus than nitrogen, it can still be the limiting nutrient because of availability. Cyanobacteria can take up more phosphates than is immediately needed (luxury uptake), which can be used as a store for times when it is scarce. This could potentially allow 3–4 cell divisions even when it is theoretically limiting in the surrounding water. Although it is common practice to measure orthophosphate in water, this may only be a fraction of what is there (Chorus and Bartram 1999). Although this is directly available for uptake by the algae at any time, other compounds of phosphorus may be present, and these can be transformed by microbial and chemical action into orthophosphates and so become available. It is thus sensible to monitor both orthophosphate and total phosphorus in the water. Phosphorus elimination through preventing soil erosion and better sewage treatment can be very effective and more cost effective than trying to curtail nitrogen inputs by preventing it from leaching from soils. Which nutrients are available to the algae at any time also depends upon any chemical reactions, and changes that take place in the water can always affect the rate of algal uptake of both N and P by the algae. It is possible that some of the transformations that take place result in more complex forms that are less available. This is the reason for measuring total nitrogen and phosphorus rather than just the most easily available forms.

Blue–green algae usually grow in the summer and autumn at a time of year when the lake/reservoir is thermally stratified. Their populations can dominate in the epilimnion (upper layers), where there is most light. If the water is artificially mixed so that the cells are circulated, at least for some of the time, into the darker hypolimnion, they will get less light and produce smaller populations. Not all algae respond to mixing in the same way. Visser et al. (2016) identified four groups of responses: (i) phytoplankton favoured by mixing (*Asterionella, Fragilaria, Staurastrum* and *Oscillatoria*), (ii) phytoplankton favoured by stability and/or reduced optical depth but whose increase stimulates cropping by zooplankton and in turn enhances loss rates by grazing (*Cryptomonas, Rhodomonas, Ankyra*), (iii) phytoplankton whose growth is merely arrested by episodes of deep-column mixing (*Volvox Anabaena, Ceratium* and *Microcystis*) and (vi) phytoplankton whose growth is abruptly reversed by episodes of deep-column mixing (*Sphaerocystis* and *Eudorina*). The authors concluded that many diatoms favoured deep mixing, but some had their growth abruptly reversed. The response is not the same for all cyanobacteria as mixing favoured *Oscillatoria agardhii*, now called *Planktothrix aghardii*. The authors also pointed out that mixing affects other factors. It can also increase nutrient concentrations and increase suspended matter, which also reduces light penetration and might improve conditions for fish and zooplankton, all of which impact different algae in different ways. Whether or not algal blooms are a problem, it is still sensible to control the flow of nutrients from the lake catchment. While the calculations of Vollenweider and Kerekes (1980) are useful guides that must be taken into account are the individual characteristics of each lake, not only their area but also their depth profiles and stratification regimes. If mixing does occur naturally, the depth to which this happens is important. This is 'the mixing depth', which together with the subsurface light regime includes the depth at which light extinction takes place, if at all. These factors can provide important control information. If artificial mixing is used, the mixing depth should be down to the lowest light zone or just below. Mixing that is too shallow could actually increase algal productivity. Cook et al. (1993) advocated that phosphorus concentrations should be kept as low as possible in the region of 30–50 μg/l of total phosphorus. These concentrations will not completely prevent growth but will substantially reduce it. If TP can be kept less than 10 μg/l, algal growth would be significantly lower, but

Table 6.11 Suggested acceptable and dangerous inputs of nitrogen and phosphorus in lakes of different depths (all units in gm^2/a).

	Acceptable inputs		**Acceptable inputs**	
Mean depth (m)	**p**	**N**	**P**	**N**
<10	<0.1	<1.5	>0.2	>3.0
<50	<0.25	<4.0	>0.5	>8.0
<100	<0.4	<6.0	>0.8	>12.0
<150	<0.5	<7.5	>1.0	>15.0
<200	<0.6	<9.0	>1.2	>18.0

Source: Adapted from Harper (1992) in Chorus and Bartram (1999).

a target of 5 μg/l should be the aim to control most species (Reynolds 1997). Different groups of algae have different strategies for survival (Chorus and Bartram 1999), so it is important to know which are dominating at any given time. Harper (1992) proposed different levels of nitrogen and phosphorus for lakes of different depths that could be acceptable or likely cause problems. These are given in Table 6.11.

The data given in this table are only guidelines, and other authors, such as Reynolds, suggest that a maximum of 5 μ/l should be a more realistic target for final lake water concentration. However, other factors such as the residence time of water in the lake and land use practices in the catchment may change either with season or overall, but if permissible nutrient inputs are exceeded, then increased biological productivity, including HABs, is extremely likely. Quite probably stronger management measures will then be needed that could involve greater costs. If such measures are needed the following questions have to be asked to avoid spending in the wrong places: (i) Is there a significant contribution of nutrients from sewage (including all inputs from septic tanks) and other more dispersed sewage inputs? It can be assumed that 2–4 g/person/day is produced and has to end up somewhere. (ii) How much does agriculture and run-off from land contribute? (iii) What is the target concentration of phosphorus that is being aimed at and how much needs to be removed from inputs to achieve this and over what time? The type of quantitative information required is not easy to obtain and is needed not just on odd occasions but also regularly over time as some of the contributing factors can change markedly with season, e.g. with tourism and crop growth. Point sources are relatively easy to monitor and should be frequently used, but diffuse sources are more difficult. Both need to be monitored and recorded as frequently as possible.

For phosphorus reduction, the use of phosphate-free detergents is beneficial where phosphorus has been substituted for less harmful alternatives, which could reduce the phosphorus load in sewage by as much as 50%. Treatment methods to remove nutrients do not have to be highly technological. High-tech nutrient stripping can be expensive. One example is using holding lagoons for liquid wastes to remove nutrients, which have been tried for many years with some success as has land treatment, but these can require as much as $10\,m^2$/person. With lagoons, care must be taken to ensure that seepage through the soil to streams and the water table does not occur. Artificial wetlands should also be considered, but these also require some maintenance to operate properly. They can be very useful for dealing with fluctuations in population numbers in tourist areas. They can be used for individual dwellings or hotels. Within the lake or river basin, area controls may be needed on land use changes such as deforestation, buildings, hard surfaces, etc. Where such changes are allowed, buffer strips around them and also along rivers and lakes that restrict fertilizer use and access to animals will also be required. Sewage works should introduce methods specifically for removing phosphorus by, for example, chemical precipitation with aluminium salts and other methods Gleisberg et al. (1995), Harremoes (1997). Biological phosphorus removal can be achieved in wastewater treatment works by alternating aerobic and anaerobic steps in biological treatment that enhances phosphorus uptake by bacteria. This has the advantage of not requiring the addition of flocculating chemicals but does require proper design and operation of the works. A significant reduction in P can be obtained (Harremoes 1997). Areas with large seasonal tourist influxes do cause particular problems because the loads may increase many fold in the tourist season. An example of this is Lake Balaton in Hungary (Somlyody and van Straten 1986). Industrial discharges can amount to 50% of the flow, and this can rise to nearly 100% in the dry season in some areas. Slow-flowing rivers such as the Thames through London and the Havel River through Berlin, because of the higher concentrations of phosphorus through a variety of inputs, are susceptible to cyanobacterial growths. All effluents and discharges into rivers like these (and all others) should be treated with the objective of lowering the phosphorus concentrations to 0.03–0.05 mg/l P or even 5 μg/l. This means that tertiary treatment for all sewage is needed. This could be achieved by sand or membrane filtration or other methods. The costs of treatment would rise but not excessively, and this could be no more than US$ 0.15–0.30/m^3 (Heinzmann and Chorus 1994).

Artificial Mixing

Artificial mixing has been widely used as a method of light limitation to algal growths in many water bodies although less for shallow bodies. Depending upon the transparency of the water, if the algae can be moved into the poor light zone, usually in the hypolimnion, their photosynthetic activity will decrease, and thus their growth and population sizes will also decrease. This is also aimed at overcoming the ability of some cyanobacteria to regulate their buoyancy and thus maintain a favoured position in the water column. A frequently used technique is the use of air pumps to produce bubbles. As the air bubbles rise to the surface, they act as airlift pumps and circulate the water (see Figure 6.6). These consist of an air generator that is bankside mounted and connected by a pipe to either a fixed diffuser or a variable depth diffuser controlled from a pontoon mounted telescopic tube at the base of which the air diffuser is mounted. This latter case is a more complex mechanism and to work properly must be properly maintained, but it does have the advantage of controlling not only the degree of mixing but also the exact mixing depth required. In all cases, regular and detailed sampling was performed at a range of depths to ascertain the numbers and species of algae present. It is important in most mixing operations to ensure, if possible, that over three quarters of the water is mixed and that the mixing rate is enough to overcome the buoyancy ability of the algae. This is a particular problem with

(a)

(b)

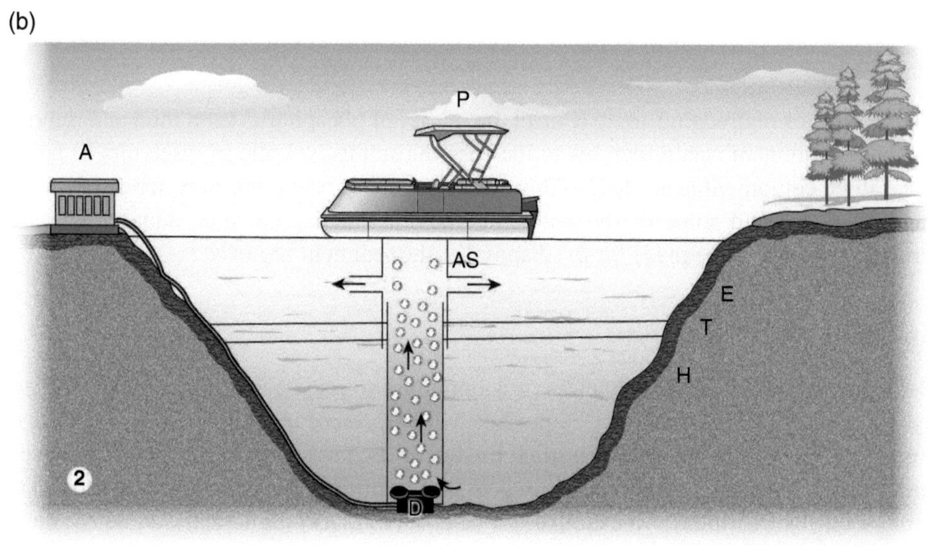

Figure 6.6 Air bubbles for destratification. (a) Simple bubble mixer or destratifier. (b) Tube destratifier with sliding adjustable outlet tube. E,epilimnion; T,thermocline; H, hypolimnion; A, air pump; D, diffuser; P, pontoon; AS, adjustable depth exit for bubble stream.

certain cyanobacteria (blue–green algae), which have gas vacuoles that enable them to float and cause scums at the surface. These gas vacuoles can be collapsed by pressure. The increased pressure can be applied in a pipe where the water is pumped from a reservoir, and in the enclosed space of the pipe, the pressure can be applied and then either recirculated into the lake or piped to treatment works. Without gas vacuoles, the algae will no longer float to the surface and form scums. Small explosive devices set off in water can produce enough pressure to collapse vacuoles (Walsby 1992). Once the buoyancy of the algae is overcome, aeration will work more effectively although the treatment may have to be repeated in time as the cells are able to eventually generate new gas vacuoles. Medium to large water bodies may require more than one aerator. Shallow lakes generally have lower circulation rates that can make mixing more difficult. As phosphorus can be released through biochemical reactions from the mud under anaerobic conditions, a further technique is to target hypolimnetic aeration where air or pure oxygen is injected into the lower hypolimnion to limit phosphorus release at the anoxic mud-water interface. In addition, NH_4^+ will be reduced. Water quality is generally improved as long as this is carried out in conjunction with controls of external nutrient loading. Keeping the lower levels of the lake oxygenated also helps the cold water fish habitat and tends to increase the zooplankton diversity and biomass, which also provides more food for fish and helps with the grazing of unwanted algae. The capital and running costs of aeration can be high, so its potential effectiveness in a particular lake or reservoir needs to be carefully assessed. Of course, the system does not need to operate continuously, as at certain times of the year, natural overturn will occur in the lake.

Dredging

The main aims of dredging are to control the release of phosphorus from mud and remove unwanted sediment accumulations to increase the storage volume in reservoirs or lakes. Specialised equipment is needed, so this can be a costly process. The very action of dredging will re-suspend some of the sediments, possibly damaging some aquatic habitats. Provision also has to be made for the disposal of the sediment removed.

Algicides

Before discussing the use of algicides, it is important to issue a warning. This is that any cyanobacterial toxins that might be present will be present in the whole algal cell. It is easier to remove an intact cell by normal treatment such as sand filtration than if the cell has lysed and disrupted so that the toxin goes into solution, which can be more difficult to remove during treatment. An example of this is with *Microcystis aeruginosa*, where more than 95% of the toxin is normally contained within the healthy cell. A similar situation is true for *Anabaena circinalis*. With *Cylindrospermopsis raciborskii*, however, only about 50% or less is contained in the cell (Newcombe et al. 2010). Algicides destroy algal cells, but their removal could be made more difficult with ruptured cells than with intact cells, so methods other than sand filtration might be needed. Copper sulphate has been a common and inexpensive algaecide for a number of years, but its use must be monitored so as not to have too much copper residue left in the water. The WHO guideline is that there should be no more than 2.0 mg/l. The USEPA (2014) does not recommend its use because of the risk of cell lysis unless there is a multi-barrier approach to remove any cyanotoxins. The best time to use an algaecide is at the early stage of a bloom's development, almost a prophylactic dose. As the cell density would be much lower than in a bloom, this minimises the possible risk of intracellular toxins and other metabolites that could cause problems in the drinking water. Copper sulphate is relatively cheap and easy to apply so has gained popularity, but its effectiveness does depend upon the pH, alkalinity and amounts of dissolved organic matter present. These should be measured before application. Also analysis for copper residuals in the water should be monitored for at least a week after a treatment. Copper sulphate use has been decreasing because of concerns about the accumulation of copper in lake sediments and its toxicity to some other organisms. An alternative is the peroxide-based algicides that are being developed and are already used in some areas. These damage cell membranes before dissociating to water and oxygen. Some of these have been registered with the USEPA as being safe to use in drinking water reservoirs.

Removal of whole cells can be achieved with chemical coagulation treatment and through slow sand filtration. Membrane filtration where the pore size is less than 1 μm should remove any intact cyanobacterial cells. A further treatment method that can be used to remove cyanobacterial cells is dissolved air flotation (DAF), which can be effective against species with gas vacuoles that can be difficult to settle in any treatment involving sedimentation of the cells. Once the water has been treated and the cells removed, there will be a sludge of removed material in which there will be cells that are disrupted and will release their toxins. Great care must therefore be taken in disposing of this sludge as it is likely to contain toxins. This should be stored away from the water for at least four weeks to allow time for the toxins to degrade and become inactive. If there are known cyanobacterial toxin-producing species in the raw water, it is always wise to send samples of water for analysis for their presence. Although it is infrequent for eutrophic waters to have harmful effects on humans, they may have some other problems. These could be taste, odour and colour. Although these may not be harmful, they are unsatisfactory and frequently give rise to complaints by the public. Taste and odour compounds and toxins can be removed using activated carbon by oxidation with chlorine or ozone, chloramines and potassium permanganate and through slow sand filtration or granular activated carbon as these filters usually have an active biological film (schmutzedecke). In all cases, it is wise to send water samples away for analysis to be sure that all toxins are absent.

Newcombe et al. (2010) in their Research Report No.74 point out that if possible, the history of HAB growths in the lake or reservoir should be examined to determine when and under what conditions they have occurred. They identified the following three degrees of probability for algal growth based upon the availability of phosphorus to the algae. *Worst case*: This assumes that 80% of the phosphorus in the water is available for algal growth and that all of it could potentially be used for algal growth. In this situation, cyanobacteria could dominate and produce toxins, tastes and odours that would be released into the water. *Most likely case*: This assumes that around 50–60% of the phosphorus is bioavailable, of which a proportion will be converted into cyanobacterial biomass. Cyanobacteria do not dominate, and toxins, tastes and odours are produced but only moderately. *Best case*: This assumes that only a small amount of phosphorus is bioavailable for cyanobacterial growth (30–40%). This means that cyanobacterial production will be low and will not dominate, so toxin, taste and odour production will be low. These guidelines will also have to take into account local conditions in the lake and will require a reasonable knowledge of the nutrients, especially phosphorus, and algal history of the lake. Even in the best-case scenario, if any cyanobacterial populations are present, it would still be prudent to take samples for analysis of toxins. Newcombe et al. (2010) also give predictions of predicted toxin concentrations of different cyanobacteria and their metabolites at different bioavailable phosphorus concentrations based upon experiences with a water supply reservoir in South Australia. They also devised a framework for a graduated response to cyanobacterial growths in the reservoir. They called this an alert level framework. They suggested four levels of alert based mainly on cell counts for *Microcystis aeruginosa* ranging from >500 cells/ml for low alert to 65 000 cells/ml for very high alert. All samples should be taken from the lake or reservoir as close as possible to the water supply intake. It must be noted that *Microcystis* occurs in colonies, and counts of individual cells can be difficult. Colony counts and their average size are simpler. The intake is often at depth, so there could be higher concentrations in the surface waters. Alert levels from all but the lowest would require notification to the relevant authorities concerning hazard potential. If the hazard level is above level 2, it is quite likely that treatment of the water in the treatment works will need activated carbon treatment as well.

Examples of European and United States Regulations and Regulatory Authorities Responsible for Water

The European environment agency monitors the concentrations of both nitrogen and phosphorus in freshwaters in Europe. For groundwaters, their results show that, due to the small overall increase in nitrates between 1992 and 1998, the concentrations have declined to about the pre-1992 level. River nitrogen concentrations have declined steadily over the period 1992–2012 almost certainly due to the EU Nitrate Directive as well as other measures put in place in some individual countries. An even greater trend has been observed in the phosphorus concentrations in rivers. The N and P concentrations have halved in European rivers between 1992 and 2012. These decreases have, in part, been due to the EU Urban Wastewater Treatment Directive that was concerned with the removal of nutrients and the wider use of phosphate-free detergents. Phosphate concentrations in lake waters have also seen a gradual decline in many lakes. This has been in part due to improvements in sewage treatment and that discharges have been diverted away from lakes. Improvements are still needed with diffuse sources being the main source of agriculture. In the European Union, the policy objective is to achieve a good ecological status of all freshwater bodies within Europe. This is the basis of the Water Framework Directive (2000/60/EC) and is a legally binding commitment for all member states. To achieve this, the water pollution must be reduced, in particular from nitrates arising from agricultural activities (Nitrates Directive; 91/676/EEC). Because a large amount of water pollution, particularly for nitrogen and phosphorus, arises from sewage work treatment effluents, these must be adequately treated before discharge and discharged in a safe manner (Urban Wastewater Treatment Directive (91/271/EEC)). Any pollutants in freshwaters can potentially harm human health, particularly when that water is abstracted for drinking water purposes. In order to protect human health, the Drinking Water Directive was introduced (98/83/EC). The Water Framework Directive also requires the quantitative status of groundwater to be assessed and that abstractions should not exceed replenishment rates. Overall water abstraction from any water body should not exceed 20% of the available renewable resources. Overall, there should be a proper integrated water resource management

Table 6.12 US Federal agencies with some responsibilities for water and the government departments in which they are situated.

Department	Federal agency	Responsibility
Dept. of Defence	US Army Corps of Engineers	Navigation, dredging, flood control
Dept. of Interior	US Geological Survey	
	Fish and Wildlife	
	National Parks	
	Bureau of Land management	
	Bureau of Reclamation	
Dept. of Agriculture	Resource conservation	
	US Forest service	
Dept. of Energy	Energy Regulatory Commission	
Dept. of Commerce	Marine Fisheries Service	

Source: Adapted from Cech (2010).

strategy in place for every country, including for transboundary waters. Countries of the European Union are obliged to incorporate these directive into their national laws and regulations.

In other countries, the approach to the regulation of water use and water pollution may be different. In the United States of America, for example, responsibilities for water are divided between certain federal agencies and state agencies (see Table 6.12). Initially, the need for some regulation and responsibility was to maintain and aid navigation and trade on two main rivers, the Ohio and Mississippi and its tributaries although the Ohio River drains the eastern catchment from the state of Pennsylvania westwards and includes many important industrial areas, whereas the Missouri drains from the north and east and a large area of agricultural lands. Any flooding on either of the rivers can also result in either industrial or agricultural pollutants being spread around.

After their confluence, as well as with several other smaller rivers and streams, the Mississippi flows to the Gulf of Mexico through Louisiana and New Orleans. The Mississippi was subject to many sand bars and other obstacles to navigation as well as a heavy sediment load, which became not only a hazard to shipping but also a barrier to trade. A major problem in the Mississippi is flooding because of the low-lying extensive floodplain zone. The river had extensive floods in 1849 and 1850. As a result of the damage to land and property as well as navigation, a commission to control floods was set up in 1879 (the Mississippi River Commission). Major floods still occurred in 1912 and 1913, and these resulted in greater federal involvement in controlling the problem. In 1917, Congress passed the flood-control act to protect the lower Mississippi and the Sacramento River in California, which also flooded. Although work on control progressed it took the great Mississippi flood of 1927 to persuade the Federal authorities to significantly increase funding of flood control activities. In this flood, the flow exceeded 1.6 trillion gallons/day. The 1927 flood resulted in the death of more than 200 people and destroyed the homes of about 600 000 people. Such was the magnitude of this disaster that Congress passed the 1928 Flood Control Act, which was designed to increase funding for improvements to the river channel and increase levee construction along vulnerable parts of the river. Although the Army Corps of Engineers was already involved in flood control activities, it was given scope to construct dams across some of the tributaries to control and hold back excessive flows to the main river. It was also charged with improving and increasing the levees along the river. The federal government would pay for these improvements. As a result, further discussions on flood control took place in Congress over the next years and resulted in a new flood Control Act in 1936 and later 1944. This acknowledged that flood control was a federal responsibility throughout the nation and instructed the Army Corps of Engineers to cost future work that was needed. It should be noted that the Army Corps of Engineers suffered for many years with underestimates of the actual and final cost of many projects. Many holding reservoirs were built on tributaries of the Mississippi, including the Garrison Dam (North Dakota), Carlyle and Shelbyville dams, the Rend Lake Dam (Muddy River) and the Clarence Cannon Dam (Salt River). Building dams and the reservoirs that result requires land, and this has to be purchased and may cause conflicts with other agencies also needing land for economic development in an area. Reservoirs can provide opportunities for fishing and other recreational activities. The Flood Control Act of 1938 greatly enhanced the St. Louis flood protection project by approving both levee and dam construction to alleviate possible flooding. Unfortunately, all of the efforts to control flooding on the Mississippi have not prevented problems, and flood damage still occurs as local governments still allowed buildings to an extent in areas with zoning restrictions. Not only was the Corps of Engineers responsible for flood control, dredging and levees but also for lock maintenance and where needed increasing their capacity. Some of the structures were beyond their life expectancies and in need of repair, rebuilding and in zones where traffic has significantly increased. One example is the Alton locks situated between the upper Mississippi and Illinois to the north and the Ohio River to the south.

There are several other federal agencies with various responsibilities for water. The Bureau of Reclamation (USBR) is part of the Department of the Interior. Their prime aim was to develop water resources west of the 100th meridian, which is where the most arid lands occur. It was responsible for developing irrigation projects in order to encourage settlement in these arid areas. During the twentieth century, the USBR initiated over 200 irrigation projects, including major dam construction projects such as Hoover and Grand Coulee. Many of the dams also served the purpose of providing hydroelectric power and water supplies to urban areas. Because of their activities, the water they provide serves about 30% of the population in the west and irrigation for about 5% of the land. This irrigation water supplies 60% of the national vegetables and 25% of the country's fruit and nuts (US Bureau of Reclamation, Homepage www.usbr.gov). The construction of dams has become less popular in recent years, particularly since the failure of the Teton dam in 1976 (Box 6.4). This, although it was the only dam failure of those made by the Bureau, led to a revision and tightening of the dam-safety program. Currently, the annual budget of the Bureau is $1.17 billion (2019 data). An overview of the water-related activities of the USBR is provided in Box 6.3

Any scheme involving major engineering structures must be carried out with proper evaluation of the risks as problems could affect many people and the economy of the area. The second and the last of the priorities above were certainly brought into focus by the Teton dam failure that occurred in 1976.

Box 6.3 Water-related Activities of the Bureau of Reclamation

Objectives

Manages, develops and protects water and related resources in an environmentally and economically sound manner in the interest of the American people.

It is the nation's largest wholesale water supplier.

Provides 1 out of 5 Western farmers with irrigation water for 10 million farmland acres that produce 6% of the nation's vegetables and one quarter of its fresh fruit and nut crops.

The second largest hydropower producer in the United States.

Delivers 10 trillion gallons of water to more than 31 million people each year.

Manages, with partners, 289 recreational sites.

Reclamation's management and recreation activities contribute $62.7 billion in economic output and support about 457 754 jobs.

Priorities

Ensure the continued delivery of water and power benefits in conformity with contracts, statutes and agreements.

Operate and maintain projects in a safe and reliable manner, protecting the health and safety of the public and USBR employees.

Honour State water rights, interstate compacts, contracts with reclamation users further the Secretary of the Interior's Indian Trust responsibilities and comply with all environmental statutes.

Plan for the future on areas in the West where conflict over water is or may exist.

Enhance the business operations of USBR in accordance with the Managing for Excellence initiative.

Also develop water strategies for the future, work in partnership with the states, Tribes, water and power customers and others. Ensuring our dams do not create unacceptable risk to the public and economy by monitoring, evaluating and, where appropriate, performing risk reduction modifications.

Adapted from http://www.usbr.gov/main/about/fact.html

Another Federal Agency with an important role in water regulations is the US Environmental Protection Agency (USERPA). This is an independent agency within the federal government. It was created in 1970 with headquarters in Washington, D.C. and has 10 regional offices and a number of laboratories. It is a regulatory agency with legislative powers to impose fines when needed. It is one of the main authorities involved in the protection of the environment but can only act within statutes that are the authority of laws passed by Congress. Many, either directly or indirectly, are involved in water quality. An outline of some of the federal laws that the USEPA is responsible for upholding is shown in Table 6.13.

Some of the areas in which the Environmental Protection Agency (EPA) has been particularly active have been initiatives for safer and less polluting surfactants, setting disposal restrictions for hazardous waste disposal, cleaning sites such as the Love Canal,

Box 6.4 Teton Dam Failure

The Teton Dam was constructed by the Bureau of Reclamation on the Teton River, Idaho. It was started in 1972 and was virtually completed in June 1976 when filling was started for the first time. It was an earth dam 93 m (305 ft.) high and 1 km (0.6 miles) long. The impounded reservoir created was estimated to be 27 km (17 miles) long. In addition to generating electricity, it was to control flooding, provide irrigation water and be recreational. The cost of the whole project was nearly $49 billion. The irrigation water was to service 40 000 ha of farmland in a relatively arid area.

The geology of the area, the Snake River Plain, where the Teton site was underlaid with basalt rocks and rhyolite, which are permeable and generally thought to be unsuitable for dam construction. As part of the preliminary investigations on site selection, numerous core drillings were performed, and these showed that the rock at the dam was highly fissured and potentially unstable. The Bureau design engineers decided that these could be filled with grout, which could be injected into the fissures to seal them. The US Geological Survey, however, had indicated that in previous surveys in the area there was potential seismic activity. Five earthquakes had occurred within 50 km of the proposed dam site within the previous five years, and they had notified the USBR in 1973. Dam construction went ahead in February 1972. The outlets from the dam were closed and filling was started in 1975, but the river outlet work tunnel and the auxiliary work tunnel stayed closed as they were not complete. The fill water rose more quickly than predicted (3 ft/day as opposed to 2 ft/day maximum input). Some small seepages were observed and increased inspections of the dam were initiated. On 5 June 1976, a major leak was spotted with a flow of 500–800 l/s. This increased to 1100–1400 l/s. A whirlpool and sinkhole were observed, and four bulldozers were sent to push riprap into the hole. Unfortunately, the hole increased rapidly in size, and two were swallowed up by the hole. Fortunately, the drivers were saved. A number of seepage points developed. Later that day, a portion of the dam fell into the whirlpool and then the entire dam collapsed.

The collapse was investigated by an independent panel (Independent Panel to Review Cause of the Teton Dam Failure 1976). The general conclusions were (i) the permeability of the loess soil used in the core and the fissured nature of the rhyolite, (ii) this allowed water to seep through leading to internal erosion, (iii) this lead erosion (piping) occurring beneath the grout curtain, (iv) the leakage water gradually filled the void spaces in the rock which then flowed along the core rock interface causing an erosion channel which gradually got larger and (v) sinkholes and whirlpools were formed leading to further erosion and collapse (see Box 6.4) (Seed and Duncan 1981, The Teton Dam Failure – A Retrospective Review).

The mass of water released by the collapse affected communities immediately downstream, e.g. Wilford, Sugar City and Rexburg, destroying thousands of homes, about 80% in some communities. One factor that exacerbated the problem was a timber yard which wen the water struck thousands of logs were washed into buildings some of them hitting a bulk gasoline storage tank. This ruptured and when the gasoline ignited, sending a flaming slick of oil on the water and subsequent fires destroying many buildings. The older American Falls Dam further downstream held back the flood water preventing further damage. By this time, many tens of thousands of hectares of land had been flooded and stripped of topsoil by the surge of water. The post-dam failure clean-up started quickly, and the USBR set up claims offices in several towns. By the end of the claims programme in 1987, 7563 claims had been paid costing $322 million. Several lessons arose from the Teton failure of earth dams. Among these are that (i) great care must be taken in the choice of material, and these must be resistant to piping of water through the structure. (ii) The materials must be compacted properly to the required density. (iii) If a grout curtain is used, it must be continuous and properly sealed to the underlying rock. (iv) There must be sufficient monitoring of the dam's performance and enough permanent instrumentation installed to provide early warning of piping.

Whether or not a dam collapses its presence on a river will affect the river ecology. In the case of the Teton dam failure, there were impacts both upstream and downstream. There was a particular problem where sediments and rocks swept downstream, altering the geomorphology and the changes to the river upstream, which not only affected the character of the river but also altered the biota.

Table 6.13 Examples of environmental laws directly concerning water for which the EPA has responsibility.

- Clean water act
- Comprehensive environmental response, compensation and liability act ('superfund')
- Federal insecticide, fungicide and rodenticide act
- Resource conservation and recovery act
- Safe drinking water act
- Toxic substances control act

There are several other acts which the EPA contribute together with other agencies. These include:

- Endangered species act
- Pollution prevention act
- Oil pollution act
- National environmental policy act
- Food quality protection act and the federal food, drug, and cosmetic act

encouraging water use efficiency, reducing the use of hazardous chemicals and creating a national inventory of all existing chemicals in US commerce. This latter initiative has been followed in Canada, Japan and the European Union. In contrast, the EPA withdrew a draft document that stated that climate change imperilled public welfare. The new EPA administrator, Scott Pruitt, under the direction of the then president Donald Trump downplayed human causes of climate change, including suppressing a mapping study on sea level rise which had $3 million awarded to it.

The USGS is located within the Department of the Interior and is a scientific agency. It has offices in every state whose duties are to monitor surface and groundwaters and their characteristics. This work is done in cooperation with local and state governments, universities and other federal agencies. The USGS monitors stream flows, for which it maintains over 7000 stream gauging stations, groundwater levels and certain other components of earth and life sciences. The stations are not just for measuring flows but are also used in predicting supplies for drought avoidance and management, flood protection, ecological management, recreation and water quality management. Groundwater abstraction is an important water source, and the USGS maps and quantifies amounts and levels of water in major aquifers and wells. They are also responsible for clean-up operations of contaminated aquifers, which includes saltwater intrusions. In general, the USGS plays a major role in water resource management.

Several other agencies have roles associated with freshwater. These include the US Fish and Wildlife Service (USFWS), the National Park Service (NPS), Bureau of Land Management (BLM), Natural Resources Conservation Service (NRCS), US Forest Service (USFS), Federal Energy Regulatory Commission (FERC) and The Federal Emergency Management Agency (FEMA). An outline of the responsibilities of these agencies is provided in Table 6.14.

It can be seen that the United States Congress has, over the years, chosen to give various aspects of water responsibility, including legal duties to different agencies. There are also local, regional and state water agencies. Many countries in the world, including Australia, Canada and the United Kingdom, do have, like the United States of America, strong protection of people's rights. This protection includes the provision of adequate and regular supplies of water and the provision of proper sanitation and protection from natural disasters, which, in the case of water, include droughts and floods. In the United States of America, as the west was colonised, the settlers dug their own wells or tapped into their nearest river or lake. Wastes were often disposed of into the nearest surface water or other unproductive sites such as ravines (Cech 2010). These areas were then sparsely populated, so the general environmental impact was small. Little attention was paid to downstream users. As the population and number of settlements increased, water pollution increased, water shortages

Table 6.14 An outline of some of the freshwater responsibilities of other federal agencies.

Agency	FW fishery	Endangered species	Habitat conservation	Nat. resource management	Dam construct	Water resources and quality	Disaster recovery
USFWS	+	+	+	+			
NPS			+	+		+	
BLM	+	+	+			+	
NRCS[a]						+	
USFS						+	
FERC	+	+	+		+		
FEMA			+				+

Acronyms: USFWS, US Fisheries and Wildlife Service; NPS, National Park Service; BLM, Bureau of Land Management; NRCS, Natural Resources Conservation Service.

[a] It indicates that this agency is within the Department of Agriculture. USFS, US Forestry Service; FERC, Federal Energy Regulatory Commission; FEMA, Federal Emergency Management Agency.

occurred and water conflicts increased. The dangers of disease also increased. These factors lead to the creation of local, regional and state water authorities to regulate water supplies and serve communities. Several agencies have been created to manage water. These include local water agencies, regional water agencies, state water agencies and multistate water agencies. All of these may, if required, benefit from Federal advice or help in providing services.

Local Water Agencies

Municipal Water Departments

At a local level, the authorities in most towns and cities operate their own systems for providing drinking water and waste disposal systems. The cost of these provisions is usually based upon charges to the customers and/or local taxes. The raw water sources are often controlled by local agencies. These may be surface waters or underground aquifers. The municipal department will have its own treatment plants and have installed its own piped delivery system. Although these may be built and operated locally, they must comply with Federal Laws regarding drinking water standards, environmental protection, etc.

Water and Sewer Districts

These are almost the same as municipal water departments except that they also treat and are responsible for sewage disposal. They are not only concerned with towns and cities but may also provide a service for some rural locations. Some derivatives of both these districts and municipal water departments (e.g. metropolitan water and wastewater districts and water and sanitation districts) can be formed to service growth areas outside city boundaries. They are created under state law and are often organised by developers.

Flood-control Districts

To a large extent, flood control is a federal responsibility, but local problems are often dealt with by local flood control districts. These often date back to the beginning of the last century and deal with relatively small local problems. They may be formed to merely control or avoid floods, but in California they are designed to not only control floods but also mitigate, capture and reuse flood water.

Mutual Ditch and Irrigation Companies

These are privately owned companies whose aim is to deliver irrigation water to their customers. They only occur west of the Mississippi. Because of the importance of irrigation to the west of the 100th meridian, the companies often had to raise funds from outside investors. Once constructed, the irrigation ditches had to be maintained and supplied with water, which often required the construction of diversion dams on appropriate rivers. To obtain irrigation water from a ditch is available to share owners. The owners of shares can portion the water available to each shareholder. This can vary with time depending upon how much water is available.

Regional Water Agencies

These agencies normally serve wider areas covering several counties or states. As such, they will control larger flood and irrigation projects. For increased efficiency, they are divided into districts targeting certain aspects of regional water problems. In some areas, they perform the function of integrating several smaller agencies into one larger agency, thus minimising the overlapping of the responsibilities of different authorities, reducing costs and improving efficiency. Their irrigation projects cover larger areas that cannot be covered by individual farmers and are partly financed by taxing properties in the area. The irrigation districts formed now act as a link between federal irrigation activities of the USBR and local projects.

Conservancy Districts

These are often referred to as river basin conservancy districts. These are government agencies funded by local property owners and are concerned with flood control and water supply issues. They often work with the US Army corps of Engineers and the USBR. They cover areas ranging from schools to cities and fire services usually within a certain watershed boundary.

Natural Resource Districts

These are often amalgamations of a number of smaller diverse water agencies, e.g. as in Nebraska. They have overall responsibility for flood control, irrigation water, wetland drainage, drinking water supplies and overall watershed management. These activities are also funded through taxes.

Groundwater Management Districts as in Kansas

Groundwater resources are important in many states, and Kansas has developed an organisation of local districts to administer groundwater resources in the western part of the state, which is quite arid, receiving only 43–53 cm of rain annually. The main groundwater source for this region is the Ogallala Aquifer. Although originally mainly for agricultural supplies with increases in population, municipal supplies are becoming more important. The districts are governed by local boards elected annually.

State Water Agencies

Waters within each state in the United States of America have a governmental department to administer them. Each one is individual, but there are many similarities between them. In general, they administer water quality, water allocation, water conservation, drought planning and flood control and protection. The power of these agencies will vary from state to state from merely advisory to being a political force. The degree of activity of these agencies depends, to a large extent, upon the geography/topography and climate of that area.

Some states are very arid, e.g. Arizona, which receives only 3–12 in. of rain annually, while others, such as Rhode Island, receive 42 in./annum. Each state will have different problems to confront and administer. Cech (2010) sums up these conflicting positions thus. 'State water agencies have a unique relationship with federal counterparts. On the one hand, states are equal partners with federal agencies, since both levels of government enforce the laws of respective state and federal governments. State agencies are often empowered through federal legislation to carry out the water quality laws of the USEPA. On the other hand, state water allocation law typically takes precedence over federal law, although endangered species and pollution laws can counter these rulings. This dichotomy of laws creates conflict where state water allocation law is pre-empted by federal regulations'.

Multistate Water Agencies

Many water issues occur across state lines, e.g. certain large watersheds and aquifers, and so need multistate management. These interstate agencies are created between states with the agreement of the appropriate federal government departments. The aim is to promote agreement in planning, communication and coordination, thus saving money and promoting efficiency. Communication between states is essential as different states may have different objectives in their management plans. An example of such a multistate agency is the Missouri River Basin Association (MRBA), which can be compared with the states encompassing the Olagalla or High Plains Aquifer.

Missouri River Basin Association

The Missouri River Basin Association (MBRA) was founded in 1981. The river, for management, resources, flood protection, etc. has had six large dams constructed in North and South Dakota that store about 90 billion m^3 of water. In all it flows through eight states, Missouri, Kansas, Iowa, Nebraska, South Dakota, North Dakota and Montana. Part of the catchment is also in Wyoming. The objective behind the formation of the MBRA was to coordinate planning activities on the Missouri River. The Board also included not only state representatives but also representatives from the basin's 28 Native Indian tribes. Representatives from interested federal agencies also acted as advisors. The MBRA works in consultation with river users and farmers to develop the economic and environmental status of the basin. The lower reaches of the river have been channelized by the Army Corps of Engineers with the aim of reducing flood risks and improving navigation. Although this has some benefits, it has adversely affected adjoining wetlands and other aquatic habitats. This management has reduced pollution of the water, and the dams have not only provided storage but also moderated the flood risk not only on the Missouri but also helped the Mississippi. The dams did alter downstream flows, which reduced and changed habitats and created a different riverine system. There were also conflicting demands from navigation, irrigation and environmental needs, especially as overall flows were reduced. The MRBA had to deal with these and other conflicting interests. As we entered a new century, other interest groups came to the fore. Prominent among these were those wishing to protect endangered species and reduce flood plain wetlands. Superimposed on these for the future is the possible impact of climate change. As the population of the region grows, more buildings, including dwellings, are being constructed on the river flood plain. How much flood protection should be provided? If more water is to be held back behind dams, river flows could back up for hundreds of kilometres, potentially causing problems upstream. Should more water be released increasing flows downstream and thus affecting nesting river bank birds? These and a myriad of others have to be sorted out. At least there is a single representative body, the MRBA, to discuss these.

The Ogallala Aquifer

The high plains area of the United States of America is a moderately arid zone but fortunately is underlain by one of the world's largest aquifers – the Ogallala (Figure 6.7). After World War II, the extraction of water from the aquifer increased considerably under pressure to increase food production and the introduction of centre pivot irrigation. Apart from the danger of contamination of the aquifer's water quality, there is a danger of overabstraction. The aquifer also provides water for cities and municipalities and runs from Wyoming through South Dakota, Nebraska, Colorado, Kansas, Oklahoma, New Mexico and Texas. The northern part of Texas is known as the Panhandle. The farmers of this region could not exist economically if it was not for groundwater pumping from the Ogallala aquifer. Within Texas, about 40% of the water used comes from the aquifer (https://www.texastribune.org/2010/06/17/how-bad-is-the-ogallala-aquifers-decline-in-texas). Although the aquifer is large and potentially has an enormous volume of water, it supplies not only Texas farmers and cities but also those in all of the other states

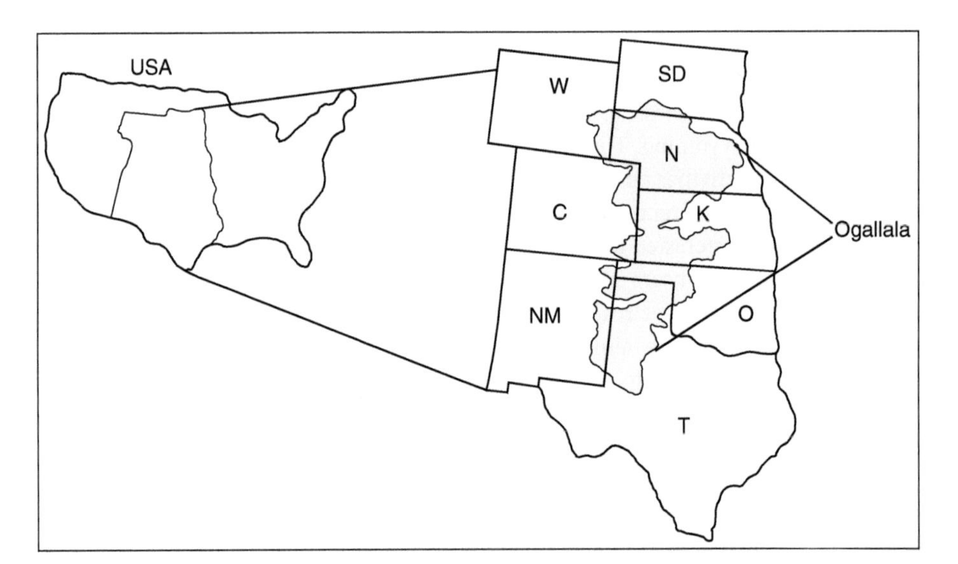

Figure 6.7 United States of America under which the Ogallala aquifer is found. SD, South Dakota; W, Wyoming; N, Nebraska; C, Colorado; K, Kansas; NM, New Mexico; O, Oklahoma; T, Texas.

mentioned above. There is, however, state by state regulatory control of its use and abstraction rates. Because of the heavy use and low recharge rate in the aquifer, water levels fall sharply. During one year in the Panhandle, the water level in the aquifer fell, on average by 1.5 ft (76 cm), but in Texas the state water plan predicts that the aquifer water volume will fall by 52% by 2060 and livelihoods would be cut back severely (Texas Tribune). The Texas groundwater conservation districts do now have some water conservation regulations, but the rules are not strongly enforced. There is a 50-year plan that allows the authorities to control permits to drill new wells, which should help with the aim being to have about 50% of the groundwater remaining in 50 years. There is, however, some resistance to these plans from oilmen who own water rights and want to sell water to large cities. The 50% remaining plan means that there may be no water available for sale in the future, so the rights become worthless. Also there is the problem of population increase in Texas, which is expected to double by 2060 from 21 to 46 million, and water demand will increase by at least 27% in that period. All this occurs at a time when the recharge rate of the aquifer is well below the abstraction rate. The State of Texas created the Texas Water Development Board (TWDB) in 1957 after a severe drought with a view to planning water management in the future and encouraging conservation measures. Because of its activities, agricultural consumption has decreased as has the overall water demand. The TWDB now uses sophisticated modelling techniques to manage the water resources with less or no unrestricted pumping replaced by permitted pumping, which is defined by the groundwater conservation districts. The needs of river and stream ecosystems and their flows are also included in the new approach. Diffuse sources of pollution are a major problem. Drainage from agricultural land can be a major source of nitrogen and phosphorus from the use of fertilizers. This applies not only to artificial fertilizers but also to the spread of animal wastes. It also applies to many land use changes. The spreading of animal wastes on land is a problem throughout Europe, especially in countries with more intensive animal production such as the Netherlands. It can also be a problem in the United States, for example in California, Vermont and Pennsylvania, where intensive livestock production can be quite common. The use of animal manure needs to be applied sparingly. Where crops are being produced, care should be taken over the tillage practices. Sediment and nutrient run-off is greatest after ploughing and seeding of the new crop. Also in the winter when the crop has been harvested. It is advisable to sow a cover crop, such as nitrogen-fixing legumes. Over this period, nitrogen is added to the soil and helps prevent soil erosion. Where possible, it is beneficial to have a band of land several metres deep around the surface water course to trap sediments and dissolved nutrients, preventing them from entering the surface water. Filtration of the run-off water through a wetland would be ideal. Tile drains are commonly used for drainage, and it is advisable that these drains discharge into a sediment basin rather than directly into a river or lake. All of the above apply to pesticides as well as nutrients (Logan 1993). Many pesticides persist in the environment for many years before being degraded, which provides ample opportunities for them to enter environmental food chains and be biomagnified. Monitoring of pesticides and other toxins in the water is essential. However, it is also important to monitor their concentrations in sediments as they may be bound onto particles that are deposited and accumulate in river, lake and marine deposits. An example is with PCBs, where it was observed that although

Table 6.15 Selected examples of some contaminants and their average concentrations in run-off from different establishments (all concentrations in mg/l).

Substance	Urban dwelling location	Commercial area	Industrial area
BOD_5	74	100	500–750
Suspended solids	200–48 000	50–830	450–1700
Total phosphorus	0.3–4.8	0.1–0.9	0.9–4.1
Total nitrogen	0.2–18.0	1.9–11.0	1.9–14.0
Lead	0.14–0.5	0.7–1.1	2.2–7.0
Copper	0.02–0.21	0.07–0.13	0.29–1.3

water concentrations might be low, those in a contaminated sediment could be quite high and pose a risk to organisms living in and around that sediment. Run-off from urban roads and hard surfaces in commercial and industrial establishments can contain a wide variety of substances, such as hydrocarbons, rubber particles, metals, nutrients and suspended solids (Bellinger et al. 1982). The suspended solid load can be as high, or even higher, than that in wastewater effluent. Ideally, all such run-off should be collected separately in a holding pond to allow settlement and, if required, treatment before disposal. Table 6.15 illustrates some of the different contaminants and their average concentrations in urban run-off from different establishments compared with raw sewage.

Although there are many problems concerning water quality, the solutions are largely available. The seven fundamental solutions are as follows:

- Prevent pollution.
- If pollution accidentally occurs, procedures must be in place to treat it immediately.
- Ecosystem restoration must be given a high priority.
- Because many freshwaters are already contaminated, better water treatment must be provided before discharge of all effluents.
- Prevent disposal of untreated wastes both human and industrial into the environment.
- Introduce effective regulatory mechanisms to protect water management and quality and provide financial incentives to achieve this.
- Educate the public about these issues and use multiple routes, such as information on best practices, promoting local public involvement and disseminating information on the benefits of proper water management.

These points have been adapted from Palaniappan et al. (2012).

Although water quality impairment can occur through natural events, the main reasons for the increases in pollution in recent decades have been through human activities. Because freshwater is a finite resource, the use of this water could cause problems. Modern technology can help with some of them but at a cost. Certainly, in developed countries, people have become accustomed to having clean water in reasonable amounts 'on tap' and at a reasonable cost. The increases in treatment costs will inevitably result in rise in price. It is thus to everyone's benefit to help reduce water pollution and help improve water quality by reducing pollution.

References

Addiscott, T.M., Whitmore, A.P., and Powison, D.S. (1991). *Farming, Fertilizers and the Nitrate Problem*. Wallingford: CAB International.

Anderson, D.M., Hoagland, P., Kaoru, Y. and White, A.W. (2000). Estimated Annual Economic Impacts from Harmful Algal Blooms (HABs) in the United States. *Technical Report WHOI-2000-11*, Woods Hole Oceanographic Institution.

Banks, D., Younger, P.L., Arnesen, R.T. et al. (1997). Mine water chemistry: the good, the bad and the ugly. *Environmental Geology* 32: 157–174.

Bellinger, E.G., Jones, A.D., and Tinker, J. (1982). The character and dispersal of motorway run-off water. *Journal of the Institute of Water Pollution Control* 81 (3): 372–390.

Bellinger, E.G. and Sigee, D.C. (2015). Freshwater algae. *Identification, Enumeration and Use as Bioindicators*. Wiley Blackwell

Billen, G., Garnier, J., and Lassaletta, L. (2013). The nitrogen cascade from agricultural soils to the sea: modelling N transfers at regional watershed and global scales. *Philosophical Transactions of the Royal Society B* 368: 20130123.

Bowling, L.C. and Baker, K.T. (1996). Major cyanobacterial bloom in the Barwon-Darwin River, Australia in 1991, and underlying limnological conditions. *Marine and Freshwater Research* 47: 643–657.

Bruinsma, J. (ed.) (2003). *World Agriculture: Towards 2015/2030. A FAO Perspective*. Earthscan, London: FAO, Food and Agriculture Organization.

Burgis, M.J. and Morris, P. (1987). The World of Lakes. Freshwater Biological Association Special Publication No. 15. UK: FBA & MPM publishing.

Burt, T.P., Worrall, P., Howden, N.J.K. et al. (2011). Nitrate in United Kingdom Rivers: policy and its outcomes since 1970. *Environmental Science & Technology* 45: 175–181.

Bussink, D.W. and Oenema, O. (1997). Ammonia volatilization from dairy farming systems in temperate areas; a review. *Nutrient Cycling in Agroecosystems* 51: 19–33.

Carr, G.M. and Neary, J.P. (2008). *Water Quality for Ecosystem and Human Health*, 2e. UNEP Global Environmental Monitoring System.

Carruthers, R.B. and Barlow, F.J. (1973). An ecological survey of the West Fork of the Obey River, Tennessee with emphasis on the effects of acid mine drainage. *Journal of the Tennessee Academy of Science* 32: 162–147.

Carlson, R.E. (1977). A trophic state index for lakes. *Limnology and Oceanography* 22 (2): 361–369.

Cattaneo, A., Asioli, A., Comoli, P., and Monca, M. (1998). Organisms response in a chronically polluted lake supports hypothesized link between stress and size. *Limnology and Oceanography* 43 (8): 1938–1941.

Cattaneo, A., Couillard, Y., Wunsam, S., and Courcelles, M. (2004). Diatom taxonomic and morphological changes as indicators of metal pollution and recovery in Lac Dufault (Quebec, Canada). *Journal of Paleolimnology* 32: 163–175.

Cech, T.V. (2010). *Principles of Water Resources. History, Development, Management and Policy*. Wiley.

Chorus, I. and Bartram, J. (eds.) (1999). *Toxic Cyanobacteria in Water*. Printed for the WHO by Hobbs printers Ltd. And transferred to Digital Printing by Spon Press, London.

Cook, G.D., Welch, E.B., Peterson, S.A., and Newroth, P.R. (ed.) (1993). Artificial circulation. In: *Restoration and Management of Lakes and Reservoirs*, 419–449. London: Lewis Publishers.

Dentener, F., Drevet, J., Lamarque, J.F. et al. (2006). Nitrogen and Sulphur deposition on regional and global scales: a multimodel evaluation. *Global Biogeochemical Cycles* 20: Gb4003.

Dobson, M. and Frid, C. (2009). *Ecology of Aquatic Systems*, 2e. Oxford University Press.

Dodds, W.K. and Welch, E.B. (2000). Establishing nutrient criteria in streams. *Journal of the North American Benthological Society* 19 (1): 186–198.

Duce, R.A., LaRoche, J., Altieri, K. et al. (2008). Impacts of atmospheric anthropogenic nitrogen on the open ocean. *Science* 320: 893–897.

Filatova, D., Jones, M., Haley, J.A. et al. (2021). Cyanobacteria and their secondary metabolites in three freshwater reservoirs in the United Kingdom. *Environmental Sciences Europe*. Springer Open. https://doi.org/10.1186/s12302-021-00472-4.

Fowler, D., Coyle, M., Skiba, U. et al. (2013). The global nitrogen cycle in the 21st century. *Philosophical Transactions of the Royal Society B* 368. https://doi.org/10.1098/rstb.2013.0164.

Galloway, N.J., Schlesinger, H., Levy, H. et al. (1995). Nitrogen fixation: anthropogenic enhancement – environmental response. *Global Biogeochemical Cycles* 9: 235–252.

Galloway, J.N., Dentener, F.J., Capone, D.G. et al. (2004). Nitrogen cycles: past. Present and future. *Biogeochemistry* 70: 53–226.

Ghosh, T.K. and Mondal, D. (2012). Eutrophication: causative factors and remedial measures. *Journal of Today Biological Sciences: Research and Review* 1 (1): 153–178.

Gleick, P.H. (2012). *The World's Water, Volume 7. The Biennial Report on Freshwater Resources*. Washington/London: Island Press.

Gleisberg, D., Erftstadt, H., and Hahn, H. (1995). Zur Entwicklung der Phospherentferentfemung aus Abwassern der Bundestepublik Deutschland. *Korrespondenz Abwasser* 42: 958–969.

Graham, J., Dubrovsky, N.M. and Eberts, S.M. (2017). Cyanobacterial Harmful Algal Blooms and U.S. Geological Survey Science Capabilities. *USGS Report 2016–1174*.

Green, P.A., Vorosmarty, C.J., Meybeck, M. et al. (2003). Pre-industrial and contemporary fluxes of nitrogen through rivers: a global assessment based on typology. *Biogeochemistry* 68: 71–105.

Greenfield, J.P. and Ireland, M.P. (1978). A survey of the macrofauna of a coal-waste polluted Lancashire system. *Environmental Pollution* 16: 105–122.

Harper, D.M. (1992). *Eutrophication of Freshwaters*. London: Chapman and Hall.

Harremoes, P. (1997). The challenge of managing water and material balances in relation to eutrophication. In: *Eutrophication Research, State of the Art. Department of Water Quality Management and Aquatic Ecology* (ed. R. Roijackers, R.H. Aalderink, and G. Blorn), 3–12. Wageningen Agricultural University.

Hartman, P.E. (1983). Nitrate/nitrite ingestion and gastric cancer mortality. *Environmental Mutagenesis* 5: 111–121.

Hartmann, M., Hediger, W., and Peter, S. (2008). Reducing nitrogen losses from agricultural systems – an integrated economic assessment. *Schriften der Gesellschaft fur Wirschaffs- und Sozialwissenschaften des Landbaues e* 43: 335–344.

Heinzmann, B. and Chorus, I. (1994). Restoration concept for Lake Tegel, a major drinking and bathing water resource in a densely populated area. *Environmental Science & Technology* 28: 1410–1416.

Howden, N.J.K., Burt, T.P., Worrall, F. et al. (2010). Nitrate concentrations and fluxes in the River Thames over 140 years (1868–2008): are changes irreversible? *Hydrological Processes* 24: 2657–2662.

Hudnell, H.K. (2010). The state of U.S. freshwater harmful algal blooms assessments, policy and legislation. *Toxicon* 55 (5): 1024–1034.

ILEC/Lake Biwa Research Institute (ed.) (1988-1993). *Survey of the State of the World's Lakes*, vol. I-IV. Nairobi: International Lake Environment Committee, Otsu and United Nations Environment Programme.

Imboden, D.M. (1992). The impact of physical processes on algal growth. In: *Eutrophication* (ed. D. Sutcliffe and J.G. Jones). Ambleside, Cumbria, U.K.: Freshwater Biological Association.

Kallqvist, T. and Berge, D. (1990). Biological availability of phosphorus in agricultural runoff compared to other phosphorus sources. *Verhandlugen der international Vereingen fur theoretische und angerandte Limnologia* 24: 214–217.

Kelly, M.G. and Whitton, B.A. (1993). The trophic diatom index: a new index for monitoring eutrophication in rivers. *Journal of Applied Phycology* 7: 4232–4244.

Lillesand, T.M., Johnson, W.L., Deuell, R.L. et al. (1983). Use of Landsat data to predict the trophic state of Minnesota Lakes. *Photogrammetric Engineering and Remote Sensing* 49 (2): 219–229.

Logan, T. (1993). Agricultural best management practices for water pollution control: current issues. *Agriculture, Ecosystems and Environment* 46: 223–231.

Lopez, C.B., Jewett, E.B., Dortch, Q. et al. (2008). Scientific Assessment of Freshwater Harmful Algal Blooms. Interagency Working Group on Harmful Algal Blooms, Hypoxia and Human Health of the joint subcommittee on Ocean Science and Technology, Washington, DC.

Meybeck, M. (1982). Carbon, nitrogen and phosphorus transport by world rivers. *American Journal of Science* 282: 401–450.

Meybeck, M., Chapman, D., and Helmer, R. (ed.) (1989). *Global Freshwater Quality: A First Assessment*. Oxford: Blackwell Reference.

Michalak, A.M., Anderson, E.J., Beletsky, D. et al. (2013). Record-setting algal bloom in Lake Erie caused by agricultural and meteorological trends consistent with expected future conditions. *Proceedings of the National Academy of Sciences of the USA* 110: 6448–6452.

Minaudo, C., Meybeck, M., Moatar, F. et al. (2015). Eutrophication mitigation in rivers: 30 years of trends in spatial and seasonal patterns of biogeochemistry of the Loir River (1980–2012). *Biogeosciences* 12: 2549–2563.

Monteny, G.J. (2001). The EU nitrates directive: a European approach to combat water pollution from agriculture. *The Scientific World* 1 (S2): 927–935.

Newcombe, G., House, J., Ho, L., et al. (2010). Management strategies for Cyanobacteria (Blue–Green algae): A guide for Water Utilities. *Research Report No 74*. Water quality research Australia, Adelaide, South Australia.

Oenema, O. and Roest, C.W.J. (1997). Nitrogen and phosphorus losses from agriculture into surface waters. *Eutrophication Research; State of the Art*. Symposium 28–29 August 1997, Wageningen, The Netherlands.

Palaniappan, M., Gleick, P.H., Allen, L. et al. (2012). *Water Quality in the World's Water Vol. 7. The Biennial Report on Freshwater Resources*. Washington/London: Island Press.

Palmer, C.M. (1962). Algae in Water Supplies. U.S. Department of Health Education, and Welfare Public Health Service. Division of Water Supply and Pollution Control, Washington D.C. Publication No. 657.

Peter, S., Hartmann, M., and Hediger, W. (2006). *Neuberechnung der landwirtschaftlichen Emissionen umweltokonomie*. ETH Zurich.

Pickering, A.D. (2001). *Windermere. Restoring the Health of Englands Largest Lake*. Ambleside, Cumbria, UK: Freshwater Biological Association.

Ravera, O. (1981). Effects of nutrient enrichment of water bodies and its correctives. *Water Industry 81*, CEP Consultants, Edinburgh; pp. 63–69.

Reynolds, C.S. (1984). *The Ecology of Freshwater Phytoplankton*. Cambridge University Press.

Reynolds, C.S. (1990). Modelling phytoplankton dynamics and its application to lake management. In: *The Ecological Bases for Lake and Reservoir Management* (ed. D. Harper et al.), 123–131. Dordrecht, The Netherlands: Kluwer Academic Publishers.

Reynolds, C.S. (1997). *Vegetation Processes in the Pelagic: A Model for Ecosystem Theory*. Nordbunte, Germany: Ecological Institute.

Ripley, E.A., Redmann, R.E., and Maxwell, J. (1979). *Environmental Impact of Mining in Canada*. Kingston, Ontario: Centre for Resources Studies, Queens University.

Round, F.E. (1965). *The Biology of the Algae*. London: Arnold.

Seed, H.B. and Duncan, J.M. (1981). The Teton Dam Failure – A retrospective review. International Society for Soil Mechanics and Geotechnical Engineering, *SIMSG ISSMGE*. https://www.issmge.org/publications/online-library (accessed December 2019).

Selman, M. and Greenhalgh, S. (2000). Eutrophication: Policies, Actions, and Strategies to Address Nutrient Pollution. World Resources Institute Policy Note No. 3. Washington.

Sigee, D.C. (2005). *Freshwater Microbiology*. Chichester, UK: Wiley.

Smil, V. (1999). Nitrogen in crop production: an account of global flows. *Global Biogeochem Cycles* 13 (2): 647–662.

Smil, V. (2001). *Enriching the Earth. Fritz Harber, Carl Bosch, and the Transformation of World Food Production*. Cambridge, MA: The MIT Press.

Smith, R.A., Alexander, R.B., and Schwartz, G.E. (2003). Natural background concentrations of nutrients in streams and rivers of the conterminous United States. *Environmental Science & Technology* 37 (14): 3039–3047.

Somlyody, L. and van Straten, G. (ed.) (1986). Background to the Lake Balaton eutrophication problem. In: *Modeling and Managing Shallow Lake Eutrophication*, 3–18. Berlin: Springer Verlag.

Sprague, L.A. and Lorenz, D.L. (2009). Regional nutrient trends in streams and rivers of the United States. *Environmental Science & Technology* 43 (10): 3430–3435.

Sutton, M.A., Bleeker, A., Howard, C.M. et al. (2013). *Our Nutrient World*. Edinburg, UK: Centre for Ecology and Hydrology.

United Nations World Water Assessment Programme (UNWWAP) (2003). The First UN Water Development Report: Water for People, Water for Life. http://www.unesco.org/water/wwap/wwdr/wwdr1

United Nations World Water Assessment Programme (UNWWAP) (2009). *The World Water Development Report 3: Water in a Changing World*. Paris: UN Educational, Scientific and Cultural Organization http://www.unesco.org/water/wwap/wwdr/wwdr3.

U.S. Environmental Protection Agency (USEPA) (2014). 810-F-11-001. Cyanobacteria and Cyanotoxins: Information for Drinking Water Systems.

United States Geological Survey (USGS) (2017). Cyanobacterial Harmful Algal Blooms and U.S. Geological Survey Science Capabilities. Graham, Dubrovsky & Eberts. *Report 2016–1174*, Ver 1.1 December 2017.

Visser, P.M., Ibelings, B.W., Bormans, M., and Huisman, J. (2016). Artificial mixing to control cyanobacterial blooms: a review. *Aquatic Ecology* 50: 423–441.

Vitousek, P.M., Aber, J.D., Howarth, R.W. et al. (1997). Human alteration of the global nitrogen cycle: sources and consequences. *Ecological Applications* 7 (3): 737–750.

Vitousek, P.M., Cassman, K., Cleveland, C. et al. (2002). Towards an ecological understanding of biological nitrogen fixation. *Biogeochemistry* 57: 1–45.

Vollenweider, R.A. and Kerekes, J. (1980). The loading concept as a basis for controlling eutrophication philosophy and preliminary results of the OECD programme on eutrophication. *Progress in Water Technology* 12: 5–38.

Voss, M., Bange, H.W., Dippner, J.W. et al. (2013). The marine nitrogen cycle: recent discoveries; uncertainties and the potential relevance of climate change. *Philosophical Transactions of the Royal Society B* 368. https://doi.org/10.1098/rstb.2013.0121.

Walsby, A.E. (1992). The control of gas-vacuolate cyanobacteria. In: *Eutrophication: Research and Application to Water Supply* (ed. D. Sutcliffe and J.G. Jones). Ambleside, Cumbria, UK: Freshwater Biological Association.

Warner, R.W. (1971). Distribution of biota in stream polluted by acid mine drainage. *Ohio Journal of Science* 71: 202–215.

Westhoek, H., Rood, T., Berg, M. et al. (2011). *The Protein Puzzle: The Consumption and Production of Meat, Dairy and Fish in the European Union*. The Hahue/Bilthoven: PBI.

World Health Organization (2011). *Guidelines for Drinking-Water Quality*, 4e. WHO, Geneva, Switzerland.

World Resources Institute (2000). Eutrophication: Policies, Actions, and Strategies to Address Nutrient Pollution. *WRI Policy Note, Water Quality and Hypoxia No. 3*. Washington.

Worrall, F., Burt, T.P., Howden, N.J.K., and Whelan, M.J. (2009). Fluvial flux of nitrogen from Great Britain 1974–2005 in the context of the terrestrial nitrogen budget of Great Britain. *Global Geochemical Cycles* 23: GB3017.

Züllig, H. (1989). Role of carotenoids in lake sediments for reconstructing trophic history during the late quaternary. *Journal of Paleolimnology* 2: 23–40.

7

Drinking Water Treatment

Pure water rarely, if ever, occurs naturally in the environment. For example, as a raindrop falls through the sky, it will dissolve gases from the atmosphere and pick up any particles that it comes into contact with. Surface waters will dissolve minerals from the geological strata over and through which they flow as will groundwaters. Surface waters will also be contaminated from fields and general land drainage, including roads. This could include, among others, fertilizers, pesticides and animal wastes. All of these may have to be removed to render water potable. In broader terms, water after treatment when delivered to the customer must meet the following criteria:

- There must be no unpleasant taste or odour.
- It must be chemically safe and free from harmful chemicals.
- It must be biologically safe, i.e. it does not contain any harmful pathogens.
- It must be clear and not contain any turbidity or suspended solids.
- It must be reasonably soft (low in calcareous compounds) to allow washing of clothes, household utensils and personal hygiene so as not to require large amounts of detergents.
- It should not contain any substances including acidity that would cause the corrosion of metal pipes, such as copper or lead.
- The organic content should be low to avoid growths of microorganisms and invertebrates in the distribution system or storage tanks.

In order to make the water safe for consumption, it should meet these criteria and in particular the microbiological standards as well as those for chemical, physical and radioactivity. These standards are usually set by the appropriate regulatory authorities in each country, but most are based on WHO Guidelines for drinking water quality (WHO 2011). In Europe, the European Union instigated two main directives that state the quality of raw water to be used for drinking water abstraction and treatment: The Surface Water Directive (75/440/EEC) and the Ground Water Directive (80/68/EEC). After suitable treatment, the water delivered to consumers should meet the requirements of two further directives: 80/778/EEC and 98/83/EEC. In the United States of America, the US Public Health Service brought in the first comprehensive drinking water standards in 1942. Later, the USEPA became the enforcement agency for drinking water standards. These were not too dissimilar to the WHO standards (see Table 7.1 for examples).

Water: Our Sustainable and Unsustainable Use, First Edition. Edward G. Bellinger.

Table 7.1 A comparison between WHO and USEPA drinking water standards for some selected substances.

Substance	WHO guideline	USEPA standard
Chemicals		
Copper	2.0	1.3
Fluoride	1.5	4.0
Nitrate nitrogen	11.0	10.0
Atrazine	0.1	0.003
2,4-D	0.03	0.07
Dioxin	0.05	0.0
Xylenes (total)	0.5	10.0
Microorganisms		
Giardia lamblia	0	0
Cryptosporidium	0	0
Viruses	0	0

Values in mg/l.

Care must be taken when comparing standards from different countries as the exact terminology can vary.

Most drinking water is obtained from surface waters such as lakes, reservoirs and rivers or from sub-surface aquifers. The latter generally provide water that is microbiologically good but can, in some regions, have contamination from metals and other chemicals. These groundwaters and most surface waters need to be treated to prevent harm to humans and for other uses and meet certain standards, most of which are based on WHO requirements.

Aims for Provision of Drinking Water and Sanitation

The WHO/UNICEF (2015) Agenda for Sustainable Development aims that by 2030 all member states will be required to take appropriate steps to achieve a pathway to sustainable and resilient development as well as human rights for all, including gender equality adopting the slogan 'no one left behind'. The task of collecting water for drinking, washing and cooking takes many hours in many developing countries and is a major contributor to poverty. The WHO/UNICEF (2006) Agenda proposes a development goal that includes many targets, including reducing inequalities related to water and sanitation (WASH). A summary of these targets pertaining to WASH is as follows:

1) By 2030, achieve universal and equitable access to safe and affordable drinking water for all.
2) By 2030, achieve access to adequate and equitable sanitation and hygiene for all and end open defecation and paying special attention to the needs of women and girls and those in vulnerable situations.

3) By 2030, ensure all men and women, in particular the poor and vulnerable, have equal rights to economic resources as well as access to basic services.
4) Achieve universal health coverage, including financial risk protection, access to quality essential healthcare services, and access to safe, effective, good quality and affordable essential medicines and vaccination for all.

Source: Abridged from UNICEF/WHO (2019).

The overall aim of the WHO guidelines, and should be for national regulations, is that drinking water should be safe not only for adults but also for infants, young children, elderly people and people with health risks. It is also important that water should be safe for drinking over a long period, e.g. there should be no substances that will bioaccumulate in the body and cause health problems. The WHO guidelines are the framework to support the development of good practice and risk management strategies that will ensure the long-term safety of drinking water. Setting standards on paper is not enough. Regular monitoring and analysis of samples is vital and requires both expertise and financial support for trained personnel and laboratories.

This approach must also form the basis for the selection of appropriate treatment methods for providing safe drinking water. The selection of the treatment techniques will vary depending upon the raw water source. Contaminated sources may require more complicated and costly techniques, so it is wise to choose as clean a raw water source as possible so that treatment can be as easy and effective as possible.

The Joint Monitoring Programme (JMP) aimed to use indicators based on the proportion of the population of a country that achieves the goals for WASH. In addition, Goals 3 and 4 of the Millennium Declaration do not specifically refer to WASH, and both play an important part in achieving them either directly or indirectly. Each country should set its own targets for rates of progress depending upon the local situation, for example their economic development and gross domestic product (GDP). It is also important to provide good education and training both in schools and for operators and administrators concerning the importance of water to livelihoods and economies as well as its sustainable use. People at all levels should understand the importance of WASH in helping healthcare. It is also important that they understand that a healthy natural environment will help human well-being. The global targets are irreversible and apply to all countries. They are aspirational as different countries have different economies and different rates of development. Generally, in the past, wealthy groups enjoyed higher standards. It follows that disadvantaged groups will need greater and faster progress to achieve international targets. This does not always happen, however. In one of the wealthiest countries in the world, the United States of America, and within that country, one of the wealthiest states, California, 88 000 people live in houses without a private indoor flush toilet. A further 120 000 people experience homelessness and are unlikely to have access to toilet facilities (Feinstein and Daiess 2019). The emphasis to date has been to expand and improve the reliability and quality of drinking water for households using domestic wells. This focus meant that sanitation and wastewater were left a little behind. For those dwellings without satisfactory wastewater collection facilities, the contamination of watercourses by faecal bacteria is a particular problem through wet weather discharges (Steele et al. 2017). Although globally there is still work to be done, there have been many improvements. A total of 2.6 billion people have

Table 7.2 % people with improved sanitation based on income and the proportion of the population having access in selected countries.

For improved sanitation comparing access for different income levels				
Cambodia	Bottom inequality	For the bottom quintile	1995, 0%	2012, 36%
		For the top quintile	1995, 89%	2012, 100%
Vietnam	Top inequality	For the bottom quintile	1995, 41%	2012, 63%
		For the top quintile	1995, 98%	2012, 99%
For improved drinking water				
Bolivia	Bottom inequality	For the bottom quintile	1995, 38%	2012, 56%
Kyrgyzstan	Top inequality	For the bottom quintile	1995, 72%	2012, 70%
		For the top quintile	1995, 100%	2012, 99%

gained access to improved drinking water supplies since 1990, and 2.1 billion have gained access to improved sanitation (UNICEF/WHO 2015). Table 7.2 summarises the changes in drinking water and sanitation connections for different regions since 1995.

Since the millennium development goals (MDG) declaration, there have been substantial improvements although some regions have fared better than others. Globally, it is estimated that there was an improvement in access to improved drinking water sources from 76% to 91%, which represents an additional 6.6 billion people with access. Unfortunately, the least developed countries, especially in Sub-Saharan Africa, have below average access although improvements have been made. UNICEF/WHO (2015) reported that in 2015 it was estimated that 663 million people worldwide were still using unimproved water sources. These included unprotected wells, springs and surface waters. Approximately 159 million people still use surface waters that are often contaminated and pose a risk to health. Most lived in Sub-Saharan Africa and Southern Asia. Some of the greatest improvements were in Eastern Asia, where MDG targets were exceeded. In China, for example, over 0.5 billion people gained access to improved drinking water. Unfortunately, some of the least developed countries have the highest population growth rates which, together with their relatively weak economies, hinder improvements. Many countries have also been affected by conflict. These countries, as well as many others, are experiencing a population shift from the countryside to towns. In 1990, 57% of the global population was rural, but by 2015 about 54% lived in urban areas. This causes problems of both infrastructure and access to services. This applies to sanitation as well as drinking water. Improvements have been made, and the use of improved sanitation rose from 54% to 68% by 2015, still 9% below the MDG target. Again, the lowest countries for access are in Sub-Saharan Africa, but Southern Asia also has access problems. The WHO has always targeted open defecation as the main factor needing elimination to improve the health and well-being of the population. UNICEF/WHO (2015) reported that more than half the population in 16 countries practiced open defecation. In further 62 countries, the figure was more than 10%. These figures did improve by 2015 with a 20% reduction in 23 countries and a 10% reduction in further 44 countries. The greatest improvements were in Ethiopia and Nepal, with the smallest in Guinea, Benin and Malawi. The largest improvements from the point of view of population numbers were in Pakistan, Bangladesh and India. For countries with low economic development, the costs of improvement in the provision of better sanitation and drinking water have difficulty in keeping up with population growth.

Some countries have made great progress in reducing open defecation. Bangladesh, for example, reduced the % of the richer population not using this from 94% to 100% between 1995 and 2012. Most dramatically, the poorest quintile not using open defecation improved from 45% in 1995 to 86% in 2012. Unfortunately, although further improvements need to be made in many countries, improved sanitation receives less attention than piped drinking water access.

Although for many years the water industry had been regarded as rather conservative in its approach to drinking water treatment in recent years, there have been many new technologies and developments of existing technologies that have been developed (wastewater treatment technology is discussed in Chapter 8). Methods are now available to treat almost any type of water from any source to drinking water standards. The main constraint, however, is financial. For public supplies in particular, cost is important, whereas in industrial and other uses, the cost may be of less importance than the final quality. For public water supplies, it is thus important to select, if possible, the cleanest supply of raw water and so minimise the treatment costs. Not only quality but also available quantity must be taken into account.

As raw waters vary greatly in their overall composition, a selection and combination of different technologies and barriers is usually needed to make the water 'safe' and able to meet the required standards. This is becoming increasingly difficult in some regions because of water contamination. Indeed, Corcoran et al. (2010) conclude that the world is facing a global water quality crisis because of widespread water contamination.

In the United Nations declaration of 2000, a number of MDG for all nations to achieve in order to improve the lives of an ever-growing world population were stated. There were a number of targets, specifically Goal 7, to ensure environmental sustainability that would involve water supply and management and improved sanitation, which is required to be achieved. Examples include food production and education. There were also specific targets within the eight main goals. One target was to halve, by 2015, the proportion of the population without sustainable access to safe drinking water and basic sanitation (United Nations 2012). Over 1.1 billion people do not have access to drinking water from improved sources, and 2.6 billion people do not have basic sanitation (WHO and UNICEF 2006). Proper access to safe drinking water and sanitation is essential to health and well-being. The benefits from supplying adequate drinking water and proper sanitation are not only in the health of the population but also in the economy and general environment of the area. Progress in achieving this was monitored by the WHO/UNICEF joint Monitoring Programme for Water Supply and Sanitation (see below). The progress report of 2012 showed some progress globally in reaching the target, but with an ever-increasing world population and faster urbanisation, many developing countries were having difficulties in making the progress necessary. Sub-Saharan Africa is still the area of greatest concern. The 2015 update reported that improvements had been made. For example, in 1990, improved drinking water was supplied to 76% of the world's population and improved sanitation to 54%. By 2015, the figures were 88% for drinking water and 77% for sanitation (UNICEF, WHO 2015). The UNICEF WHO (2019), however, provides a slightly more pessimistic picture of an improvement for drinking water of 61% in 2000 rising to 71% in 2017. For sanitation, the improvement for the same period was from 26% to 45%. Although great improvements had been made, the improvement was not evenly spread over all countries, and those where poverty and population growth were high often showed little improvement. Others did not have political will, and conflict hampered others. In addition, there had been a global financial crisis affecting many during this period. Progress was showed by the UNICEF/WHO report (2015, 2019).

The Joint Monitoring Program of the WHO developed definitions and criteria for various levels of access to drinking water and sanitation (http://www.wssinfo.org/pdf/Policies_Procedures_04.pdf), which are outlined as follows:

Improved drinking water sources
- Piped water into dwelling, plot or yard
- Public tap or standpipe
- Tubewell/borehole
- Protected dug well
- Protected spring
- Rainwater collection

Unimproved drinking water sources
- Unprotected dug well
- Unprotected spring
- Cart with small tank/drum
- Surface water (river, dam, lake, pond, stream, canal and irrigation channels)

Improved sanitation facilities
- Flush or pour flush to a piped sewer system, septic tank and pit latrine
- Ventilated improved pit latrine
- Pit latrine with slab
- Composting toilet

Unimproved sanitation facilities
- Flush or pour flush to elsewhere
- Pit latrine without slab or open pit
- Bucket
- Hanging toilet or hanging latrine
- No facilities or bush or field

A summary of the later findings from the 2019 UNICEF WHO report is provided in Table 7.3.

If we do not come to grips with these problems, around 4000 people will die each day simply because of preventable diarrhoea and other water-borne diseases (Department for International Development [DFID] 2008). In addition, millions of working hours costing billions of $US will be lost each year, slowing down the advancement of many economies.

Bottled water is only considered improved when the household uses water from an improved source for personal hygiene and cooking. Improved sanitation only applies when the facility is not shared or is public. Unimproved flush or pour-flush toilets should not be flushed into a street, yard or plot, open sewer, ditch or other open location. Not only does

Table 7.3 A summary of the findings of the UNICEF WHO (2019) report.

Drinking water	Sanitation	Hygiene
Coverage of safely managed supplies rose from 61% to 71%	Coverage of safely managed services rose from 26% to 45%	60% of global population had washing facilities with soap and water at home
Rural cover, which was often neglected Rose from 39% to 53%	Rural cover for safely managed services rose from 22% to 43%	3 billion people still did not have basic washing facilities at home
Global access to basic water supplies Fell from 1.1 billion to 785 million	The number of people practicing open defecation fell from 1.3 billion to 67 million	Three out of five people had basic handwashing facilities in 2017
Eight out of 10 people who lacked Basic services lived in rural areas and half of these were in least developed countries	Seven out of 10 people who had no access to proper facilities lived in rural areas and one third were in least developed countries	Three quarters of people in least developed countries lacked handwashing with soap and water

Source: Data taken from UNICEF WHO (2019).

the provision of WASH facilities need to be provided, but the maintenance of the systems must be allowed and the upgrading of existing systems, especially in urban areas, should be included. It is hoped that the JMP data on progress will help identify any barriers to progress and point the way to solving them.

At the present time, a range of treatment processes are available for the 'clean' and the 'dirty' water sides. Generally, the provision of both improved drinking water and sanitation lags far behind in rural areas compared with urban areas.

Treatment processes for drinking water and, to a lesser extent, wastewater have developed over many centuries. Drinking water treatment was used by the ancient Egyptians, Hindus in India and the ancient Greeks and Romans. Early treatment often involved simple settlement to remove suspended solids, thus clarifying the water and improving its appearance and taste. Boiling was also used to improve appearance and taste. This also helped remove any pathogens although the existence of both bacteria and viruses was unknown at that time. As long ago as the fourth century BCE, the Greek Hippocrates promoted the idea of 'healthy drinking water', a concept that was subsequently followed in the Roman Empire. The Romans were expert builders and used this skill to transport water over large distances in open channels into cities. These open channels allowed UV light to penetrate the water and help, to an extent, to disinfect it. The Romans also used sand filters to further clarify the water for drinking with systems not unlike those currently used. As Roman cities grew, they became aware of the need to remove wastewater and make provisions for the safe removal of storm water. Gutters were provided and cleaned on a regular basis to allow good flows. The wastewater from this source was washed into the nearest convenient stream. In Rome, human and other waste was collected in a network of sewers that fed into the main sewer, the Cloaca Maxima, which discharged into the River Tiber. Unfortunately, with the collapse of the Roman Empire, attention to healthy drinking water and efforts to dispose of wastewater in an environmentally friendly way were decreased, and washing, bathing and cleaning clothes were not high on the priority list of most people. This resulted in the spread of disease and ultimately death. In the sixteenth and seventeenth centuries, more attention was paid to the quality of drinking water with simple experiments on purification of water for drinking, and later with the use of the microscope by van Leeuwenhoek (1632–1723), the presence of small organisms, including larger groups of bacteria, was seen in the water. The loss of labour due to waterborne disease was a problem in many large estates and some, in the sixteenth and seventeenth centuries, issued beer to their workers while working in the fields, rather than water, as this was seen to be safer. Later, more attention was started to be paid to clean drinking water, and in 1746, the first patent for a water filter was given to Joseph Amy. Although towns and cities existed at this time, they were not generally very large, and the majority of people lived in rural communities where reasonably clean sources of freshwater were often available. With the industrial revolution, however, towns and cities grew rapidly as people migrated from the countryside to provide labour in the new factories. Unfortunately, the increased demand for housing led to 'slum' dwellings being rapidly built with inadequate sanitation and waste disposal. Wastes were usually disposed of into the streets, which often also had open sewers flowing along them. At some locations, the sewage and waste deposits were more than 6 ft deep. This resulted in outbreaks of cholera and typhoid as well as other waterborne infectious diseases. The first major cholera outbreak in Britain (1831–1832) killed more than 6000 people. The second (1848–1849) killed 14 000 people, and a third from 1853 to 1854 killed another 10 000 people. Although cholera probably occurred in ancient times, it was from the start of the nineteenth century that most recorded outbreaks occurred in developed countries. Seven pandemics from 1816 to 1975 are recognised. The first started in India and spread to China, Indonesia and to the Caspian Sea area. The second reached Russia, Hungary, Germany and Egypt before spreading to Paris and London. It also spread to North America. The subsequent three pandemics, between 1852 and 1896, showed similar patterns. The sixth and seventh pandemics mainly affected the far east. The spread of cholera in the early nineteenth century in Europe was thought to have been spread by a mist or miasma. At that time, waste of all kinds was simply disposed of into the gutters. In addition, there was no motorised transport only horse driven. There were often many thousands of horses used in major cities for transport, and their waste was also deposited in the streets and gutters and rarely cleared away. As a result, there was a drive to provide clean safe drinking water to all houses, especially those of worker houses that were closely packed, and the first public water treatment process was provided in Paisley in Scotland in 1804, followed by 1806, when parts of Paris, France, were also provided with treatment processes. During the third pandemic, the physician John Snow studied an outbreak in London in 1854. Snow was not convinced by the miasma theory concerning the spread of the disease. At that time, the microbial cause of cholera was unknown, and hence remedies were usually ineffective. Snow had previously published an article, in 1849, that was contrary to popular belief, and reported that cholera was not a disease of the blood but a digestive disease. This, he concluded, was because of the symptoms of vomiting and diarrhoea. He also concluded that it was caught through ingestion not by breathing in bad odours. These ideas were not widely accepted. The serious outbreak in Albion Terrace, an area of London, gave him a chance to investigate in detail what was happening. Water was supplied to domestic premises in that area by two companies. Snow found that there were six times as many deaths due to cholera in houses where the water was provided by the Southwark and Vauxhall Company as compared to those supplied by the Lambeth Company (The Times 1856). This occurred even though water for both companies came from the River Thames. The epidemic continued for a little over 10 weeks, and during that time, Snow obtained records of cholera deaths and water supplies. His data clearly showed that six times as many deaths occurred in the Southwark and Vauxhall area than in the Lambeth area. The Lambeth Company drew its water from further upstream where it was less polluted. He identified the main source of the outbreak to be a public water pump on Broad Street (now Broadwick Street), which subsequently was discovered to have been dug only 3 ft from an old cesspit that was leaking. Not only people living near the pump, but also businesses in the area using the water were affected. Snow also identified people in the area who did not contract cholera and tried to ascertain why. A nearby workhouse, with over 500 inmates, did not have any cases, and Snow found that they used water from the Grand Junction Water works. Men working in a local brewery also escaped the disease as they drank water from the brewery well. Snow concluded that the cholera cases could be directly linked to the Broad Street pump. His results and conclusions related to the cause of the outbreak started to change popularly held views about miasma, and as a first measure, he persuaded the authorities to remove the pump handle preventing its use. This greatly reduced the number of new cases occurring in the area. This change of view did not happen quickly, and Snow's views were initially rejected, and only in the following decades, these were started to be accepted. At around this time, the Chelsea Water Company installed slow sand filters (SSFs) to treat water for drinking, and this helped reduce or prevent cholera in the area supplied by them, which was adjacent to an area not provided with such filters where cholera was rife.

Around this time, other prominent workers in the field of sanitation in cities arose. The summer of 1858 was very hot, and because of the human waste that was continually being disposed of into it, strong smell or stench arose from the River Thames was made worse by the very warm summer weather (34–36 °C in the shade and 48 °C in the sun). This smell permeated the houses of Parliament, causing an abandonment of proceedings. Members were forced to flee to areas of fresher air in the countryside. The Thames, according to Michael Faraday of the Royal Society in 1855, had become 'an opaque pale brown fluid'. The authorities agreed that urgent action was required to rid London of the 'evil smell', which by many was believed to be the cause of disease and death. This crisis was called the 'Great Stink' and gave rise to a major advance in urban planning. London needed an overall organisation, rather than a number of separate authorities, to be created to oversee and raise the money required to solve the sewage disposal problem. This power was given to the Metropolitan Board of Works, who was given the authority to raise £3 million (late nineteenth century price value) to start work. A chief engineer was appointed in 1852, Joseph Bazalgette, who had previously produced plans for a sewage disposal system and was then given this task. His plan was to build a network of main sewers that would collect sewage and surface water and run parallel to the Thames transporting the waste by pumping to outfalls at Barking and Crossness, one on the north side and the other on the south side of the river and both east of London. These combined sewers would transport the waste downstream of the city and flow into the sea. The system was opened in 1865 although it was not completed until nearly a decade later and involved the construction of some 82 miles (132 km) of underground main sewers built with bricks. One very important feature of the system was due to Bazalgette's forward planning. He made generous calculations for the volumes of sewage per person and an upper estimate of the population at that time. He calculated, on the basis of this, the diameter of the pipes, but then to allow for unforeseen changes, he doubled the diameter of the pipes again, saying that this project will only be done once allowance must be made for future increases. This foresight proved invaluable as, with the rapid population increase over the next century, the system would have overflowed without the larger pipes. This new system greatly improved London's sanitation, virtually eliminated cholera and typhoid and gradually improved the river, especially after the additional construction of extensive sewage treatment facilities at the turn of the century. A further benefit to London was that the scheme involved the creation of the Victoria, Albert and Chelsea embankments along the Thames. Bazalgette's project and foresight marked a major step forward in the modernisation of London as it grew with the industrial revolution and was also followed by other major cities in both Great Britain and the world. Without this project, the development and growth of London would not have been possible. For more details of Bazalgette's work, see Halliday (2009). Another important reformer at that time was Sir Edwin Chadwick, who was responsible for the inquiry into the operation of the Poor Law in 1832 and later, together with T.S. Smith, extended this into looking at sanitation and public health. He published his report on *The Sanitary Condition of the Labouring Population* in 1842 (Chadwick 1842) and a supplementary report in 1843. Table 7.4 shows examples of deaths from water-related diseases in some English counties at that time. He is acknowledged as one of the important health reformers of this period. He was also appointed as a commissioner of the Metropolitan Commission of Sewers in London from

Table 7.4 Recorded deaths from water-related deaths in some English counties during the year ending 31 December 1838.

County	Deaths	% of deaths from all causes
Cornwall	443	11.6
Cumberland	165	9.8
Devon	615	10.4
Essex	417	10.6
Hereford	84	8.7
Lancaster	2866	9.6
Middlesex	4422	14.4
Northumberland	366	12.2
Surrey	1348	13.7
Westmorland	41	6.3

Source: Data taken from Chadwick (1842). Note that the counties with lower % deaths through water-related diseases had more rural populations. Those with higher % deaths from this cause had populations associated with industrialisation and a non-rural workforce.

1848 to 1849 and as a commissioner of the General Board of Health from 1848 to 1854 when he continued to work on sanitation issues until his death in 1890.

Much of our water infrastructure in major towns and cities has been developed over a period of time with changes and additions introduced as needed. It is important to realise that many of the decisions being made at the present time have to take into account what has happened in the past and may often be constrained by these previous actions. This applies not only to drinking water and wastewater provision and treatment but also to all other aspects of water management.

Depending upon population numbers and their density and distribution, the problems of drinking water supply and wastewater treatment will vary. Where most of the population could be regarded as rural, the requirements will be different for a rapidly growing city. Katko et al. (2006) illustrate some of these differences for Finland. Initially, the demand for better water supplies was not only for health reasons but also the need for water for firefighting. In rural areas, the primary need was for watering cattle. Drinking water services started earlier than sewerage systems in many other European countries, whereas in Finland the two were often developed simultaneously (Juuti and Katko 2005). Helsinki established Finland's first water system in 1876 operated by a private company, and by 1890, it had introduced a metered billing system for water. At the start of the twentieth century, water closets for toilets started to become accepted. There was, for example, a marked increase in the rate of wastewater treatment plants in Finnish cities from 1900 to 1993, which indicates the large increases in construction that occurred after the water act of 1962 and the sewage surcharge act of 1974. Many other countries have expanded and improved their drinking and wastewater facilities over the past century.

The development of drinking water and wastewater systems followed a similar pattern in the United States. As the population grew and cities developed, local water sources gradually became contaminated as wastes were disposed into them. Similarly, as industries developed, local lakes and rivers were used as the easiest route for the disposal of their wastes. In addition, because wood was used as a major building material, water for firefighting was of great importance. Private companies sprang up in cities to provide water from sources outside by building pipelines and aqueducts. The water quality was not always good, and outbreaks of cholera and typhoid were not uncommon, so private companies were often forced out of business, and the control of drinking water provision and eventually wastewater treatment was taken over by municipal authorities. The first public water works in the United States of America were built by Hans Christopher Christiansen in Bethlehem, Pennsylvania, in 1772. Soon after New York, in 1799, steps were taken to overcome the problem of pollution of the well supplies, which were then used by the growing population and commercial/industrial activities. The Manhattan Company was formed to provide a new source of clean water. The Croton River was eventually chosen for this purpose. In 1870, New York's first filtration plant was constructed at Poughkeepsie to treat Hudson River water. This greatly improved drinking water quality and decreased the occurrence of typhoid fever. Around this time, other cities were developing their own schemes following this lead. The Baltimore Water Company was formed in 1805, but as the city grew, its water sources, mainly springs and wells, became polluted. Only a small portion of the city around the centre was connected to a piped water supply, and as with many cities at that time, cholera outbreaks became a serious health problem, so improvements in water supply became urgent. Connections to clean piped water were increased to about 30% of the population. In 1853, the Baltimore Water Company was taken over by the City Authorities in order to provide the finances required for further improvements and expansion. Similarly, Boston needed to develop a new adequate water system after an extensive fire in 1825 that destroyed many homes and commercial properties in the city centre. There was a debate as to whether the task should be placed in private or public hands. Contamination of wells was also a problem, and a more extensive network of sewers was introduced in the mid-nineteenth century. Several small companies were created to supply drinking water. In 1836, it was proposed that the city authorities should take over supply and build and operate the waterworks. There was still much discussion for the next 10 years regarding the preferred raw water source, and finally, it was agreed to use Long Pond, outside the city. Currently, both drinking water and wastewater treatments are the responsibility of the Massachusetts Regional Water Authority. In 1909, chlorination was introduced for the first time in New Jersey, resulting in dramatic decreases in bacterial contamination. Nationally, in the United States of America, some drinking water standards were first adopted in 1914, and chlorination as a disinfection process was widely adopted in 1915. By the 1940s, the incidence of waterborne diseases was considerably reduced compared with 1910 (National Research Council 2002). After the Second World War, many new contaminants became problematic. Some of these were highlighted by Rachael Carson in her book 'Silent Spring' (Carson 1962). Although many well-established methods of drinking water and wastewater treatment continue to be used and upgraded, there have been many new developments in both areas during the past few decades of the twentieth century and more recent years. The standards for drinking water quality and regulations for wastewater treatment and the disposal of treated wastewater are now much stricter and more comprehensive. These aim to not only protect human health but also, in the case of wastewater discharge, the environment. Reducing the burden of waterborne disease not only improves health but also the economic power of a community. Individual countries may have their own drinking water standards, but most are based on those outlined by the WHO (WHO 2011). To achieve the targets outlined by the MDG on WASH, all countries will require to invest more money. It is equally important to ensure that the money is spent effectively and fairly not only on equipment and technology but also on water resource management. These finances must be available not just for capital investment but also for long-term maintenance, operation and training for operatives (DFID 2008). Care must also be taken at the international level to better coordinate funding programmes ensuring that they are properly and fairly targeted. There should also be continuing monitoring of progress together with follow-up advice where needed.

This chapter first examines drinking water treatment. Wastewater treatment is discussed in Chapter 8. Within these categories, it is then divided into small-scale systems suitable for individual dwellings or groups of a few dwellings and larger scale systems as would be used for larger villages, towns or cities. Not all of the population, even in developed countries, is connected to public water supplies. For example, although Cyprus and the Netherlands have 100% connection, Lithuania has only 75% and Romania has 52% (Eurostat 2007).

The basic approach used in drinking water treatment, and to a certain extent wastewater treatment, is to place multiple barriers to prevent the passage of pathogens and toxins from the source water to the consumer. In the case of wastewater, this would be a barrier between the wastewater and the receiving environment.

Drinking Water Treatment

Drinking water treatment processes usually function by using physical, biological or chemical techniques for removing contaminants. Normally, there are several stages of treatment, so this is termed the multiple barrier principle. In most surface waters and some groundwaters, especially in developed countries, there are a large and ever-increasing number of potential chemical and microbiological contaminants that might be present. The fact that there are usually multiple treatment barriers means that even if there is a malfunctioning of one stage, the end consumer is still protected by the proper functioning of the other stages. It is important to understand the nature, possible impact on human health and the source of these contaminants. Unfortunately, especially in the case of microorganisms, their detection requires highly specialised laboratories and techniques rarely available to many communities and is very costly. Because of this, reliance is placed upon indicators of contamination, both microbiological and chemical (Dufou et al. 2003). This approach provides much easier monitoring and analysis and, although not specific, gives good guidance as to the quality of water. The type and number of individual stages of the management plan will depend upon the geology and nature of the source water. The system of multiple barriers also means that even if one part of the treatment chain malfunctions, the rest will still operate to provide a continuous supply of treated water to the consumer. Cost has always been a factor in deciding on the treatment used, but of paramount importance is the protection of the final recipient. The barriers should occur at many stages, including the protection of the source water, one or more treatment stages of the water, final disinfection before the water leaves the treatment processes, the protection of the distribution system and multiple testing at all stages, including at the testing at the consumers tap. For drinking water, the main consideration must be formed around health-based targets. These targets are based on national and international standards such as the WHO Guidelines for Drinking Water Quality (WHO 2011). It must be recognised from the outset that standards and targets change with time, so provisions must be made to upgrade both the management plan and the process chain to accommodate these changes (Davison et al. 2005). Because both chemical and microbiological contaminants can cause both short- and long-term health problems, Davison et al. (2005) identified the following main types of health-based targets, which are indicated below.

Depending upon the source water quality, the number of treatment stages may increase. This is particularly so as international standards for drinking water become stricter and include new substances. In order to comply with these standards, the whole system needs to be regularly monitored, and this requires an overall water safety plan, the objectives of which are (i) to protect all source waters from contamination, (ii) to provide effective treatment of water used for drinking, etc. to meet the required quality standards and (iii) to distribute and store this water without contamination to consumers. As there are a number of different stakeholders involved in the use and treatment of water, for example public health authorities, water supply utilities and municipal authorities, their responsibilities and roles must be clearly defined and agreed. This can be a difficult task, especially for smaller supplies, as the expertise may not be available, nor may there be adequate sampling equipment or laboratory facilities. One approach is to use the 'hazard analysis critical control point' (HACCP) concept. This has been used for many years by the food industry and is now, with the encouragement of the WHO, being introduced into the water industry (Hellier 2000). A new approach was seen to be needed as even large municipal supplies in developed countries sometimes suffer from microbiological pathogen outbreaks. The UK Department of the Environment reported over 25 outbreaks of cryptosporidiosis since 1988. In HACCP, it is important to identify the potential hazards, identify critical control points, establish critical limits at these points and establish a robust system of monitoring at these points. Then, if any parameters exceed the critical levels, establish what remedial action must be taken. It is also necessary to initiate regular tests on whether the HACCP system is working properly, and to aid this, full documentation of all procedures and records of results must be kept (FAO/WHO 1996; Hellier 2000). In assessing hazardous substances that might be present, a toxicity evaluation for both acute and chronic effects needs to be taken into account and also the probability of a consumer receiving that amount. It may be necessary to involve the area health authority as to whether there are any epidemiological signs of a problem. In supplies for small communities, the handling and storage of the water should not be considered once treated as this can lead to additional contamination. Any identified hazard could impact not only humans but also livestock and economic activities. As sources of waters often vary with time, it is essential to understand and quantify these variations. These could be on an hourly, daily or seasonal basis. Often, individual hazards occur in groups at certain times and may be dealt with as a group. For example, cyanotoxins produced by blue–green algae usually occur during the peak growing season and mid to late summer. Bacterial pathogens transmitted by birds, e.g. *Campylobacter* and *Salmonella*, could be monitored together. If these occur, grouping, with a suitable indicator from that group, can lessen the demands on analytical facilities (Hellier 2000). Monitoring of the whole water chain, including the distribution system, is needed, as if a failure at one or more critical control points occurs, sampling must be intensified at all points down the chain for any increased risk.

It is essential to have the full support of management and staff before implementation of an HACCP plan. The team involved must have the appropriate amount of training and time to carry out their duties. Indeed, specialised training in some aspects may be required. It is also useful to have regular meetings to discuss the data obtained. Regular monitoring within the distribution system and at the consumer's tap should verify the effectiveness of the treatment being given and the HACCP approach. This approach is being more widely implemented, including in Melbourne, Australia and Iran (Tavasolifar et al. 2012). In the latter case, the source water, the Zayandehroud River, exhibited both seasonal and yearly variations in quality. They identified eight critical control points. These with an outline of some of the hazards identified together with control measures and critical limits are provided in Tables 7.5 and 7.6, which gives the steps in structure as used in Iran.

The first key factor in the multiple barrier approach is protection of the choice of source water and its protection from contamination. Wherever possible, as clean as possible source water should be chosen. Usually, some treatment will be required even if it is only boiling for individual use. The treatment sequence should be designed based on the chemicals and materials that have to be removed to make the water safe. The designer of the treatment sequence needs to take into account (i) the chemicals to be removed and whether they have to have their state altered to make them easier to remove, e.g. by aeration to remove unwanted dissolved gases or change soluble salts to insoluble ones such as soluble ferrous

Table 7.5 Water quality treatment based on HACCP.

This outline, based on Hazard Analysis and Critical Control Points (HACCP), is based on various adaptations of the FAO/WHO (1996) control approach to food hygiene for drinking water treatment processes. The process can be divided into a number of steps.

Step 1: Firstly, identify all possible hazards that could occur in the water at all stages of its collection and treatment.

Step 2: Assess the potential severity of each of these hazards.

Step 3: Examine the treatment chain for the water and ascertain where critical points occur (CCP) and practices that are in place to control the processes.

Step 4: Determine the limits of parameters for water quality and process efficiency, which are used at each CCP.

Step 5: Organise a robust system of monitoring at each CCP. This must include the ability of each monitoring technique and analytical procedure to determine each parameter at the analytical limits laid down by the water quality standards.

Step 7: Establish actions that must be taken should any of the hazards at any critical point exceed the specified level.

Step 8: Have regular trial procedures to verify the correct operation of the overall system. This should include sampling and analytical procedures. This should ensure that the programme is working correctly.

Step 9: Establish a comprehensive recordkeeping procedure documenting all of the above activities so that full documentation is available for discussion and regular review.

Table 7.6 HACCP structure applied to Isfahan drinking water treatment system, Iran.

CCP	Hazard Identification	Controls	Critical limits
Works intake	Chemical and microbiological deterioration source	pH, TOC Conductivity	4.5–11.5 <10 mg/l <3000 μS/cm
Coagulation and flocculation	THM precursors, organic pollutants, algae, Heavy metals, pesticides	Final TOC Conductivity Heavy Metals Plankton	<2 mg/l 750–2000 μS/cm National standard <2000 cells/ml
Sedimentation	Turbidity, metals Bacterial numbers Organic matter	Turbidity TOC Colour Smell	<3 NTU <2 mg/l <20TCU <5TON
Filtration		Turbidity Bacteria Aluminium	<0.5 NTU <100/100 ml <0.2 mg/l
Final disinfection	Chlorine residual	Free chlorine Turbidity	1–2 mg/l <1 NTU
Distribution pipes and reservoirs	Bacterial growth	Chlorine Colour and smell minimal	0.5–0.8 mg/l

Source: Modified from Tavasolifar et al. (2012).

to insoluble ferric (ii) remove as much suspended matter, both inorganic and organic (living or dead) (iii) disinfect the water. When treatment methods are being chosen, they should be environmentally sound technologies (ESTs). This, in general terms, means that any techniques chosen must not only be fit for purpose but also be durable and can be maintained with the finances available for maintenance and trained staff to operate the system. This will help focus on water use efficiency both for urban and individual use and help balance supply and demand. In most cities, domestic demand for water is either the largest consumer or second only to industry, whereas in the rural environment, agriculture is usually the greatest consumer with domestic consumption often far behind. Many cities are situated on rivers that have been used both for water supply and for the disposal of sanitation wastes. Only about 10% of wastewater globally is adequately treated (UNEP 2008), which results in its disposal, causing both surface water and groundwater contamination. This is often exacerbated by inadequate provision for dealing with storm-water run-off. In year 2000, over 25% of people living in towns and cities in the developing world did not have adequate sanitation. Over 95% of the world's population had some sort of drinking water supply, but only two-thirds of these had tap water supplies. Less than one-third of the global urban population had a water tap within their homes, 10% had access to public taps and 8% relied upon hand pumped water from protected wells (UNEP 2008). There is also a large variation in the provision of sanitation depending upon the size of the conurbation, with cities with less than 100 000 inhabitants having less than 40% provided with piped or well water on the premises and with flush toilets and sewer systems. In cities with 1–5 million inhabitants, 70% have flush toilets, and in cities with 5+ million inhabitants, more than 80% have flush toilets. As domestic sewage, if not properly disposed of, is a major source of contamination and is likely to spread disease, there is an urgent problem to solve this issue and thus improve both the human health and economic status of an area.

The provision of safe drinking water and proper sanitation in both cities and rural areas requires a properly organised infrastructure and institutional framework. This framework can be divided into three categories (UNEP 2008): (i) public and centralised systems with good technological backup, (ii) collective and semi-decentralised systems that have a degree of technology and institutional backup and (iii) decentralised systems where individuals organise their own systems. These are usually low technology and do not rely on any institutional support. In all cases where ESTs are used, international standards must always be met. The choice of ESTs must take into account local conditions, including water quality, variations in supply, environmental needs and possible changes in demand in the future. All categories of use should be incorporated into a regional programme for integrated water resource management.

Where possible, water supplies should include a range of source waters, such as rainwater harvesting, desalination and even inter-basin transfer schemes taking care, especially in the latter option to avoid the movement of alien species and damage to existing ecosystems. Any sustainable use of water must be embedded in an integrated plan involving all users. It is important to have adequate finances in place, and this involves the cost of providing the facility and the overall economics of running and maintaining it. International standards should be met for both drinking water and sanitation. Concerning the overall running of the scheme, especially when it is being provided for large villages, towns or cities, a decision must be made whether it should be a public supply of water and treatment of waste or privately owned and run. These discussions concerning decisions should involve all users to ensure an equitable allocation of resources between all sectors. Consideration should also be made of both current and future population needs and the proper training of workers and administrators to meet the technical capacity needed. Finally, all legal aspects, local, national and international, must be taken into account.

Once a programme has been agreed, there are several steps for preserving both the quantity and quality of the resource and supply system. The following steps must be incorporated as an integral part of any plan.

- Control the disposal and discharge of wastewaters in the region and enforce the rules. These rules should include setting standards for any discharges so that sources and ecosystems are protected.
- Seasonal variations in supply and demand.
- List all industrial and agricultural inputs and a note of diffuse pollution sources and whether there should be restrictions on land use around raw water sources.
- Clearly, state the role of relevant authorities in the management of all aspects of the water cycle.

This is usually more easily done for groundwaters than for surface waters although problems can occur with both.

In addition to the requirements for humans and society, provisions must also be met for environmental needs. Again, this means that all actors with interests in sustaining the environment need to be engaged. An assessment must be made as to the resource requirement of the environment that needs to be quantified, and this should include both the quantity and quality of the water. Any new developments, such as factories and land use changes including the expansion of housing, should have an environmental impact assessment made with special regard to water impacts. Existing and new developments should be monitored regarding water use and environmental impacts. A note must be taken of any particularly sensitive areas, such as wetlands, forests, inland and coastal deltas. Although this may be taken to only apply to waste discharge, it can also apply to the amounts of water abstracted for drinking and other economic uses as if the amounts are large ecosystems can also be affected.

With all measures that need to be taken always involving public participation, especially in developing countries, the women of the households, as in rural communities, are usually the main water collectors. As many of the measures taken involve some expenditure, the idea/concept of the costs of providing water and the economic gains that can be obtained must all be included. For any treatment systems being considered, a budget must be drawn up that includes construction, use and future maintenance. All prices and tariff arrangements must be discussed and agreed with users and kept within their ability to pay. There should be an agreed allocation of the resource, especially if supplies vary with season, and this should take into account future changes in population and possible expansion of other economic activities such as industry and agriculture. Future changes in climate and potential changes to the water cycle should also be recognised so that as access to water improves to the users, the volumes used in all sectors will invariably increase. In some climates, there is also a risk of extreme events such as seasonal floods, and these events must be allowed in the design and construction of structures and by having reserve funding if such events occur.

When choosing the best environmentally sustainable treatment or the location, be it for the supply of water or disposal of waste, it should be designed as an integral part of all other water objectives for that area. These objectives must be part of an overall sustainable Integrated Water Resources Management Plan. In many parts of the world, demand for water has already exceeded supply, and this has been exacerbated, especially in some developing countries, by drought and climate change. To cope with this water storage, the augmentation of supplies is needed. To implement such plans, a proper water balance for the whole area or even the country is required and should include an assessment of all natural inputs and outgoings and take into account seasonal and geographical variations. In addition, detailed information on water quality from all sources is needed. If there are records over a number of years note should be taken of any multi-year cycles, for example 10 or even 100 year cycles. Where abstractions are from either surface or groundwaters, consideration must be given to potential impacts upstream and downstream or to other parts of the aquifer at other locations. This is where it is important to involve public participation. It is also important when controlling abstraction volumes to have regard to environmental flows, i.e. the amounts of water needed to maintain ecosystems. This can require detailed specialist knowledge, but if ignored, it can lead to environmental degradation and possible adverse changes to the hydrological cycle in that area. As the demand for water varies seasonally both for human use and by ecosystems, the augmentation of supply may be needed in peak periods. Rivers, for example, are subject to seasonal variations and may have long-term variations in flow due to factors such as climate change. One way of dealing with these variations is to dam the river in order to store water when it is abundant. These dams are usually built across valleys and at some distance from cities. The reservoirs created behind the dam will have a larger surface area than the river and, as a result, will be subject to greater evaporation losses. Although small dams can be constructed and operated by small communities such as villages, larger dams are usually constructed and operated by regional or even national authorities. The dam can also provide additional income to the authorities if a hydroelectric scheme is incorporated in its construction. In addition, where water is used in irrigation schemes, agricultural charges can also provide significant income. Although dams can be beneficial, they can have severe effects on river ecosystems. The reservoir created will inundate land, and sometimes human dwellings and sediment that would normally carry down the river will be trapped behind the dam, which will gradually build up initially, making the reservoir shallower and even eventually filling it up and rendering it useless. In rivers subject to seasonal floodwaters, a dam can help prevent and control flooding downstream. The disadvantage is, however, that controlled flooding can deposit sediments and nutrients onto agricultural, land improving their fertility, and is relied upon in some agricultural societies. An example of this is the construction of the Aswan High dam on the River Nile. If the river water forming the reservoir drains fertile land, then that in the reservoir will also be rich in nutrients, which could result in adverse eutrophication effects and decreased water quality. It can also lead to the proliferation of parasites such as *Bilharzia* and provide breeding areas for malaria mosquitos. Another type of reservoir that can be used in flatter terrain is a pump-storage reservoir, such as that used in the Thames valley, UK. In these areas, a raised embankment is created around a designated stretch of land, lined with an impermeable material such as clay or plastic (in the case of small reservoirs) and then filled with water by pumping from a nearby river.

Table 7.7 Examples of vegetation indicating the presence of water with root depths.

Botanical name	Common name	Depth of root penetration in metres
Cyperus rotundus	Java grass	3–7
Vangueria tomentosa	Wild medlar	5–10
Delonix elata	Creamy peacock flower	5–10
Grewia occidentalis	Crossberry	7–10
Markhamia hildebrandtii		8–15
Hyphaene thebacia	Duom palm	9–15
Borassus flabellierfer	Palmyra palm	9–15
Ficus natalensis	Natal fig	9–15
Kigelia aethiopica	Sausage tree	9–20
Newtonia hildenbrandtii	Lebombo wattle	9–20
Acacia seyal	Red acacia	9–20

Source: Adapted from Nissen-Petersen (2006).

In developing countries where there are seasons with substantial rain interspersed with arid dry seasons, water capture is very important. One increasingly popular method is the use of sand dams (see Chapter 4).

Sandy riverbeds can supply water when many other sources have dried up. Places that have the best reserves can often be identified by observing vegetation, especially trees that have tap roots penetrating deep into the ground (Table 7.7).

There are a wide variety of sources of drinking water for towns and cities, each of which can have different qualities. Table 7.8 shows some examples of these sources. One prime feature in producing clean safe drinking water is to select the cleanest source and then protect it from future contamination as much as possible. This could mean placing restrictions on land use in the catchment, possible planning restrictions on buildings and development and restrictions on recreational activities. Intakes to the treatment processes should be carefully selected to avoid other contaminated discharges.

An important preliminary point is that, regardless of the source of water being used, the quality should be protected not only from the point of view of the end user but also to lessen the costs of treatment.

If raw water is abstracted from a surface source such as a slow flowing river or a lake/reservoir, especially if it is thermally stratified, a number of different depth draw-off points should be provided so that the best water quality depth can be selected as water quality can vary with time of year, for example if the lake/reservoir is stratified or there is a possible polluted layer to be avoided. In the latter case, the siting of the abstraction point is extremely important.

The basin of proper selection of source water and its treatment to make it suitable for drinking and comply with the appropriate standards is to provide multiple barriers in the treatment process to eliminate unwanted materials and pathogens.

Table 7.8 Examples of drinking water sources for cities.

City	Main drinking water source
Edmonton Alberta (Canada)	North Saskatchewan River (Seasonal glacial Meltwater)
Montreal, Quebec (Canada)	St. Lawrence River
Paris (France)	Surface water from the R. Seine and R. Marne watersheds and groundwater
London (UK)	R. Thames, R. Lea filling pump-storage reservoirs and groundwater
Reykjavik (Iceland)	Groundwater
Al-Jubail (Saudi Arabia)	Desalinated salt water from Red Sea
Amman (Jordan)	River and groundwater
Sydney (Australia)	Various reservoirs in river catchments
Washington, DC (USA)	Potomac River
New York (USA)	Delaware and Croton River watersheds

Source: Taken from Cech (2010), Northcliff et al. (2008) and Berliner Wasserbetriebe (2016.)

The concept of multiple barriers is illustrated by considering the reduction in bacterial numbers in water abstracted from an urban river, stored in a reservoir and then passed through several stages of treatment prior to being used for drinking as shown in Table 7.1.

Using *Escherichia coli* (Faecal coliform bacterium, FC) as a marker of abundance:

Human FC sewage discharge	195 000 000/person/d	
	8 260 000/100 ml	
↓		
Reduction through traditional sewage treatment	Cumulative (%)	Reduction
	50	4 130 000 FC
	80	1652000 FC
	98	165 200 FC
	99.9	800
↓		
Self-purification and dilution either through discharge to river or storage facility reduces FC by a further 10–15%		
↓		
Self-cleaning in river	50	200–350
Storage in a reservoir	80	80–140
Water supply treatment	99.99	0.000–0.0001

The concept of multiple barriers to eliminate contamination in drinking water treatment.

Whatever the raw water source being used, it must be protected from contamination. This frequently means that restrictions are placed upon the activities that can be carried out in the water catchment, for example the use of fertilizers on agricultural land, the discharge of industrial wastes or the disposal of solid wastes into unlined disposal sites. If surface waters are being used, depending upon their depth and size, a range of abstraction points may be needed regarding position and depth. This is particularly true of reservoirs and lakes that may become thermally stratified, which can produce different water qualities at different depths. These standing bodies of water and rivers can also have floating debris, plants or algal blooms on the surface, and these should either be avoided by suitable subsurface abstraction or screens provided to prevent the debris from entering the treatment processes. Similarly, in lakes and reservoirs, bottom water abstraction should be avoided as this could suck in silt and sediments as well as the water possibly being anoxic. Intake pipes can be clogged by biological growths (biofouling) on the inside of pipes. An example of this is the possible growth of freshwater mussels, where growths can significantly reduce the internal diameter of the pipe. The use of many biocides for their control should be avoided as the chemicals can pass through the treatment process and into supply. One safe method is to isolate the section of the pipe and chlorinate it to kill the biofouling. This can then be removed, the pipe flushed to waste, and then put back into service.

In developing countries, especially in villages and rural areas where wells are used for water supply. Protection around a well, e.g. fencing to prevent access by animals, and a stone paved area to prevent excessive erosion around the wellhead are required.

The main objectives, in general terms, for drinking water, apart from meeting international standards, can be summarised as follows:

- There must be no unpleasant taste or odour.
- It must be safe and free from pathogens.
- There should be no colour or suspended solids. The water should be clear.
- It should be reasonably soft to allow for good personal hygiene and clothes washing.
- It should not be corrosive to protect pipes and pumps, etc.

In order to meet these criteria, the source of raw water should be as clean as possible. The cleaner the source is, the fewer treatment steps will be needed and the lower the costs. Although pollutants should be avoided, there may also be natural substances that may need to be reduced, such as particulate matter, floating detritus, dissolved materials, oil, iron, hardness and carbon dioxide.

As treatment systems are usually based on a multiple barrier principle, some initial stages of treatment are often required to protect more vulnerable later stages. These are termed Pre-treatment stages.

Pre-treatment

A range of options are available for removing potentially nuisance substances before the water passes on to other stages, including Screens.

If the source is from surface water such as a river, lake or reservoir, the intake is usually protected by screens. These are designed to protect against large floating objects, pieces of vegetation including leaves and grass cuttings. Sometimes protection against oil spills is needed in which case floating booms can be deployed. For solid objects, bar screens are used. These are vertical bars set at an angle of about 60–80°. The gaps between the bars will vary depending upon the nature of the material being removed. For coarse matter, the aperture between bars is about 50 and 100 mm; for medium matter, it is 20–50 mm; for fine matter, it is 5–20 mm. The screen bars can be cleaned manually by raking or, with fine bar screens, with air or water backwash. Automatically cleaned screens are available. Two examples are band screens and drum screens (see Figure 7.1).

Microstrainers Microstrainers, often called rotary microstrainers (RMs), are a more complex and sophisticated version of drum screens. They are used to remove finer-grained organic

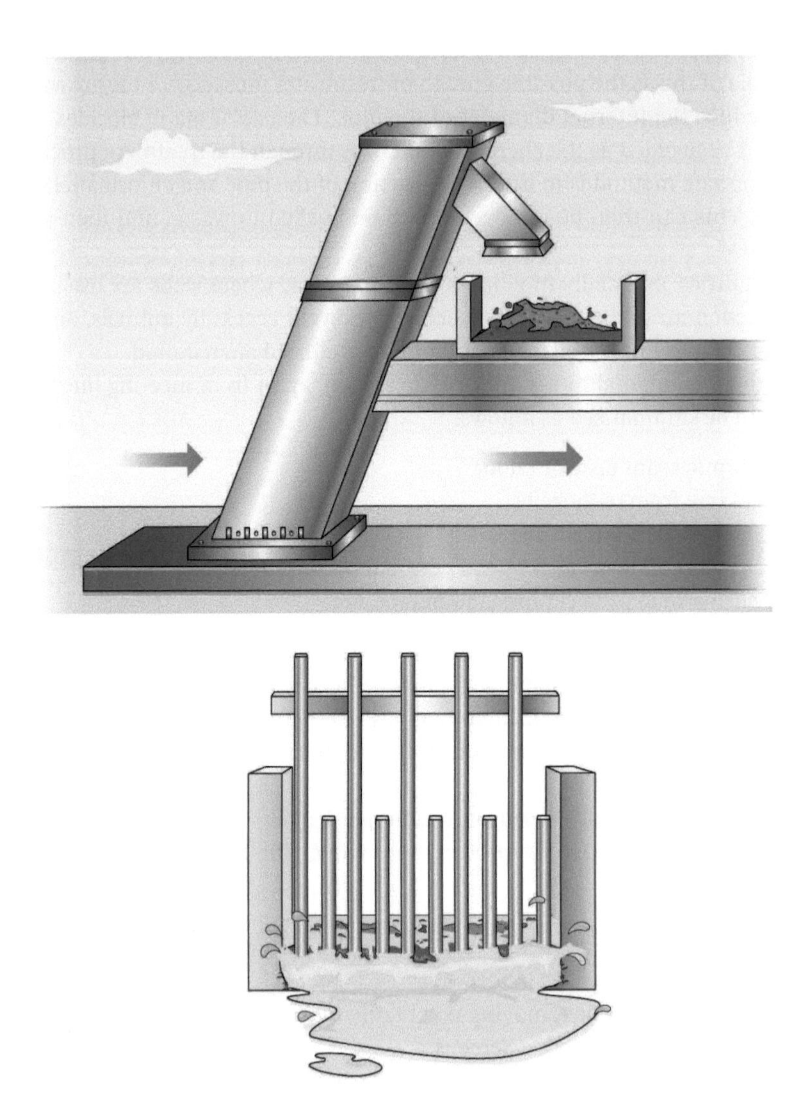

Figure 7.1 A bar screen curved for hand raking to clean.

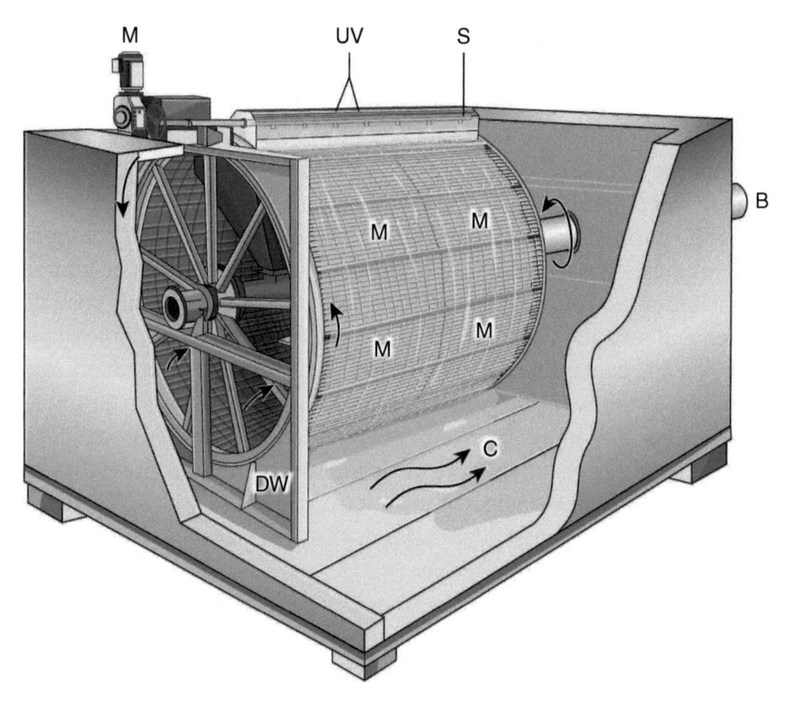

Figure 7.2 Exploded diagram of a rotary microstrainer. M, motor rotating drum; M on face of drum, straining mesh making the sidewall of drum; UV, ultra violet lamp to disinfect the water; S, spray of clean water to wash the mesh; DW, dirty water entering through one end of the drum; C, clean water which exits through the other end; B, drum axle. *Note:* The drum illustrated is held in place by a sealed rim within the solid wall at each end.

and inorganic particles from water. Figure 7.2 shows the construction of a typical RM. The microstrainer drums are usually 3 m long and 1.5–3.0 m diameter and are generally housed in open concrete tanks. Water passes into the drum through its centre and out through the fine mesh walls. These mesh walls are made of woven stainless-steel wire or a finely woven supported fabric. The aperture sizes in the mesh range from 15 to 64 μm. These will not only remove small inorganic particles but also many planktonic algae and invertebrates, which can be important if lake or reservoir water is being used (Bellinger 1968). If a mild eutrophic reservoir is the source water, this will probably contain, at certain times of the year, large diatom populations. These will be trapped on the screen, forming a secondary mesh that, while allowing the flow of water, will retain smaller particles than the microstrainer screen alone. As the mesh becomes clogged by the particles, it is retained as the drum continuously rotates clean water is sprayed onto the outside at the top, dislodging the collected material, which falls into a collecting trough below and is then piped away to waste. To prevent microbial growths developing on the mesh, UV radiation can also be provided shining onto the mesh as it passes during rotation, keeping it clean. If the waters are hard, calcium carbonate scale can develop clogging of the mesh apertures, in which case the drum is removed from operation and washed in dilute acid solution to remove the scale. Depending upon the particulate load, if clogging of the mesh increases, the speed of rotation of the drum can be increased above its normal speed of rotation of 5 rpm. The normal headloss across the drum mesh is low at about 0.3 m, and the consumption of washwater is also generally low, below 3% of the total volume treated. Microstrainers can be regarded as an actual treatment stage that removes quite fine particles rather than a simple screen that removes only coarse particles. Their disadvantage is that they are relatively expensive to install and run.

Groundwaters, although normally cleaner than surface waters, are also at risk from contamination, especially in areas that have intensive agriculture, large numbers of human dwellings or industrial activities. Information must be gathered concerning all possible sources of groundwater contamination in the catchment and, if needed, control measures introduced to prevent possible contamination. In addition, the depth of the borehole into the aquifer must be known as this will give guidance as to the possible ingress of any contaminants. Shallow aquifers are more prone to contamination than deep aquifers. Depending upon the quality of the water, the type of treatment and number of stages will vary. Groundwaters usually require fewer stages of treatment than surface waters. Whatever the method of treatment selected, it is essential that proper maintenance is provided and that all operatives are properly trained.

Water abstracted from aquifers is usually low in dissolved oxygen. This water needs to be aerated to replenish the dissolved oxygen as this improves its taste. This can be done mechanically using either fountains or passing the water over a series of cascades. Aeration will also drive off excess CO_2 and increase the pH of the water. There may be corrosive substances in the water initially, and these substances may not be adequately reduced by aeration alone. In this case, neutralisation using lime is often needed. Depending upon the geological formations housing the aquifer, the groundwater may contain dissolved iron and manganese, which can not only affect the taste of the water but also stain fabric garments washed in the water. Dissolved iron is naturally in the ferrous state. When exposed to oxygen, it changes its state to ferric. Aeration will help raise the pH of the water above 6.5 when the iron is converted into insoluble iron hydroxide, which forms reddish-brown particles that then need to be removed by sedimentation or additional treatment processes. Manganese occurs in groundwater in a soluble form, which, upon exposure to air, forms brown or black deposits that can stain fittings and clothes. If organic matter has contaminated, the groundwater its decomposition could lead to the production of hydrogen sulphide that is also stripped out by aeration. If the aquifer is in an area of agricultural land where fertilizers have been frequently used, there is a danger that nitrates could have migrated through the soil and rock into the groundwater. If concentrations are high, the nitrates may have to be reduced using other unit processes.

A typical treatment scheme for groundwater from a medium to deep borehole that is of good quality and has no significant contamination would be

Borehole → Aeration → Disinfection → Service Reservoir → Distribution → Consumer

If the aquifer is in a limestone-rich area, the water may be regarded as too hard. Hardness in the water is caused by dissolved calcium and magnesium salts, which affect the use of water by consumers as excessive hardness affects the consumption of soap and washing detergents. Hardness also causes scale in kettles, pipes and water heaters and any other parts of the water system. Hard water also affects all types of domestic washing, often making the appearance of clothes look dull. Because the amount of hardness in natural waters varies from location to location, residents in a particular area may well become accustomed to a certain degree of hardness. For example, someone living in an area with 300–350 mg/l (relatively hard) would find water in a different area with 150 mg/l too soft, although in either case, they might get used to it after a while. Although hardness is not a health hazard, large amounts are a nuisance, and the amount may have to be reduced. Waters with 0–75 mg/l are usually considered soft, those with 75–125 mg/l fairly soft, 125–250 mg/l moderately hard, 250–400 mg/l hard and > 400 mg/l very hard (Source: From Ministry of Health, New Zealand, 2007, N.Z. Ministry of Health, PO Box 5013, Wellington, New Zealand).

In rural areas, shallow aquifers are in danger of being contaminated by pathogenic microorganisms from both animals and humans, e.g. farm animals and overflowing septic tanks. In villages and towns there may also be a risk of chemical contamination from industrial activities and road run-off. If the aquifer is deep and the borehole is encased, then the risk of contamination is less. Although disinfection of the water would be required before consumption this is not normally carried out for wells in developing countries. If possible, a comprehensive analysis of the water chemistry and microbiology should be carried out, and any treatment should be recommend if needed. In rural areas, the well/borehole must be securely fenced with a ring fence at least 50 m radius around the borehole. This would be the minimum, and with shallow aquifers, a larger safety zone would be needed. All boreholes and wells should be protected against surface water ingress. Communities that are lucky enough to use deep aquifer waters that yield excellent quality water will only need to invest in minimal treatment, perhaps only aeration and disinfection.

The most common type of disinfection used is chlorination, which is effective against most harmful bacteria, viruses and many protozoan parasites such as *Giardia*. It is not as effective against *Cryptosporidium*. Ultraviolet (UV) light is also an effective disinfectant, especially against most bacteria, viruses and protozoa. UV systems have the disadvantage of needing an electricity supply but have the advantage of needing a relatively low level of maintenance. The efficiency of UV disinfection can be impaired if there is significant colour or suspended solids in the water. The latter should be removed by prefiltration if UV radiation is to be used. If analysis shows that unwanted chemicals were present in the aquifer, additional treatment will be required.

A typical treatment sequence might be

Bore hole – aeration – filtration – (possible removal of unwanted chemicals) – disinfection – service reservoir – distribution. Typical drinking water process sequences are provided in Table 7.9.

Traditional methods of water treatment are based on filtration, and currently, many developments of filter technologies are still being used as well as many others. Filtration is mainly used for removing particles, both mineral and biological, from water. Examples of the range of particles to be found are shown in Table 7.10.

Removal of particles, and to a lesser effect, colour, taste and some chemicals during filtration involves physical, chemical and biological processes. Within a filter, a number of

Table 7.9 Examples of typical drinking water treatment process sequences.

Process treatment	Purpose of action
Bar screening	Removes leaves, branches and large objects
Chemicals for coagulation	Flocculates smaller particles. May need pH adjustment
Coagulation and flocculation	Small particles join together to form larger ones for ease of removal
Sedimentation/flotation	Larger particles may now settle to the bottom or with flotation made to rise to the surface with dissolved air
Filtration	
Roughing filters	Can remove debris and large particles
Rapid gravity	
Sand filters	For removing excess turbidity and any larger suspended particles
Slow sand	Can remove fine particles, small algae and bacteria
Filters	Have biological as well as physico-chemical action
Disinfection	Kills or inactivates potential pathogenic organisms If chlorine is used can provide a residual in the distribution system
pH correction	Helps prevent corrosion in the distribution system
Storage of clean water For distribution	Usually stored in a service reservoir

Source: Modified from EPA (1995).

Table 7.10 Types of particles commonly found in water.

Particle type	Group category	Size (μm)
Inorganic minerals	Clays (colloidal)	0.001–1.0
	Silicates	No data
	Non-silicates	No Data
Biological	Viruses	0.001–0.1
	Bacteria	0.25–10.0
Algae[a]	Unicellular chlorophyta	2.0–50.0
	Unicellular diatoms	1.0–200.0
	Unicellular cyanobacteria	1.0–20.0
Other biological	*Giardia cysts*	10.0
	Cryptosporidium oocysts	4.0–5.0
	Parasite eggs	10.0–50.0
	Nematode eggs	10.0
Miscellaneous particles	Small amorphous debris	1.0–5.0
	Organic colloids	No data

[a] These are single-cell representatives of the common groups of algae, but different sizes may occur. Colonial and filamentous forms will be much larger.
Source: Table modified from Guchi (2015).

mechanisms act to achieve the required end. Straining removes particles by passing the liquid through an inert sieve whose apertures are smaller in diameter than the particle. The particles are then retained, and the water passes through. In a litre, however, particles of smaller diameter than the aperture may be removed as other mechanisms come into action, which include the following:

- Sieving
- Adsorption
- Absorption
- Biological action
- Chemical and biological reactions

As a filter matures, more of these mechanisms come into action, but the rate at which this happens depends to an extent upon the nature of the raw water. Particles are removed until a stage is reached when the filter does not allow enough water to meet its design requirements. The filter is then said to be blocked and will need cleaning to reinstate it to its full flow capacity. Traditionally, filters are classified into two groups partly based on the rate of flow of water through them and the size of particles needed to be removed. These filter types are rapid gravity sand filters (RGSFs) and SSFs. As technologies have developed, many modifications to the basic designs have been made. The following section outlines the main types of filters together with some developments.

Filtration

Filtration can be by sand, a manufactured medium such as a ceramic or polypropylene cartridge, diatomite or activated carbon (AC). Whichever medium is chosen is determined by the size and nature of the particles to be removed. Although filtration through the above media can be quite effective for removing inorganic particles and biological particles, even as small as many bacteria, they are only partially successful at removing viruses. Filtration plants, especially when serving towns or cities, are usually built to last for many decades and so need to be maintained and operated properly to achieve the required performance. Various sizes of sand grains are used in the different categories of sand filtration. These are indicated in Table 7.11.

When cleaning water, if there are only particles of a known size to remove, this can be achieved by using a two-dimensional screen. This action is termed sieving. If the removal medium is a three-dimensional structure such as a tank filled with sand, then particles and

Table 7.11 Some of the typical media sizes used in filters.

Material	Size range	Specific gravity
Conventional sand	0.5–0.6 mm	2.6
Coarse sand	0.7–3.0 mm	2.6
Anthracite/coal	1.0–3.0 mm	1.5–1.8
Gravel	1.0–50.0 mm	2.6

Source: Data from EPA (1995).

other materials are removed by a number of mechanisms, not just screening. Screening will also take place in a three-dimensional filter, and in these particles of larger diameter, the aperture size between the grains of the medium is removed at the surface. The smallest pore/apertures are approximately one sixth the size of the filter medium grains. Some of the other mechanisms that may operate in three-dimensional filters are sedimentation, interception, electrostatic forces, biological activity and chemical activity.

Sand Pre-filters

These filters are not usually used as a sole method of filtration but as an additional barrier and protection for later treatment stages in a multi-stage treatment sequence such as slow sand filtration (SSF). These filters are designed to have a relatively fast throughput and remove mostly coarser particles. As a treatment stage before SSFs, they decrease the rate of clogging on the slow filters and thus increase the filter runs before cleaning is needed. In many countries, although SSFs were operated without problems a few decades ago, population growth, increased agriculture and industrial activity in the raw water catchment area have caused a deterioration in water quality, putting a strain on the main treatment processes and hence the need to protect them with pre-filters in order to maintain their efficiency.

Sand pre-filters or roughing filters may be configured either horizontally or vertically, the latter being popular as they require less land. If source waters have a high turbidity, as much of the particulate matter as possible needs to be removed in order to protect any subsequent treatment stages against excessive contaminant loading. Although rapid sand filters may be used for this purpose, roughing filters can also be used. A roughing filter will remove floating debris such as leaves and small vegetation as well as relatively coarse suspended solids. Roughing filters often have various sized filter media arranged in layers. The coarseness of the media in these layers decreases with the direction of flow. This means that coarser solids are removed near the inlet and less coarse solids are removed near the outlet. The filter media will have grain sizes of approximately 4–25 mm. The flow of water through the filter is relatively fast at 0.25–1.5 m/h. Cleaning the filters is hydraulic but can be manual, especially in vertical flow configurations. Manual cleaning often involves digging out the filter material, cleaning it and then replacing it in layers. Because of the relative coarseness of the medium, it is suggested that roughing filters act as a multi-storage sedimentation basin, providing a large surface area for the particles to accumulate.

The medium used in RGSFs is usually coarse sand on a gravel bed and a system for pumping either air or water at the bottom for backwashing in order to clean the filter. RGSFs have a vertical flow-through system. They are frequently used as a treatment stage before SSF or after chemical coagulation/flocculation. Figure 7.3 shows the typical layout of an RGSF. Although the filter medium is commonly sand with a grain size of 0.5–1.0 mm in diameter, additional media layers may be added, e.g. anthracite. The sand depth is usually about 600 mm, overlaying 300 mm of gravel. There must be a sufficient depth of water above the sand to allow the water to have enough hydraulic pressure to pass through the medium relatively rapidly, about 5–10 m/h. When the filter accumulates particulate material, it progressively blocks and the headloss across the filter increases. When this reaches about 2.5 m, the filter needs to be washed as the flow-though is greatly reduced. These filters are cleaned by air scouring and hydraulic backwashing. The procedure is outlined in Table 7.12. Backwashing will expand the medium, thus allowing the trapped inorganic and

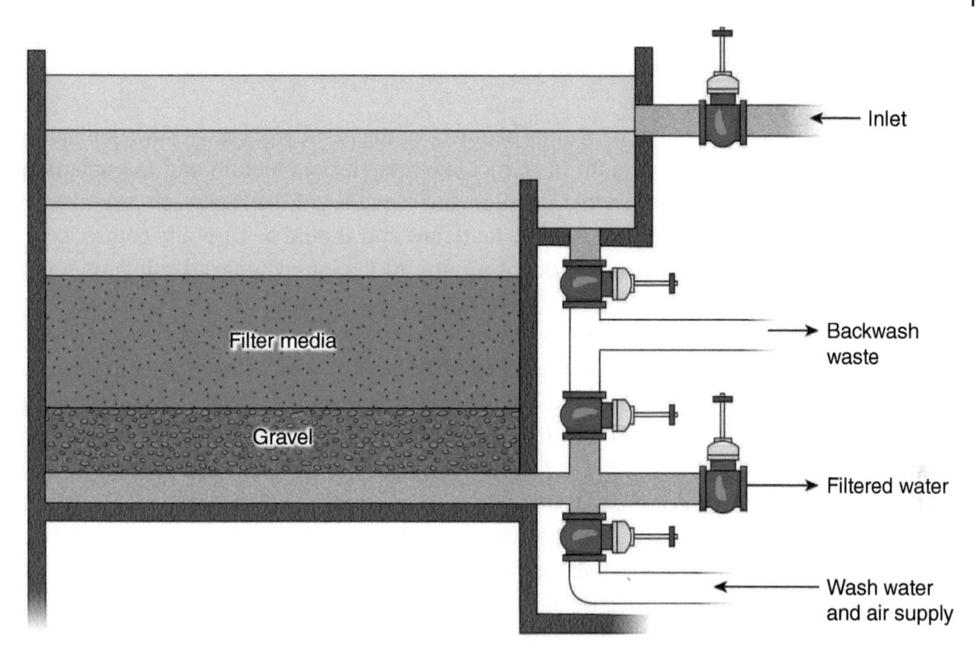

Figure 7.3 Diagram of a rapid gravity sand filter.

Table 7.12 The wash sequence for a rapid gravity sand filter (see Figure 7.3 for valve numbering).

1) Ensure that valves 1 and 3 are closed.
2) Close the inlet valve (no. 4) and allow the filter to drain down until about 10 cm of water are left above the medium.
3) Close the outlet valve no. 2.
4) Open valve 1 and apply air scour at a rate of approximately 20 m^3/m^2 of bed area. Depending upon the type of clogging matter the air scour, which agitates the medium and dislodges the clogging solids, can operate for 5+ min. The optimum time for the air scour must be determined according to the local conditions.
5) Close off the supply of air and apply clean backwash water through valve 1 at about 20 m^3/h at the same time opening valve 3. This should expand the volume of the media by about 10–20% ensuring proper cleaning. This would take at least five minutes. Or until the backwash water runs relatively clear. The dirty backwash water is collected in the overflow troughs and passed out through valve 3 to waste.
6) Close valve 2, open valve 4 allowing the bed to refill to its operating height. Valve 2 can now be opened and the filtered water can be drawn off and the filter restarted.
7) Initially, it may be necessary to run a small amount of water to waste until it runs clean in case any unwanted particulate matter was trapped in the medium.

Extra monitoring of the filtrate should be carried out after cleaning to make sure that the backwash went according to plan.

organic particles to be washed upwards and out of the medium into collecting troughs and then to waste. If the filter is not cleaned, it could result in a breakthrough of unwanted materials into the filtered water.

The length of time for a filter to operate before cleaning is, on average, 24–48 h depending upon the source water, the amount of particulate matter, the medium and the operating conditions. The correct and efficient operation and sequence of the backwash procedure is essential to remove unwanted accumulate materials and should be carefully controlled. It is also advisable not to take the filter out of service for too long and avoid using too much clean water, but importantly, to clean for enough time to ensure that the filter is cleaned properly before putting it back into service.

It is important to ensure that both the air scour and the backwash water rates are correct for the particular filter. Where possible, the surface of the media should be observed during backwashing to ensure that both the air scour and the water backwash are working evenly across the bed. If the air scour is too low, the medium will not be properly cleaned, and mudballing (see below) could result. If the scour is too high, some medium may be lost in the outflow. Similarly, if the backwash water rate is too low, the medium will not be properly cleaned. The dirty washwater should be allowed to settle for a while in a suitable holding tank, and the clear supernatant should be either discharged to waste or recycled into the treatment process. The settled solids are treated as waste sludge and disposed of as waste solids, taking into account that they may contain hazardous microorganisms and unsafe solids. When the bed is put back into operation it should initially be at a slower rate to allow it to settle before increasing to its normal operating rate. Checks on the turbidity of the inflow water and the filtrate should be carried out frequently as this will provide useful information regarding the filter's performance (see Table 7.13).

Mudballing can occur at too low air scour or backwash rates, causing unwanted solids to clump together forming round masses, sometimes to the size of tennis balls. These may block parts of the filter surface, impeding the flow of water and eventually greatly reducing the operating capacity of the filter.

Rapid gravity filters (RGFs) can operate with dual media. In this case, in addition to the sand, a layer of anthracite may be added to the surface. This can extend the range of particles and chemicals in solution removed by the filter. In some cases, other layers can be incorporated, e.g. a layer of fine garnet sand at the base of the normal sand layer for removal of some finer particles. In the case of dual or multimedia filters, great care must be taken during backwashing to maintain the structure of the different layers. Careful selection of

Table 7.13 Potential effects of different turbidity on filter performance (turbidity readings, mg/l SiO_2 scale).

Turbidity	Impact on filter
100–1000	Filter will clog quickly
<100–65	Filters operate with difficulty
30–20	Filters will operate but only with great care
10	Maximum desirable limit

Source: Data modified from EPA (1995).

the materials in the different layers according to their particle size and weight ensures that they settle out at the end of the backwashing process into their proper layers.

Rapid gravity and roughing filters are usually built as open tanks, but where a sufficient head of water is a distribution system if used for groundwater treatment, the filters can be constructed in either vertical or horizontal closed, usually steel, cylindrical tanks (see Figure 7.4). These enclosed filters can be operated with low or medium pressures depending upon construction and need. Lower pressures can often be obtained using a natural head of water that has a low to zero energy requirement as it does not require pumping. There is also the advantage of these tank filters, as they are above ground and can easily be installed inside a suitable housing unit. They do, however, have the disadvantage that the operators cannot easily see what is happening inside and thus may miss any malfunctioning. The backwash programme for cleaning is the same as that for open filters. These enclosed or pressure filters can also be operated as vertical or horizontal units.

The main purpose of RGFs is to reduce turbidity by removing particles from the raw water when granular activated carbon (GAC) is added or used as the filter medium for trace organic compounds, and some taste and odour can also be eliminated. Not all compounds can be removed equally. In addition, activated carbon can be costly, and the operation of filters becomes more sophisticated. Activated carbon can be added to existing filters or used as an entirely separate process. As GAC has a lower density than sand, flow rates through the filter need to be adjusted to avoid washout. Depending upon the amount of organic matter to be removed, frequent regeneration after about 6–20 months of use can be required.

Activated Carbon

Controlled combustion of carbon produces a highly porous material that has a high affinity for organic compounds. It can be found in two forms: powdered AC and granular AC.

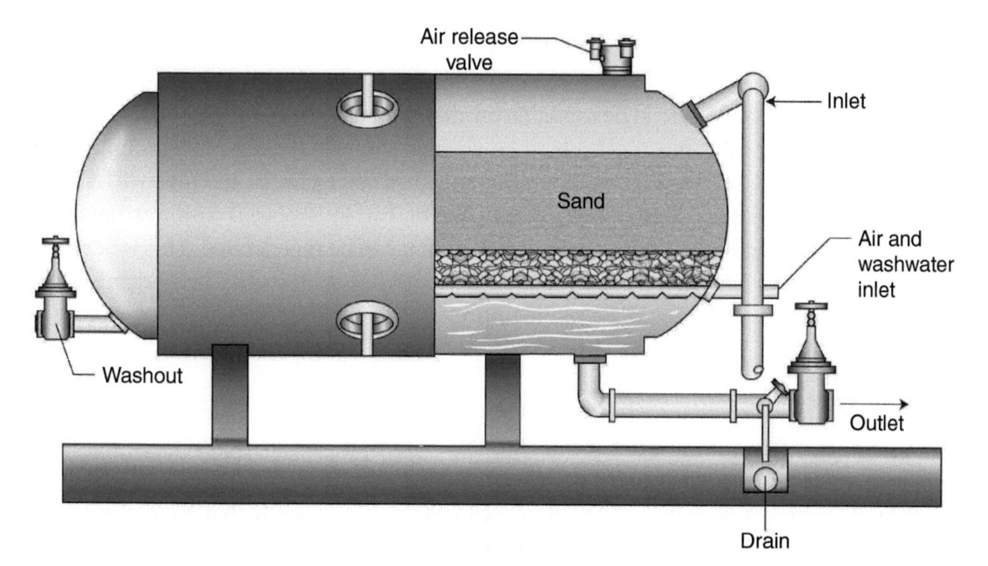

Figure 7.4 Diagram of a fully enclosed rapid gravity sand filter which allows the filter to be operated under pressure if required.

Activated carbon filters can also be used and are effective at removing colour, taste and odour and certain chemicals, including iron and hydrogen sulphide, from the water. They are not particularly good, however, at removing bacteria and viruses.

Powdered activated carbon (PAC) is dosed into the water as a slurry, which is then removed in later treatment processes where it is disposed of with waste matter. It is mainly used to remove taste and odour, which occur especially in lowland rivers. Typical dose rates would be about 20 mg/l and a contact time in the filter of only six minutes or a little more (Stevenson 1997). If algal toxins are being removed, higher doses may be required. An example could be the removal of microcystin-LR toxin present at 10 μg/l. This could require a PAC dose of between 20 and 100 mg/l to remove (Hall and Croll 1997). GAC can also be used as an additional layer in a filter or as the main medium in a filter. Because GAC beds can usually be operated for at least a year before cleaning, there is enough time for a biological film to develop over the granule surface. The bacteria in this film are able to break down geosmin, iso-borneol and other organic compounds as well as many breakdown products that are absorbed onto the GAC, reducing taste and odour in the water. GAC can be used to absorb some pesticides, but usually slower flow rates are needed and the bed life may be less (Stevenson 1997).

GAC is usually used in fixed filter beds where the GAC either replaces the sand layer or is used as an additional layer. It must be understood, however, that GAC is considerably more expensive than sand. For small undertakings, cylindrical vessels can be used to house the GAC in a dedicated filter. The depth of carbon in a dedicated filter should be at least 2.5 m. Culp and Culp report continuous GAC absorbers being used on Lake Tahoe water for organic molecule removal although in this case for wastewater treatment (Culp and Culp 1971). The filters can be operated as either downwards or upwards flow. At the end of a filter run, as measured by either headloss increase or potential particle breakthrough, cleaning and regeneration must take place. Smaller filters can have the carbon removed manually or suction devices are available to vacuum the carbon out. As carbon is saturated with absorbed materials, all of these should be removed to prevent any molecules from being released back into the filtrate. The downtime of the bed needs to be kept to a minimum, so a store of GAC should be available on site. It must be remembered that GAC beds absorbing mainly pesticides may have to be cleaned more frequently, for example every six months. Once the carbon has been removed, the bed casing/walls and pipework should be flushed out with clean water and then the carbon can be replaced with fresh material and the bed restored to action. Some grades of carbon can be regenerated. This is achieved by dewatering and heating to a high temperature driving off the absorbed organics. The carbon can then be stored for reuse in the future.

Ion Exchange

Ion exchange columns are relatively new in the water industry, perhaps becoming more widely used in the last 50 years of the nineteenth century although they have long been used in chemistry. Ion exchange mostly uses insoluble resin beads that have, depending upon the type of resin, molecules in their structure that are able to exchange either cations or anions from the water onto the bead surface. Ions on the resin are able to exchange ions in solution in the water that is in contact with it. The resins are usually in the form of synthetic beads 0.5–1.0 mm in diameter. Although most resins in use are synthetic polymers, there are naturally occurring polymers, e.g. zeolites. The process is mainly used for water softening when Ca^{2+} and Mg^{2+} are exchanged from solution by Na+. Some iron may also be removed. When the exchange sites on the resin have been used, the resin must be regenerated. This is achieved by first cleaning the column with clean water and then flushing the column with 5–10% NaCl solution. Other solutions for flushing and regeneration, such as strong acids (hydrochloric and sulphuric acids) or strong bases (sodium hydroxide), can also be used. The regeneration fluid can be passed either in the direction of treatment flow, usually downwards, or in reverse, upwards flow (counterflow). The ion exchange column can be used in a batch or continuous operation mode. It is important that the water being treated is reasonably clean and free from suspended matter; otherwise, the resin column can clog and fail to work as required. Because the resin beads have a low specific gravity, it is important to ensure that they are not washed out of the column so that inlet and outlet design are important. Competition for exchange sites on the resin occurs between ions in the water. Some examples are as follows:

$$Ba^{+2} > Sr^{+2} > Ca^{+2} > Mg^{+2}$$

$$Ag^{+} > Cs^{+} > K^{+} > Na^{+} > Li^{+}$$

$$NO_3 > CN^{-} > HSO^{4} > NO_2 > Cl > HCO_3^{-}$$

After regeneration, there is a liquid that is very high in dissolved solids containing all of the ions removed by the resin released during regeneration. This must be properly treated before disposal.

Different resins are available for the removal of a range of cations and anions. In particular, zeolites have been used for colour removal (Hill and Lorch 1987) and nitrate removal (Hall and Croll 1997). Nitrate removal is affected by sulphates in the water, as they have a higher affinity for the resin. Nitrate-selective resins are available which largely overcome this problem. In all cases, care must be taken with the materials used for the construction of the ion exchange filters as some of the solutions can be very corrosive.

Membrane Filters

Drinking water quality regulations are becoming stricter globally, acceptable limits on many substances are being lowered and many new substances are being placed on the list (see WHO 2011). Examples from the USEPA include the surface water treatment rule, the enhanced surface water treatment rule, the groundwater disinfection rule, the coliform rule and the disinfectant/disinfection by-product (DBP) rule. The latter aims at reducing the amount and number of chemicals used in drinking water treatment. Concern has arisen in recent years concerning the removal of certain pathogens from water, in particular the cysts of *Giardia* and the oocysts of *Cryptosporidium*, both protozoans. These are both quite resistant to disinfectants such as chlorine. A high level of disinfectant and longer contact times needed to kill these can run the risk of creating high levels of DBPs that themselves can be harmful to human health. This would also breach the US Disinfectant/Disinfection By-Products Rule. This has led to the encouragement of alternative technologies, such as

Table 7.14 Examples of size ranges of cysts and oocysts compared with bacteria and virus particles.

Organism	Size range (μm)
Giardia muris	7–14
Cryptosporidium parvum	4–7
Escherichia coli	1–3
Pseudomonas aeruginosa	0.5–0.8
MS2 virus	0.027
Clay particles	0.1–10.0
Planktonic algae	1.0–200.0
Sand	100–1000+

membrane filtration, that can reduce the amounts of chemicals used. To remove these protozoan pathogens from water requires membranes with small enough pore sizes. Table 7.14 shows approximate sizes of the cysts and oocysts compared with typical bacterial and virus particles.

Membrane filters are sheets of a semipermeable material as either a flat sheet or tubular shape. The membrane, depending upon pore size, is capable of removing a wide range of particles from water (Jacangelo et al. 1997). In order to pass water through the membrane, two approaches can be used. The first uses pressure or suction, and the second is electrically driven. Pressure-driven processes are used for microfiltration (MF), ultrafiltration (UF), nanofiltration (NF) and reverse osmosis (RO; for descriptions see later) membrane techniques. NF and RO are becoming more frequently used in drinking water treatment, and MF and UF require less pressure and are less expensive to operate. Electrically driven processes are an improved version of electrodialysis called electrodialysis reversal, in which the direct current is periodically reversed in order to minimise fouling of the membrane. The membranes used are ion exchange in which pairs of membranes, cation exchange and anion exchange, are packed in alternating layers through which the water flows. When a current is applied, positive ions migrate to the cathode and negative ions to the anode. Salt is removed from every other layer and collected in the intervening layer for discharge as a brine concentrate. The desalted water was passed into a pure water system.

In the past few decades, membrane filters have become more widely used, especially for groundwater sources needing little or no pretreatment, and they are simple to install. They are different grades with specific sized pores in their surface (see Figure 7.5).

They are efficient at removing particles from larger suspended matter to viral particles and even larger molecules depending upon the membrane grade. To obtain a suitable yield of clean water, filtrate pressure (or suction) must be applied, and the amount needed increases as the particle size to be removed decreases so that MF membranes capable of removing particle sizes of 0.05–5.0 μm require relatively low pressures, whereas ultrafilters removing particle sizes of as small as 1 nm require much higher pressures. The latter are consequently more expensive to install and operate, but as it is possible to have quite small

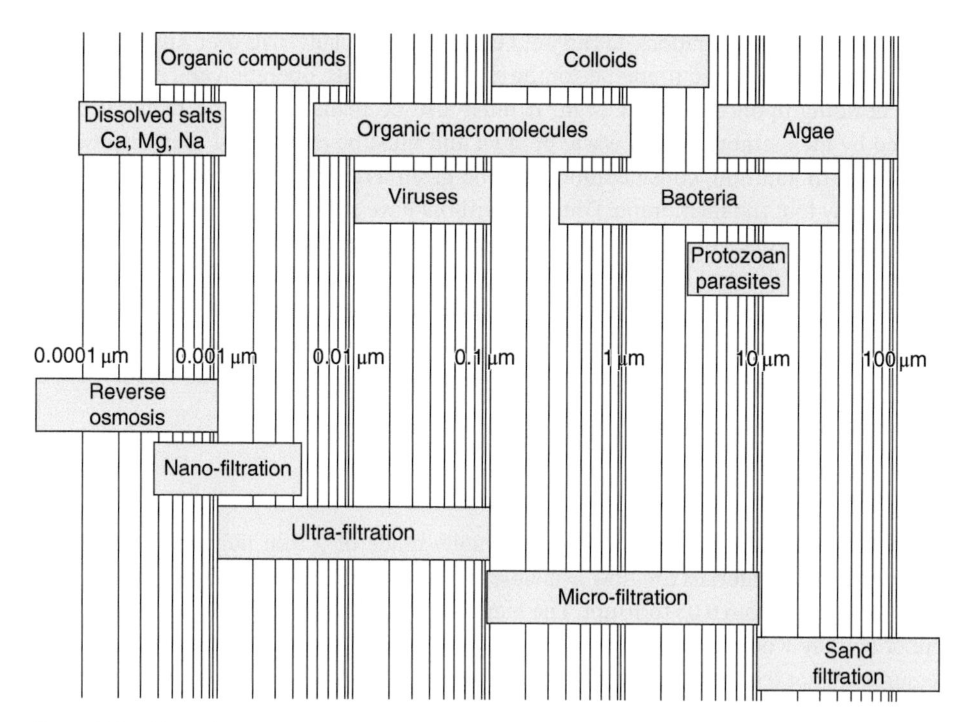

Figure 7.5 Chart showing the sizes of various particles and the types of filter that are required to remove them.

membrane units at modest cost, they are becoming quite popular for small-scale operations such as individual houses. In addition, because of the increasing demands of drinking water regulations, the range of membrane particle removal capacities makes them a more attractive proposition for both small- and large-scale works.

Before embarking upon any treatment system, but particularly before considering membrane technologies, a thorough analysis of the type and amount of contaminants throughout the year is required. If the wrong type of membrane is used, it is either unlikely to meet the required standards for that particular operation or too expensive to install and maintain.

If only larger particles of size greater than 0.2 μm are to be removed, then MF is adequate. These have the advantage of requiring relatively low pressures to operate (15–60 psi). If it is possible to chemically coagulate or precipitate smaller particles, then microfilters or nanofilters can be used. If dissolved organic molecules need to be removed, then, depending upon the size of the molecules, NF, UF or even RO may be required. These latter technologies require much higher pressures to operate (30–190 psi). If organic molecules are not a problem but inorganic molecules such as may be present if softening of the water is needed, then NF could be used. If all inorganic contaminants need to be removed or the amount of total dissolved solids is high, then RO may be required. The relative costs of these different types of filtration are $MF < UF < NF < RO$.

Although membrane systems tend to be expensive to install, costs are decreasing, and they do have the advantage that the cost does not increase markedly per unit of water treated for small installations compared to larger ones. However, it is beneficial, especially with medium to large operations, to carry out extensive pilot plant trials over a representative period of time to assess requirements for the capacity of plants, operation and maintenance before deciding upon the final design. It must also be realised that the material being removed by the membranes is a waste product and must be treated and then disposed of accordingly. In addition, consideration must be given as to whether the materials being removed may foul the membrane. Different authors have slightly different size ranges for the different categories of membrane filters. The size ranges given here are averaged from several sources. Different categories of membrane can be described according to various criteria, including aperture or pore size; the molecular weight cut-off size of the molecules that can pass through the pores: membrane construction material and its construction in the filter; the nature and type of materials to be removed and the final water standards that have to be met.

Microfiltration

Microfiltration separation membranes are usually made of a thin polymer sheet. Many other membrane materials are now available, especially UF, NF and RO filters. The pore sizes in MF range from 0.05 to 10 μm. The membrane configuration can be tubular, hollow fibre or spirally wound sheets. The hydraulic resistance of the membrane is low, so high pressures are not required. The types of particles that microfilters would be used to remove would be fine sand and clays, small algae, some bacteria and, importantly, *Giardia lamblia* cysts and *Cryptosporidium parvum* oocysts. Both of these protozoans can be severe problems in many parts of the world (see Table 7.15).

Pretreatment may be needed for high turbidity waters to remove large particles that may block the apertures to the hollow fibres in some membrane configurations. Some membrane materials need to operate within a certain pH range, so some chemical dosing may be needed to adjust the pH of the inflow water to avoid damaging the membrane.

Table 7.15 Distribution of Cryptosporidium and Giardia in surface and drinking water supplies.

Country	Source
India	Human stools
China	Contaminated surface water
Portugal	Animal and human stools, contaminated surface water
Brazil	Human stools
Poland	Human stools
Finland	Groundwater
Japan	River water
USA	Surface waters contaminated with sewage and animal waste
UK	Human stools and contaminated surface waters
Russia	Contaminated surface water

Sources: Putigeni and Menichelle (2010), Betancourt and Rose (2004), Hashimoto et al. (2002), USEPA (1999), Karanis et al. (2006), Pitkanen et al. (2015) and Clancy and Hunter (2008)

As the filter operates it retains particles and gradually clogs. They therefore require regular cleaning, usually by backwashing with either water or air/gas. The configuration can be hollow fibre, flat sheets or spiral wound. The hollow fibres can operate as inside-out or outside-in using a pressure vessel. The untreated water can be introduced either at right angles or tangentially to the membrane, but where it is from the inside to the outside, either the hollow fibres must be closed at one end, or if open, the fibres are held in an enclosed pressure vessel. The microfilters can be supplied in closed pressure containers that can be added retrospectively to the treatment chain and are particularly good at removing protozoan cysts and oocysts as well as planktonic algae. Microfilters are not able to significantly remove colour, taste and odour from water as the pore size is too large. This may be partially overcome by adding PAC to the inflow water. This retains some of the organic molecules, causing colour as well as some taste and odour to be reduced. The carbon powder is removed by the membrane and does not pass into the permeate.

MF is more frequently being considered for use as an alternative to some conventional treatment stages as they are easy to operate and take up less space. The downside is that they are more expensive and, if higher pressures are used, consume more energy. For smaller plants (<1 million gal/day; 4.6 million l/day), pre-constructed 'package plants' are available.

Ultrafiltration

Ultrafiltration is similar to MF, but the pores are much smaller, 0.002–0.1 μm. As a result, much greater pressure (30–100 psi) is required to force the water through. These membranes are generally made of polyacrylonitrile, polyimide and cellulose acetate and are commonly fabricated in a sheet or tubular form. Because the membranes are very thin, often less than 1.0 μm, they need to be supported by a stronger, porous layer. They may be tubular or spiral wounds. As with MF, all protozoan species are removed together with most bacteria and some viruses. In addition, some humic materials can be removed. If used instead of certain conventional treatment methods (coagulation and flocculation), the use of chemicals is greatly reduced. As long as the membranes are regularly cleaned and maintained, they produce a consistent quality filtrate. They are also easy to automate. Large molecules can be retained, and smaller molecules are allowed to pass through. They are frequently used in industry for this purpose. Virus and bacterial removal is another application by the food industry. The water to be treated may need pre-treatment to remove larger particles to reduce rapid fouling of the UF membrane. This is also true if the water contains dissolved iron and manganese. In this case, the water needs to be treated with suitable oxidants to precipitate them prior to UF to avoid precipitation in/on the membrane causing blockage or worse still precipitation after membrane treatment in the treated water.

Nanofiltration

Nanofiltration membranes have a nominal pore size of 0.001 μm, so much greater pressures are needed to force the water through. Operating pressures are usually around 90 psi but can be up to 150 psi, which will require a greater energy input. NF filters will remove all protozoan cysts, most bacteria and most viruses and colour and other humic materials.

NF filters also remove alkalinity and hardness and are sometimes called 'softening filters'. Because of the alkalinity removal the residual waste can be corrosive. In addition, if treating hard waters, scaling can be precipitated on the membrane, which needs removal. Membrane fouling can be a problem with all membranes but particularly with those having smaller pore sizes.

Reverse Osmosis

RO will remove most inorganic contaminants in water. It will also remove natural organic compounds and pesticides as microbiological pathogens. Although it is less affected by solids in suspension, RO is frequently used after some other pre-treatments. RO has a high capital cost, and the membranes are subject to fouling. The wastewater produced is a strong brine solution and can present difficulties in disposal. The amount of wastewater produced is also quite high, between 25% and 50% of the feed water.

Membrane Integrity

Because membranes, by their nature, are fairly delicate, damage can occur, producing a hole allowing untreated water to pass through. Continuous monitoring of the filtrate should be carried out as a routine. Any undue rise in turbidity should be noted. Other tests for breaks in the membrane include air pressure testing, air bubble testing, sonic wave sensing and biological monitoring. Jacangelo et al. (1997) found that air pressure testing was the most sensitive test.

Backwashing

Some hollow fibre membranes are arranged as suspended strands closed at the lower end. As an array is suspended in a tank, the particles are deposited on the surface. As the fibres move gently with the flow of water, there is a tendency for the particles to be displaced and fall to the tank bottom, thus reducing the rate of clogging. For other configurations of membranes, some other form of cleaning is required. As the membranes are usually arranged in discrete units, backwashing is usually carried out separately unit by unit and takes only a few minutes. This allows the overall treatment process to continue uninterrupted. The backwash frequency depends upon local conditions but could take place every hour or less. About 5–10% of water on average goes for backwashing. Some systems use pressurised air in conjunction with water and, depending upon the membrane material, low doses of chlorine to prevent biofouling. If there are scaling or other such problems, additional chemicals, such as acids, alkalis or surfactants, may be used in the backwash (see Table 7.16).

After cleaning, the membrane unit is washed with clean water to remove debris and cleaning materials before putting back into use. Chemical cleaning needs to be carried out only when required, not every time.

Disposal of the wastewater after cleaning must be considered from the outset, as not only may the volumes be larger than with most traditional treatments, but it can also be more concentrated. With care, and if no other cleaning agents have been used, it may be disposed

Table 7.16 Examples of additional cleaning agents for membranes that may be used.

Cleaning agent	Type of fouling removed
Citric acid	Inorganic scale
Hydrochloric acid	Inorganic scale
Sodium hydroxide	Certain organic fouling
Sodium hypochlorite	Organic matter, biofilms
Chlorine	Organic matter, biofilms
Surfactants	Organic and other inert particles

of to the source water. Otherwise, some additional treatment may be required, and alternative approved methods of disposal may be used, e.g. disposal to sewers, dilution and deep well injection.

Membrane fouling is very likely to occur from a range of substances found in natural waters. As they foul, the membrane water pressure needs to increase to achieve the same flow. Depending upon the nature and turbidity of the source water, a layer or cake will build up over the membrane surface. Although this cake will have an adverse effect on flow, it can help with the removal of finer particles, especially virus particles that would, to an extent, pass through the membrane.

Membrane technologies are now being more widely used, especially the low-pressure MF and UF. In the United States, for example, use increased from <10 million gallons per day (MGD) in 1994 to 140 MGD in 1999 and >750 MGD by 2005. Globally, the largest installer and user is in the United States of America at 44% of the global total. The Pacific Rim countries had 23%, Europe had 19% and the Middle East/Africa had 14% (Furukawa 2008). MF and UF use is growing as is RO, especially in countries where droughts are common and water is in short supply, such as Australia and parts of the United States of America, where many desalinations of seawater installations are projected. The greatest use of membrane technologies is for drinking water treatment although wastewater treatment use is growing rapidly. As low-pressure membranes are very effective at removing particulate matter but are not good at removing dissolved matter, a combination of low pressure followed by higher pressure NF or RO membranes could provide comprehensive treatment. This combination is called two-stage membrane filtration. This has been found to work (Vickers et al. 1997); however, the costs are still quite high, about three or four times higher than conventional treatment.

Coagulation and Flocculation

Coagulation and flocculation is one of the most widespread and common chemical methods of treatment. In it, coagulant chemicals are added to the water in order to remove any particulate matter, especially those that are difficult to sediment on their own. When considering which treatment option to use, one of the most common biological groups of particles to remove and cause problems is algae. Not only can they produce large growths in slow flowing or still surface waters in the actual treatment processes themselves they can

Table 7.17 Size and settling characteristics of different particles.

Particle size (μm)	Particle type	Settling time
10 000	Gravel	0.3 s
1000	Coarse sand	3.0 s
100	Fine sand	38 s
10	Silt	30 min
0.1	Bacteria	55 h
0.01	Colloid particles	230 d
0.001	Colloid particles	63 yr

produce nuisance growths on the walls of filters, clogging the surface of filters and producing colour, tastes and odours in the water. As algae can be present in most, if not all, surface waters, it is advisable to monitor the source water as they usually exhibit seasonal growth cycles. Knowledge of the peak growth times can allow the plant operator to take additional measures to combat them. If possible, however, it is preferable to control their growth in the raw water and prevent shock growths getting to the works. Pre-chlorination is often used as a preventative measure for raw water intake. Particles in water can be of many sizes, which affects their settling times (see Table 7.17).

The settlement times based on a sphere of specific gravity 2.65 to settle 30.48 cm in distilled water in absolutely calm conditions.

Particles of bacterial size and less do not settle conveniently for normal water treatment. These particles each carry a surface electrical charge and, as they have the same charge, repel each other. Even when in suspension they naturally bump into one another by Brownian movement or van der Waal forces they still do not join together. Coagulants, such as salts of aluminium and iron, help neutralise the charges, allowing the particles to reduce the repulsion effect and form larger groups of solid precipitates called flocs. These flocs are able to trap colloids and other smaller particles, allowing them to be incorporated into larger and easier to settle groups. This floc can then be easily separated using conventional methods.

When the metal salts are added to water, the metal ions hydrolyse quickly, forming a series of metal hydrolysis species. The process is controlled by the pH of the water. The coagulation process involving the mixing of the coagulant chemicals is carried out quickly with maximum turbulence, whereas the flocculation part is carried out with gentle mixing before passing on to a clarifier to allow completion of the floc formation and settlement. Coagulation and flocculation are effective at removing otherwise difficult fine particles and reduce the time needed for particles to settle. A range of coagulants are available, and a selection is outlined in Table 7.18.

The dose rate of the coagulant used will depend upon the nature of the water to be treated and the chemical used. Examples of typical dose rates are for 8% Al_2O_3 23.8 mg/l and for ferric sulphate 8.9 mg/l. It is important that for any given water, a laboratory 'jar test' be carried out (Gray 1999) not just once but at different seasons representing different conditions. pH optimisation should also be carried out in a series of jar tests at the same time. When the coagulant is added, it neutralises the negative charges on the particle surfaces, so

Table 7.18 Some examples of coagulation and flocculation chemicals and the form in which they are usually used.

Chemical	Form used	Comments
Aluminium sulphate (alum)	8% in liquid as Al_2O_3 or also in solid form	Very widely used in the UK and other countries
Ferric sulphate (iron(III) sulphate	Liquid as 40% as $Fe_2(SO_4)_3$	Commonly used
Poly-aluminium silicate sulphate (PASS)	Liquid at 8.3% as Al_2O_3	Effective over a wide range of pH end even at low temperatures. Doses lower than for Alum so lower sludge volumes. Less corrosive than some

Source: Data abstracted from Hall and Croll (1997).

they do not repel each other. They are then said to be destabilised colloids and associated fine particles that are then able to join together to form larger, heavier masses. This is encouraged by gentle agitation of the water, making them collide, and flocs in the flocculation process. Sometimes additional coagulant aids can be added. Simple substances such as starch or silicates can be used for this purpose or specifically manufactured polyelectrolytes that are able to attach themselves to the flocs and form bonds or bridges that link together other flocs, making even larger particles. The process of coagulation and flocculation together with the action of polyelectrolytes is illustrated in Figure 7.6. The flocculation process is carried out in a flocculator, the main types of which are shown in Figure 7.7. Whichever system or combination of systems is used, it must be ensured that (i) there are no dead spots in the tank that could allow unwanted settlement, (ii) short circuiting of the water plus chemicals should be avoided and (iii) agitation should be gentle enough not to break up the flocs. After coagulation and flocculation the solids and floc particles must be removed in a clarifier. This can be achieved by either settlement or flotation depending on whether the flocs produced are heavier or lighter than water. In either case, the solids are removed as a chemical sludge, which is then dewatered and concentrated before disposal. Some clarifiers create a floc or sludge blanket by passing the coagulated water upwards through a tank slowly, allowing flocculation to occur. The flow gradually reduces, holding the floc particles in suspension that then forms a dense blanket, the floc blanket. As the flow continues, water passes upwards through the floc blanket, which now retains all of the particles. The resulting clean water is now allowed to flow out to the next stage of treatment. Too rapid a flow will cause some floc to be carried over, and this will then have to be removed. A typical clarifier is shown in Figure 7.8.

Dissolved Air Flotation (DAF)

Dissolved air flotation acts in reverse in that the solids are encouraged to float to the surface. This works particularly well for particles that do not settle well, such as microscopic algae and light flocs, or have a neutral to negative density. The system works by introducing very small air bubbles into the dosed water between the DAF cell and the flocculator. As these bubbles rise to the surface, they attach themselves to the floc, raising it to the surface.

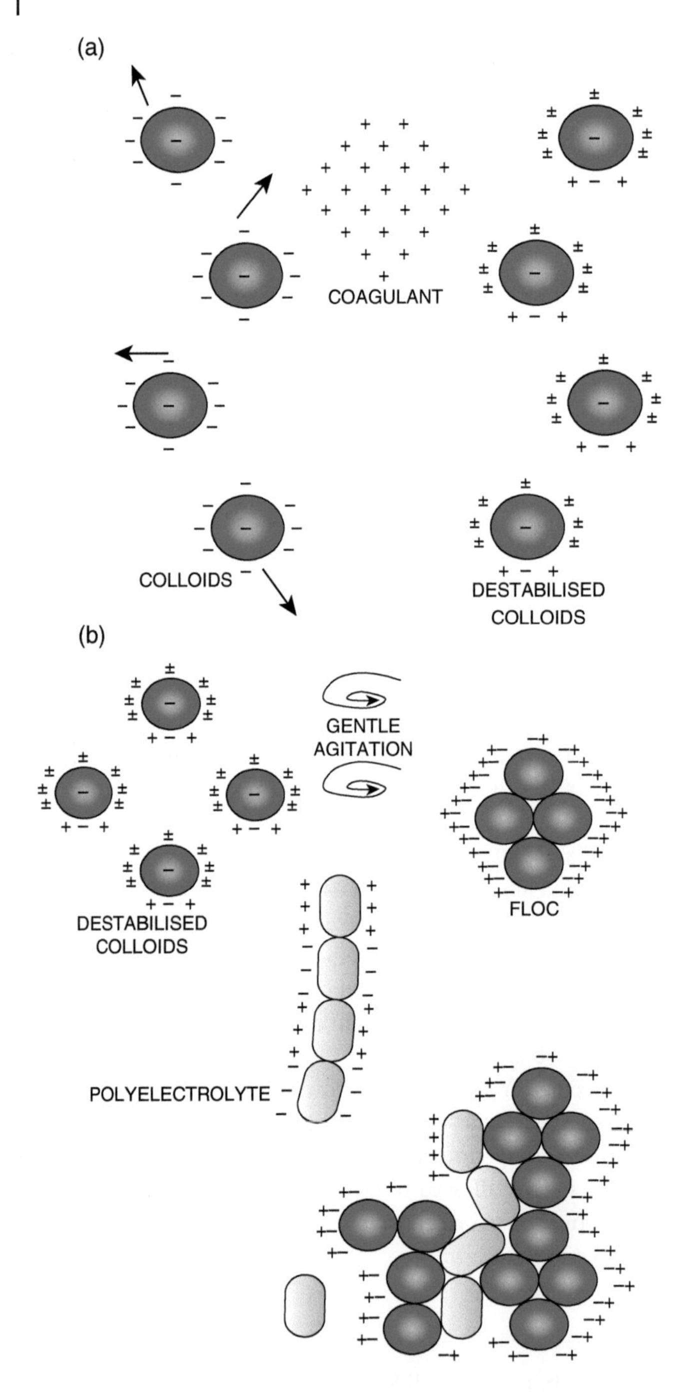

Figure 7.6 (a) Negatively charged particles need to be neutralised with a coagulant to create a destabilized colloid. (b) Flocculation of the destabilised colloid with the aid of a polyelectrolyte to form larger particle groups.

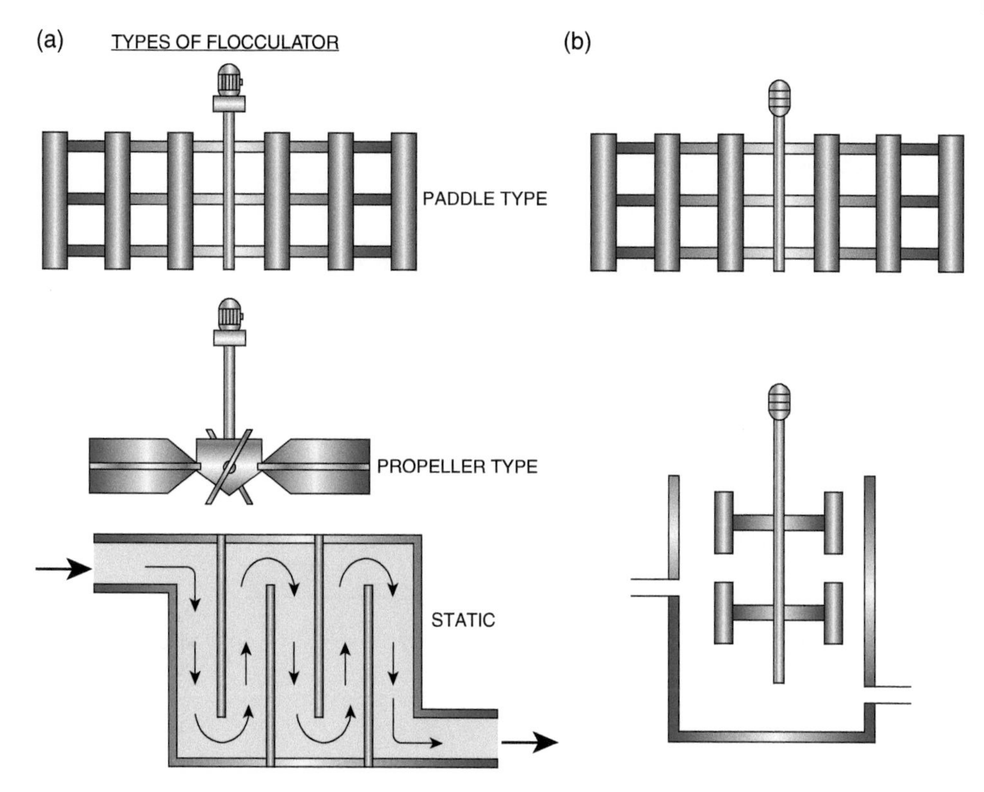

Figure 7.7 (a, b) Examples of different types of flocculator used to encourage the process of flocculation. *Source:* Reproduced by kind permission of Severn Trent Water Authority, UK.

The bubbles are produced by dissolving air in water under pressure. This water is then released into the flocculator, and the pressure is released. The dissolved air then comes out of solution as microbubbles, which, together with the floc, rise to the surface. Surface scrapers then remove this sludge floc, known as float, to waste. A typical DAF unit design is shown in Figure 7.9. A variation in this process is the adsorption clarifier. This also relies on upwards flow of the floc and particles, but here a porous plastic medium floats in the water and is held in place by a stainless steel mesh. This acts as a filter trapping the flocs that accumulate and can be removed by backwashing. DAF plants are more expensive to operate, but they do have higher treatment rates and therefore are smaller to construct.

If greater flows and improved settlement of flocs are required after coagulation and flocculation, inclined or horizontal plates or tubes can be installed. It has long been observed that particles settle more rapidly in inclined tubes. These lamella clarifiers accelerate settlement and reduce the size required for the sedimentation unit. Lamella clarifiers can be horizontal or inclined. Details of various types of plate and tube settlers can be found in Stevenson (1997).

Many treatment processes, not only in developing countries but also in all countries, could be improved in their design, operation and maintenance. This is often because the raw water analysis was incomplete or with changing circumstances is in need of updating. In many developing countries, insufficient testing of technologies at a bench scale has

(a)

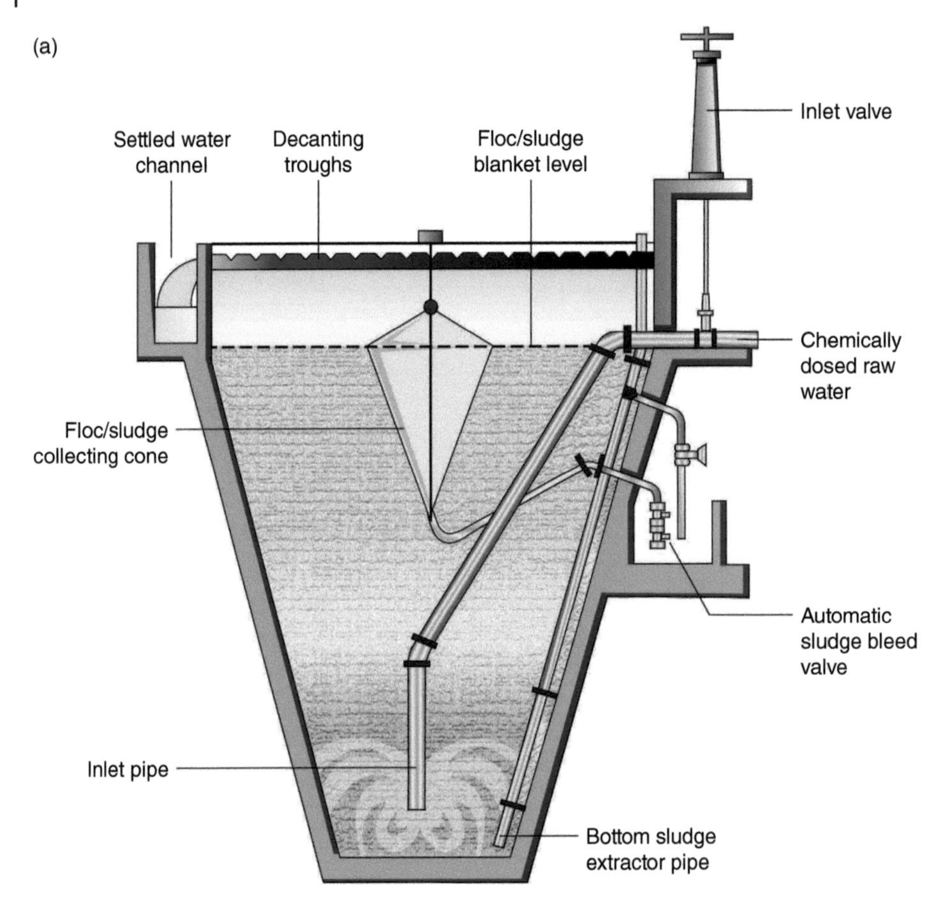

(b)

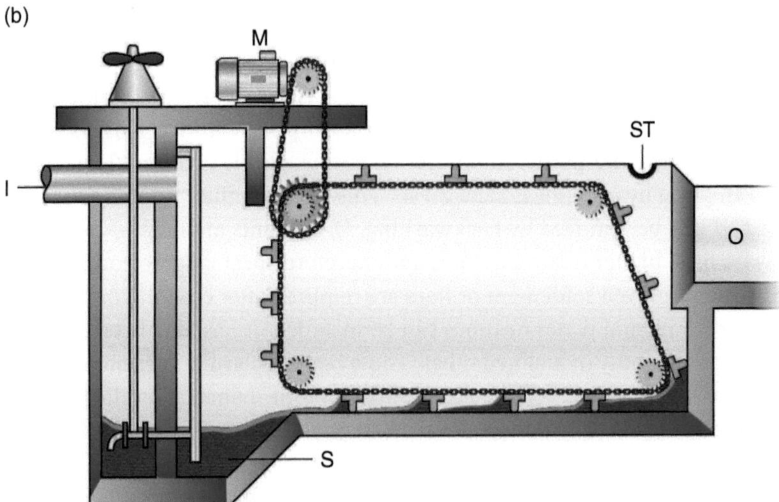

Figure 7.8 (a) Hopper type clarifier which has an upward flow. *Source:* Reproduced from Severn Trent Water Authority UK with permission. (b) Horizontal flow clarifier. M, motor to drive the scrapers; ST, scum trough; I, chemically dosed raw water inlet; O, clarified water outlet; S, sludge collecting hopper. *Source:* Reproduced by permission of Severn Trent Water, UK.

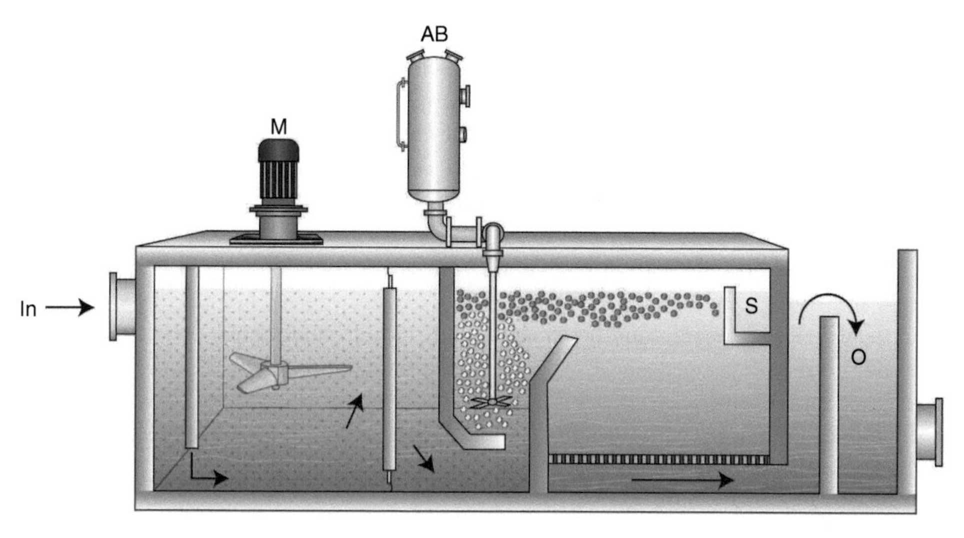

Figure 7.9 Dissolved air flotation (DAF) unit. In, inlet for chemically dosed raw water; M, motor to drive the mixer paddles in the flocculator section; AB, air injection unit; S, sludge collection trough; O, outlet of clarified water. *Source:* Reproduced with permission of Severn Trent Water Co. UK.

taken place. Bench scale testing using the 'Jar Test' should be undertaken before the process design is undertaken in order to understand the nature and potential problems that could exist with that particular water. This should include both seasonal and special variations in the source water. It should also be recognised that (i) changes may be required with the original design and its operation because of changes in demand and source water, (ii) changes in the drinking water quality standards through greater understanding and new substances becoming more common and (iii) errors and miscalculations in the original design, possibly because the design engineer did not have sufficient knowledge of the local situation. Examples include poor siting of the water intake and lack of flexibility in the positioning of the intake to take into account changing conditions. In addition, lack of knowledge concerning the volumes of the treatment tanks, their flow patterns correct timing of the chemical reactions taking place. Other examples of design problems are coagulant dose required and how this could vary with season; dosing mechanism not set up properly or old and worn; agitation incorrect for flocculation; time in flocculation unit incorrect and short circuiting of flow. After flocculation, the water with flocs is transferred to settling basins. This must be done with care so that the floc does not break up. The exit to the settlement basin must be correctly designed so as not to carry over too much energy into the settlement, which could result in short circuiting and stagnant zones. Poor settlement would result in placing a heavier load on any filters because of excess floc carryover. Again, an entrance baffle would help overcome this problem. In a normal horizontal and rectangular settling tank and under correct operating conditions, most of the flocs will sediment out in the first third. If the design or operation is sub-optimal, then settlement could reach the last third and even the tank exit. To keep the velocities of flow low, the outlet weirs should be quite large, often two or more times as long as the original design. This will ensure that floc carry-over is minimised.

A further development of an upwards flow clarifier is the Pulsator Clarifier. These consist of pairs of tanks with a sealed tower between them. The overall flow pattern is as with a flat bottomed clarifier, but the flocculated water is passed into the tower, where it then flows into the bottom of each tank. A floc blanket is formed in each tank. The air is now partially pumped out of the tower, allowing the water to rise by approximately 0.5–1.5 m over a period of 30–60 s. The partial vacuum in the tower is now released, resulting in a surge of water passing into the tanks, causing the floc blanket to pulsate and become denser, improving treatment and particle removal. This in turn allows a higher rate of treatment than other methods.

Other types of flocculators, clarifiers and DAFs can be found in Gray (1999), Stevenson (1997) and Wagner and Pinheiro (2001).

In the developing world, sand filters are usually used to capture and prevent floc carry-over only, but in developed countries, dual media filters of sand and anthracite are quite often used for the removal of specific substances. A properly designed and operated system will produce final water with turbidity less than 0.5 NTU. If this rises above 1.0 NTU, then problems in the treatment process are indicated.

Slow Sand Filtration

RGFs and roughing filters are designed to remove coarser particles from the water. SSFs act more slowly, up to 50 times slower than RGF, and have a physical, chemical and biological action hence their other name of Biological Filters. SSFs are usually constructed as rectangular concrete tanks with an impervious base. They are provided with a controlled inflow at the surface, a fine sand layer of depth 1.2–1.4 m overlaying a 0.6 m gravel layer and a controlled drainage system at the bottom for collecting the filtered water (Figure 7.10; Hendricks 1991). While RGFs can operate at filtration speeds of between 5 and 15 m/h, SSFs are operated at between 30 and 50 times slower. For this reason a much greater area of land is required. For example, a 50 Mld plant requires about 2 ha if it operates at a flow rate of 0.1 m/h. As land becomes more expensive, SSFs become less attractive, especially near towns and cities although they do yield an excellent quality filtrate. With SSFs, not only physical straining occurs, but also other important mechanisms act. As SSFs have a

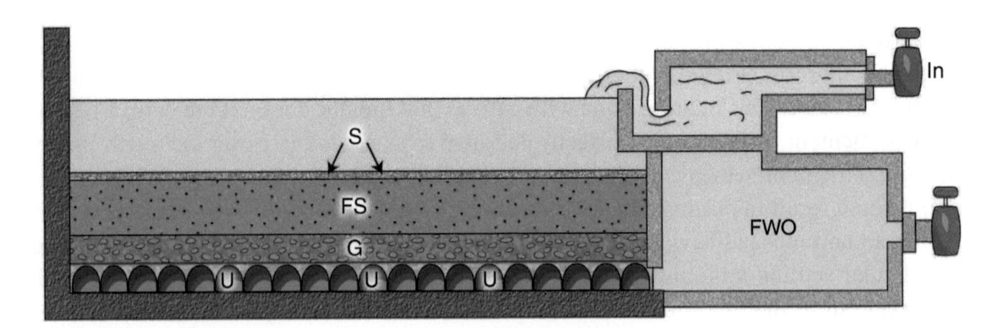

Figure 7.10 A slow sand filter. In, inflow water; S, schmutzdeke layer; FS, fine sand; G, gravel layer; U, underdrain; FWO, freshwater outlet Note: Slow sand filters are much larger in the area that is represented in the diagram. When used in treatment processes, they are often 0.25 to 0.5 ha in area but will still work on a smaller scale.

relatively slow flow-through, this allows more microbial communities to develop. They are able to remove a range of much smaller particles from the water as well as being capable of performing a number of beneficial biological and biochemical reactions. These include removal by oxidation or reduction of natural organic matter (NOM) in solution, including taste and odour compounds, ammonia, nitrates, iron, manganese and arsenic. In addition the efficient removal of organic compounds reduces the risk of trihalomethane (THM) production during disinfection by chlorination and reduces the amount of organic matter passing into the distribution system, which can promote bacterial and other growth in pipes (Scholz and Martin 1997). SSF has been the main method for treating drinking water for over 200 years. The first recorded SSF was in Paisley, Scotland in 1804 and was designed by John Gibb. It was built for his bleach factory, and he sold excess water to the public. The details of the method were refined, and in 1829, James Simpson adopted it for public supply. He installed SSFs for the Chelsea Water Company in London. By 1852, SSFs had become well established, and the Metropolis Water Act was passed, requiring all water taken from the R. Thames within five miles of St Paul's Cathedral be filtered before passing to the public. Further compelling proof of the effectiveness of filtration occurred in 1892 in Germany, where two neighbouring cities, Hamburg and Altona, both drew their drinking water from the River Elbe. The water treatment in Hamburg was just to allow settlement of medium to large particles before passing out into supply. Altona filtered its water before supplying it to the public. The river became infected, and Hamburg had a severe cholera outbreak in which more than 7500 people died. Altona did not have a major outbreak, and very few people died. In subsequent years in other parts of the world outbreaks of cholera have been linked to drinking unfiltered water. SSFs have been constructed in many countries, including the United States of America, Holland, France, England, Sweden and Japan (Huisman and Wood 1974). Since their inception, many improvements have been made, mostly concerning their construction, including reductions in the amount of land required and automation of many of the controls. Although, in general, sand filters can be divided into two categories, gravity and pressure, SSFs have mainly been of the gravity type. Gravity filters are simple in their construction and operation even though the processes taking place within them are quite complex. The purification of the water consists of a mixture of physical, chemical and biological reactions (see later). RGFs, because of their coarser sand, 0.6–2.0 mm as opposed to 0.15–0.3 mm for SSFs, operate some 20–50 times faster than SSFs. Water passes through them at 5–15 $m^3/m^2/h$, whereas in SSFs the flow rate through the medium is only 0.1–0.4 $m^3/m^2/h$. Although this means that they require less space, the water has such a short residence time in the medium that many of the cleansing mechanisms are unable to occur, so these filters rely mainly on physical screening.

Particle removal and cleaning mechanisms operating in an SSF:

There are a number of forces and reactions that take place as water passes through the sand medium (Huisman and Wood 1974). Although these will be outlined separately below, it must be recognised that many of them interact together.

Physical and physico-chemical actions:

These include (i) screening, (ii) sedimentation, (iii) centrifugal forces, (iv) diffusion, (v) van der Waal forces and (vi) electrokinetic attraction.

Screening: This is the simple removal of particles that are too large to pass through the apertures between the sand grains. This occurs mainly at or near the surface of the filter.

Although there is a minimum size that can initially be retained as a layer of material called the schmutzdecke builds up, the apertures available become smaller so that smaller particles can be retained. The schmutzdecke layer is composed of organic and inorganic particles that were in suspension in the inflow water together with algal, invertebrate, protozoan and bacterial populations that develop and help with the purification process (Bellinger 1979; Duncan 1988). As smaller particles move between the sand grains, the flow is continually changing direction, which causes the particles to collide with each other. This allows some of them to aggregate into clumps large enough to be retained. The overall effect is that the filtering efficiency improves with time, but at the same time, the flow-through the bed gradually decreases, i.e. the bed is getting blocked. This will continue until the flow of water is lower, i.e. the headloss is high and the bed needs cleaning.

Sedimentation: As the water flows past the sand grains and repeatedly changes direction, some still zones are formed, which allow the particles to sediment onto the grains under the influence of gravity. In the cumulative grain area of one cubic metre of sand, the grain size of 0.25 mm is 15 000 m^2, and there will be many upwards facing still zones into which some sedimentation can take place (Huisman and Wood 1974).

Centrifugal forces: These also act upon particles throwing them out of the stream into still zones as the direction of flow changes. This enables particles to enter sedimentation zones, coalesce or aggregate and enter the immediate zone around sand grains.

Diffusion: The natural random movement of molecules, also known as Brownian movement, brings particles into contact with sand grain surfaces. This force acts throughout the whole depth of the medium, not just in the surface layers. This effect is relatively small and is more significant for small particles of less than 1 μ.

van der Waal forces (mass attraction): These forces are also quite weak and will only act if a particle is within a micron or so of the sand grain surface. It is probably only significant in conjunction with other forces and where the flow velocity is low.

Attachment: For removal of a particle to occur, it must attach itself to either the filter medium or to something permanent within the medium. In addition to the forces mentioned above, the development of a biofilm over the sand grain surface provides an important adsorptive surface for particle attachment. In addition, particles can be held on the surface of the sand by electrostatic forces. These forces hold particles having opposite electrical charges on their surface to the sand grains. Particles having a similar electrical charge will be repelled. Clean sand has a negative charge on its surface and will thus attract positively charged colloidal particles such as cations of iron and manganese. Bacteria usually have a negative change on their surface and are thus likely to be repelled. As a new or just cleaned filter is brought back into operation, the negatively charged sites are neutralised by positively charged particles and may oversaturate the surface, making it then able to attract negatively charged bacteria. This charge reversal of the surface can occur a number of times during the life of the bed.

Adhesion: Organic molecules are deposited on the sand grain surfaces, particularly in the upper layers of the bed. These become a food source for bacteria and other microorganisms. This builds up into the biofilm, the zoogleal layer, which is gelatinous in nature and allows particles to stick to it. The build-up of organic molecules and microorganisms is greatest at the surface where it forms an active layer, the schmutzdecke or dirt layer. Biological activity involving different organisms also takes place throughout all sand layers in the bed.

Chemical and Microbiological Actions

Although physical and electrochemical actions help remove particles from water, chemical, microbiological and biological reactions have a greater influence and help purify water. SSFs are also bioreactors. As organic matter and microorganisms collect on and in the surface layers of an active layer or schmutzdecke, the dirt layer builds up. This will consist of bacteria, algae and meiofauna (nematode and oligochaete worms) as well as protozoans, rotifers, flatworms and copepods, all of which thrive in this layer. All of these factors play an important role in purifying water (Bellinger 1968, 1979).

On the surface of the sand, where larger particles, both living and dead, accumulate, the schmutzdecke develops. Depending upon the biology of the raw inflow water, different organisms may dominate. For example, Bellinger (1979) reported a range of unicellular and filamentous algae growing in irregular patches on the surface of SSFs. Different dominant algal species could affect the overall composition of the schmutzdecke (Bellinger 1979). Adhering to the surface of the sand grains both at the surface and throughout the filter medium, a jelly-like layer forms, a zoogleal film. This consists of various species of bacteria and other microorganisms. Bacteria use some of the organic matter as food material, which they oxidise to provide energy that they need. This breakdown of compounds is called dissimilation. Some of the material and energy produced is used to produce new cells. This is assimilation. The materials that are not used in the schmutzdecke together with the brake-down substances are carried down lower into the bed where many are acted upon by bacteria in the zoogleal film. Any degradable organic matter, including that in the incoming water and produced as dissimilation products by the bacteria and other parts of the schmutzdecke flora and fauna, will be gradually broken down into their inorganic components, i.e. water, carbon dioxide and simple inorganic salts. These may pass out in the filtered water. As most of this bacterial activity occurs in but is not confined to, the uppermost layers, when the top two centimetres or so are skimmed off during cleaning the whole of the ecosystem in the upper layers, including bacteria, contained in that layer will be removed. This means that when the bed is brought back into operation after cleaning, it needs time for these ecosystems to redevelop, hence the need for a ripening period. Bacteria living in the lower depths will not be affected by cleaning. At these lower depths, below 30–40 cm, biochemical reactions still occur. Bacteria such as *Rhodococcus ruber* are well known for breaking down many organic pollutants and can act in a beneficial way for cleansing water (Calvo-Bado et al. 2003). They also reported the supressing effect of some bacteria on other, potentially unwanted, microorganisms. Another important function of the zoogleal film is that it adsorbs colloidal material from water. In addition, the extracellular polymeric substances (EPS) produced by the bacteria in the film act to flocculate organisms and destabilise clay, facilitating removal. Larger predatory organisms such as protozoa and even worms remove bacteria both in the schmutzdecke layer at the surface and lower in the bed (Haarhoff and Cleasby 1991). The removal of pathogenic microorganisms and their cysts, such as *E. coli, Giardia* and *Cryptosporidium*, also occur. For this diversity of organisms to develop, it is essential to create a 'steady state' within the bed. This, together with the longer retention time, allows the relevant populations to develop and are often in the lower depths of the bed. Some are slower growing, e.g. those involved in

Table 7.19 Biochemical contaminant removal mechanisms in slow sand filters.

Contaminant type	Contaminant
Natural organic substances	Colour, taste, bio-fouling substrates
Trace organics	Geosmin, pesticides, endocrine disruptors, 2-methylisoborneol (MIB), methyl tert-butyl ether (MTBE), cyanobacterial toxins
Inorganic chemicals	Nitrates, ammonia, iron, manganese, chlorates and perchlorates

This list is not comprehensive.

the nitrogen cycle, and need more stable conditions to flourish. Examples of the types of biological and biochemical reactions that can occur in SSFs are provided in Table 7.19.

It is important to remove natural and other organic matter from the water as this reduces the chance of producing disinfection by-products (DBPs) at this stage of disinfection.

The removal of cyanobacterial toxins has also been demonstrated in SSFs (Ho et al. 2006; WHO 2011). The WHO also suggests PAC with a contact time of >30 mins as being very effective. As outlined in Table 7.19, other groups of compounds can also be removed. These are discussed in more detail below.

Natural Organic Matter

The term NOM covers a wide range of naturally occurring compounds, including humic and fulvic acids as well as carbohydrates. The problem with them is that if sufficient amounts are present, they can react with chlorine and ozone disinfectants to form unwanted by products (DBPs). These are unwanted products formed by the action of strong oxidants, such as chlorine or ozone, that react with some of the organic molecules present in the water.

Some, such as THMs, are related to problems such as some cancers. These are usually the most frequent by products of chlorine reacting with organic molecules.

Another commonly occurring DBP is haloacetic acid (HAA). The formation of DPBs is a function of temperature, concentration of organic molecules and concentration of disinfectant (WHO 2011). Most THMs are volatile and can eventually be transferred to the air where they form another possible route of exposure to drinking or taking a shower. The WHO and USEPA have recommended guideline concentrations for these compounds, which range from 60 to 300 μg/l depending upon the compound (WHO 2011). The USEPA Disinfectants by-products Rule requires annual averages at every point in the distribution system to comply with the predefined maximum contamination levels (MCLs); increased sampling frequency for larger communities and their works and any hotspots in the distribution system showing higher concentrations of DBP need to be identified (Zu and Bates 2013). DBP precursors can be removed in a variety of ways, including the use of activated carbon and pre-ozonation. Rittmann et al. (2002), for example, showed that GAC/sand filters gave better performance than anthracite sand.

Removal of geosmin and other taste and odour compounds.

Taste and odours in drinking water are an increasing problem in many waters worldwide (Zu and Bates 2013). Some of the compounds responsible are products of certain algae and bacteria, e.g. geosmin (trans-1, 10-dimethyl-trans-9-decalol) and MIB (2-methylisoborneol). Some algal blooms of cyanobacteria and certain other groups in reservoirs can give rise to these tastes and odours (T&Os) (Palmer 1962). These can be detected by many humans at very low concentrations (5–10 pp trillion) and can give rise to customer complaints. Some of the main offending algae are listed in Table 7.20. These tastes and odours can be wholly or partially removed by the biological processes found in SSFs. Many bench-scale studies have confirmed the ability of sand filters to remove this T&O from water.

Some of these are listed in Tables 7.20 and 7.21.

From the results shown in Table 7.21, the best removal results were from filters where stable mature microbiological populations had developed. Ho et al. (2006) identified four species of bacteria, and *Pseudomonas* sp., *Alphaproteoacterium, Sphingomonas* sp. and a species of *Acidobacteriaceae* sp. were the most likely species involved. The most likely processes were

Table 7.20 T&O-producing algae.

Algal genus	MIB producer	Geosmin producer
Cyanobacteria		+
Anabaena		+
Aphanizomenon		+
Microcystis	+	+
Oscillatoria	+	+
Phormidium	+	+
Phormidium	+	+

Other algal genera reported to produce taste and/or odour in water:
Diatoms: *Asteerionella, Synedra* and *Tabellaria*
Green algae: *Gomphosphaeria, Hydrodictyon, Staurastrum* and *Volvox*
Charophyta: *Chara* (Stonewort)
Chrysophyta: *Dinobryon* and *Synura*
Cinophyta: *Ceratium* and *Peridinium*
Source: Data from Chorus and Bartram (1999), Ho et al. (2006), Elhadi et al. (2006) and Palmer (1962).

Table 7.21 Geosmin, MIB and taste and odour removal by biological filters.

Percentage removal	Reference source
60% geosmin and 40% MIB at 20 °C	
36% geosmin and 16% MIB at 8 °C	Elhadi et al. (2006)
60% geosmin and 40% MIB with 5-month-old sand	
Below detection limit with 26-year-old filter	McDowell et al. (2007)
95% geosmin and 95% MIB with 30-year-old filter	Ho et al. (2006)

various biodegradation pathways. Ho et al. (2006) also concluded that biodegradation rates increased when the biofilm had been pre-exposed to MIB and geosmin.

Removal of iron and manganese. Iron is one of the most abundant metals in the earth's crust and usually occurs with iron. In the lower zones of lakes, the water is usually low in dissolved oxygen and as a result often contains raised concentrations of iron and manganese. Similarly, groundwaters often contain elevated concentrations of both elements. Iron can be found in natural freshwaters at concentrations ranging from 0.5 to 50 mg/l (WHO 2011). It can also be derived from water treatment processes using iron coagulants or in the treated water distribution system when iron pipes are used. Manganese also occurs naturally in freshwaters, often along with iron and especially when dissolved oxygen concentrations are low. Both iron and manganese are essential elements in human diets. They both can, however, give rise to problems concerning colouration, taste in water and staining even at fairly low concentrations. Manganese concentrations above 0.1 mg/l can cause staining of laundry and black deposits in pipes, which can dislodge, giving rise to black flakes in the water. They do not usually cause a health risk.

Both iron and manganese can exist as two valences, Fe(II) and Fe(III) and Mn(II) and Mn(IV). Of these, Fe(II) and Mn(II) are much more soluble than the other two. The result is that removal by filtration media is small for the lower valence compounds of these two unless a strong oxidant, such as ozone, is added prior to filtration to oxidise the lower valences to higher ones (Zu and Bates 2013):

$$2Fe^{2+} + 3/2O_2 = Fe_2O_3$$

$$Mn^{2+} + O_2 = MnO_2$$

There are two main mechanisms that are involved in the removal of iron from water: physicochemical and biological. Iron is widely used industrially and found in most soils, rocks and waters. Manganese is also found widely in the environment. In surface waters in the United States of America, the median level was found to be 16 μg/l (US Geological Survey, National Water Quality Assessment Program). It is also used industrially as well as in an organic form (methylcyclopentadienyl manganese tricarbonyl; MMT) as an octane enhancing agent in unleaded petrol in Canada, the United States and the Americas. A survey in Germany indicated that 90% of drinking water supplies to households had less than 20 μg/l of manganese (Bundesgesundheitsamt 1991). Both manganese and iron can bioaccumulate in biota, especially in lower organisms such as phytoplankton, molluscs and some fish (ATSDR 2000). Manganese intake by humans is usually much lower from water than from food, and oral intake is regarded as one of the least toxic metals.

Iron can impart a taste to drinking water but usually only at concentrations above 0.3 mg/l. Concentrations between 0.3 and 3 mg/l are considered acceptable. Staining of laundry can occur at concentrations above 0.3 mg/l. Food is the main human intake of iron, with an intake of 10–14 mg, whereas water will only contribute 0.6 mg daily. As an essential element in the human diet, the minimum daily requirement for iron is about 10–50 mg/day (FAO 1988). The average lethal dose of iron is in the range 200–250 mg/kg of body weight (National Research Council 1979). The guidelines for Canadian drinking water are 0.3 mg/l for iron with a target concentration of 0.05 mg/l and 0.05 mg/l for manganese with a target concentration of 0.01 mg/l. For the European Union, the permitted levels are 0.2 for iron and 0.05 for manganese.

Iron can be removed from water during the treatment process by oxidation by two main processes, either directly by aeration or ozone injection or by oxidation of iron adsorbed onto the sand grains in the SSF or by biological mediation. The oxidation of ferrous to ferric iron reaction by iron bacteria is shown below:

$$4Fe^{2+} + O_2 + 10H_2O \leftrightarrow 4Fe(OH)_3 + Energy$$

Several types of iron oxidising bacteria may be involved in the process (Ankrah and Sogaard 2009), including *Leptothrix, Gallionella, Sphaerotilus* and *Crenothrix.* There are still unanswered questions regarding the exact involvement of iron bacteria, but the types mentioned above are also responsible for the oxidation of manganese. The oxidation of manganese and iron does, however, require different conditions. From the figure, it can be seen that if both metals need to be removed, this is best done separately. As with most treatment methods involving biological reactions, it is essential to give the filter time to seed with appropriate bacteria and for an active population to develop.

In addition, because a number of species of iron bacteria may be involved, there is a good chance that at least one of them will thrive in the conditions provided. The results for various pilot studies are given by Gage et al. (2001), and case studies of actual works removing iron and manganese are provided in Mouchet (1992). The general conclusion was that biological removal treatment for these metals offered a good alternative to conventional treatment and is used in many countries.

Ammonia and nitrate removal nitrates are significant pollutants in surface and groundwaters in many countries, including Saudi Arabia, North America and Australia (Mohseni-Bandpi et al. 2013). Ammonium is also a common pollutant in surface waters and aquifers largely through the increasing use of fertilizers in agriculture (Tekerlekopoulou and Vayenas 2007). Natural levels of ammonia in both surface and groundwater are below 0.2 mg/l although these may rise to 3.0 + mg/l in anaerobic groundwaters (WHO 2011). Where animal wastes from agriculture pass into surface waters, concentrations may be much higher. Elevated levels of ammonia are usually taken to be an indication of sewage or animal waste contamination. Normal environmental concentrations are not a problem in drinking water, and the WHO has not issued a guideline value for them.

Although not normally a health risk, it can adversely affect disinfection efficiency in drinking water treatment as well as affecting the removal of manganese and cause taste and odour problems. Nitrates are found naturally in water and are a vital plant nutrient. Nitrates are an integral part of the nitrogen cycle, as are nitrites, although these are not usually present in raised concentrations unless anaerobic conditions prevail. Higher nitrates in surface waters and groundwaters can be a problem through the use of fertilizers in intensive agriculture. In surface waters, promotion of the growth of algae and other aquatic plants can be a problem and together with increased concentrations of phosphates can give rise to severe eutrophication. Surface water concentrations can fluctuate rapidly because of uptake by algae, denitrification by bacteria and runoff from fields. Groundwater concentrations tend to be much more constant. Drinking water may provide a significant intake of nitrates for humans, especially with bottle-fed infants (WHO 2011). The guideline value for nitrate–nitrogen is 11 mg/l or for nitrate (NO_3) 50 mg/l. For nitrite–nitrogen, the guideline is 0.9 mg/l. This is to protect bottle-fed infants against methemoglobinemia. The best way to prevent nitrate and nitrite contamination is to control sources such as the use of fertilizers and control sewage discharges of all types into surface waters. For drinking waters, the simplest control if elevated concentrations are present is to blend the source water with other low-concentration waters, thus lowering the final concentration to below guideline levels. However, the removal of both ammonia and nitrates does occur in sand filters and biological removal still seems to be an effective method of removal as other physical and chemical methods, such as ion exchange, RO and nanofiltration, which show poor selectivity for removing nitrate specifically and produce waste rich in nitrate and brine that can be difficult to dispose of (Mohseni-Bandpi et al. 2013). SSF offers a less expensive and efficient method of removal. It was also found that most NO_3 was removed in the top 10 cm of the filter and corresponded to 98% removal. NO_2 concentrations decreased beyond 10 cm and were below guideline limits in the filtrate. Biological filtration produces a harmless by product – nitrogen gas, which is dispersed in the atmosphere. The bacterial transformations of the nitrogen cycle in a filter require a carbon source, which may be in short supply in the raw water feed. Organic compounds such as sugar, methanol, ethanol and acetic acid in controlled concentrations have been used as additives.

In addition, several inorganic compounds, such as sulphur and hydrogen, may be used. The use of organic electron donors is required by heterotrophic nitrogen cycle bacteria. Examples of the types of energy reactions are as follows:

$$6NO_3^- + 2CH_3OH - 3NO_2 + 2CO_2 + 4H_2O$$

$$6NO_2 + 5CH_3OH - 3N_2 + 3CO_2 + 3H_2O + 6OH^-$$

Methanol, which is commonly used as a carbon source, should only be used in small quantities.

Mohseni-Bandpi et al. (2013) suggest that the ratio of methanol to nitrate needed to meet the requirement for denitrification can be calculated using the following equation:

$$C_m = 2.47(NO_3 - N) + 1.53(NO_2 - N) + 0.87(DO)$$

where

C_m = mg/methanol required for denitrification
NO_3–N = Initial nitrate–nitrogen concentration in mg/l
NO_2–N = Initial nitrite–nitrogen concentration in mg/l
DO = Initial dissolved oxygen concentration in mg/l

The problem with using methanol is its potential health hazards. Hence, the residual concentration must be kept below 100 mg/kg of body mass of the consumer. Many have concluded that the quantity of methanol required for denitrification is only about 10% of that from other sources (Liessens et al. 1993).

Arsenic removal

Arsenic is present in many groundwaters and is also used in many processes, such as electronics, wood preservation and agriculture, and can thus be a pollutant in surface waters. Arsenic(III) is more toxic than arsenic(V), both of which are more toxic than methylated arsenic compounds. In addition, metals, unlike organic compounds, cannot be degraded and thus remain a permanent toxic hazard.

Arsenic is present in many waters, especially groundwaters, and is used for domestic supplies in Canada, the United States of America, China and many other countries (Nordstrom 2002) often exceeding national and international guidelines.

The problem is greater in small community supplies rather than supplies for large municipalities.

There are a number of physico-chemical treatment processes available for reducing arsenic concentrations, and they vary greatly in the use of chemical additives, cost, waste products and ease of operation. Removal of arsenic by biological means can be an important mechanism involving naturally occurring bacteria.

These bacteria are able to reduce, oxidise or absorb a range of metal contaminants (Zouboulis and Katsoyiannis 2002). As some arsenic compounds are very soluble, they are bioavailable and can thus be absorbed and transformed by certain bacteria. A number of metals can be removed by bioreduction, including chromium, uranium and selenium (Lovley 1995). In the reduced form, these metals are easier to remove. Metals such as iron, manganese and arsenic are more easily removed in an oxidised state as this produces insoluble compounds that can then be removed by filtration (Mouchet 1992). Iron and manganese can then be effectively bio-removed from water, and in addition arsenic can be removed at the same time (Karsoyiannis et al. 2002).

The biogeochemical cycle of arsenic can involve several physico-chemical transformation processes, such as oxidation–reduction, adsorption–desorption and precipitation (Duarte et al. 2009). There are also biological actions, such as As(V) reduction, As(III) oxidation and a number of methylation reactions (Lievremont et al. 2009). In mildly anaerobic waters, the predominant arsenic species is As(III), which is soluble in water and more difficult to remove by conventional treatment methods such as coagulation/filtration, ion exchange, RO and lime softening (Jekel 1994; Zouboulis and Katsoyiannis 2002). The As(III) needs to be oxidised and made insoluble to be removed, and it can be co-removed at the same time as Fe(II) with an effectiveness of 90%+. Zouboulis and Katsoyiannis (2002) also demonstrated that live bacteria were essential for removal and that the removal process was greatly reduced when they inactivated the biota in a filter using chlorine disinfection. Because of the adverse health effects, the limit for arsenic in drinking water has been lowered by WHO (2011), European Commission (b) (1998) and the USEPA (2002) as well as many other countries from 50 to 10 μg/l. It is also true that arsenic removal is low without the presence of iron in the water.

Of the techniques available, sand filtration has been shown to give satisfactory results. Arsenic can be removed by SSF, especially in systems containing layers of activated carbon interspersed with sand. Removals of over 90% have been reported (Pokhrel et al. 2005). In these tests, pre-oxidation was not found to give any greater benefits. Pokhrel et al. reported that the biological oxidation of associated iron compounds was many times faster than chemical oxidation reactions. Although other treatment removal technologies are in use, biological treatment processes are gaining interest as they do not require the use of added chemicals. A recent development by Manz (2011) is to use an SSF with a smaller depth of sand and a smaller head of water above the sand. The filter is still designed to be a biological filter, but cleaning the filter is easier as the shallower depth of sand allows the bed to be backwashed rather than the surface layers needing to be skimmed.

Cleaning a Slow Sand Filter

SSFs, as with any other filter, will gradually clog up as more and more particles are retained and the through flow is reduced. The rate of flow reduction is measured as headloss (see also RGFrs). When the headloss reaches a pre-determined value, meaning that the flow-through the bed is not acceptable, the bed is taken out of service and cleaned. The first stage is to shut off the incoming water. Then, the water level is reduced to about 10 cm below the sand surface level. The beds should not be left for longer than is necessary before skimming to avoid scavenging birds that could, with their droppings, pollute the sand and in searching for invertebrates disturb the sand surface. As soon as the smutzdecke and top sand are dry enough, the surface layers are skimmed off to a depth of about 1–10 cm. Skimming can be carried out manually or mechanically. Manual skimming with rectangular flat-ended shovels could be used with smaller beds, but mechanical skimmers are more likely to be used for larger beds. The first fully mechanised skimming was used by the then Metropolitan Water Board London (now Thames Water) (Lewin 1961). Hand cleaning is relatively labour intensive, but mechanical cleaning requires specialised equipment. Care should be taken in either case to minimise disturbance of the lower sand layers. All of the dirty skimmed sand is removed and transported away for cleaning. It is important to recognise that the remaining sand layers are undisturbed as they will contain an established biomass. Any vehicles used in mechanical skimming must have a low surface loading (less than 33 kN/m^2: Huisman and Wood 1974) with the sand so as not to compact the remaining layers. Manual skimmers may need to be provided with protective walkways. Depending upon climatic conditions and the composition of the raw water, varying amounts of algal growth can occur in the supernatant water and across the sand surface. If this is largely of filamentous algae these form a mat which can be rolled up into sections to be removed before skimming proper takes place. If, however, a substantial layer of more gelatinous algae occurs, cleaning them off can be more arduous. Once the dirty sand and schmutzdecke have been removed for cleaning, the bed can be refilled to its operating level and left to ripen for a couple of days, allowing the biomass to recover before putting the bed back into use. After the ripening period, the bed was run, and the filtrate was tested to ensure that water was of a high enough quality before putting it back into service.

The removed sand and schmutzdecke layer are now washed to remove all dirt. The resulting clean sand is then stored for reuse. The dirty washwater must now be treated before disposal. Cleaning is typically only required every one or two months, but this depends on the quality of the supernatant inflow water. For illustrations of mechanical cleaning equipment, see Huisman and Wood (1974). After a much longer period of use and skimming, the bed may need to be re-sanded. After each skimming, the depth of sand will be reduced, the time will come when its minimum working depth (usually about 30–50 cm) will be reached, and new sand must be added to restore its proper depth. To do this, the bed is first cleaned as described above, and then the most common approach is to use a trenching method. In this, the bed is divided into strips. Each strip is dug out taking care not to disturb the underdrains, and sand is removed. The removed sand can either be taken away for cleaning or placed aside to be added as a top layer when the lower layer has been replaced with fresh clean sand. Digging out the lower layers and replacing them with clean sand is continued strip by strip until the whole bed as had the sand replaced to the correct depth either with wholly clean sand or with a combination of clean and a layer of older sand removed from above the underdrains (Environmental Protection Agency, Ireland 1995;. Huisman and Wood 1974). The advantage of reusing the bottom layer of sand and placing it on top is that it will contain and establish biota and thus reduce the time needed to ripen the bed. Once the bed has been re-sanded, the surface needs to be smoothed and level. The bed can then be refilled with clean water backwards through the underdrains. A period of ripening of the resanded bed will be require to allow the microorganisms to recolonise the added clean sand. Once the bed has been refilled and ripened, raw water can be allowed to flow in at the top and down through the sand. The filtrate should be checked for quality, and if satisfactory, the bed can be put back into use.

Several skimming and in situ sand cleaning machines have been tried, e.g. in Holland and France, with varying success (Huisman and Wood 1974), and although they may potentially save on labour costs, the cost of the equipment is often quite large. Many works still use more traditional methods.

Disinfection

After the various methods of treatment indicated previously, the water must be disinfected. Disinfection is an essential barrier to the passage of pathogenic organisms from treatment processes to consumers' tap. Although disinfection cannot guarantee the complete destruction of all pathogens, it will greatly reduce the risk of disease and aims to at least inactivate most potential pathogens, and certain disinfectants offer some protection throughout the distribution system. For disinfection to be effective, flocs and particulate matter should have already been removed from the water prior to the disinfection process (WHO 2011). A number of disinfection approaches are available, but whichever one is chosen, it is essential that it is operated correctly and continuously to provide an effective barrier and meet the required drinking water standard. There are a number of disinfection products in common use. The main ones are listed below;

- Ultraviolet radiation: UV light of sufficient intensity and duration can destroy organic molecules and disrupt organisms. It shows no residual action.
- Ozone: It is a powerful oxidant. It can oxidise organic matter and destroy bacteria. It shows no residual action. It is not long lasting and must be generated as required.
- Chlorine and chlorine compounds: These are powerful oxidants, but not as strong as ozone. They show a residual effect. They can be manufactured and stored and are relatively easy to handle.

Each of these is considered individually.

Ultraviolet Radiation

UV radiation was first used for disinfection in the United States of America in 1916, but it was expensive. Over the decades costs have decreased, and it is now an effective method of disinfection, especially in small and household supplies as it does not require the addition of any chemicals. UV light, at the correct intensity and duration of exposure, disrupts

Table 7.22 UV dose requirements to give a 4.0 log inactivation.

Organism	Dose
Giardia cysts	22
Cryptosporidium oocysts	22
Hepatitis A	21
Rotavirus SA11	36
E. coli	6–8
Vibrio cholerae	9
Legionella pneumophila	9.4
Streptococcus faecalis	11.2

Source: Taken from USEPA UV Manual (2006); Environmental Protection Agency of Ireland (2011).

microorganism cells and destroys a wide range of various types (see Table 7.22). The optimum wavelength of UV light is between 250 and 270 nm (National Drinking Water Clearinghouse 2000). Most UV reactors will deliver a dose of about 40 mJ/cm^2 which is adequate to inactivate most of the frequently occurring microorganisms.

A typical UV reactor consists of a UV-generating lamp embedded in a tube with the water to be disinfected flowing around and past (Figure 7.11). UV radiation is rapidly absorbed by water, so it is important not to exceed the required depth passing over the lamp. UV penetration is affected by a number of impurities in the water, including turbidity, iron and manganese in the water and colour. Knowing the amount of impurities present, the UV transmittance can be calculated (see EPA Ireland 2011). Another potential problem is that the quartz glass tube of the UV lamp can be fouled, particularly by inorganic compounds, which will partially block the light. If this is likely to occur a procedure for regular cleaning should be installed. This problem is usually greater in hard water areas. It is also essential to remove as much colour, organic particles and colloidal matter as possible as these could shield pathogens from radiation. Colour and dissolved organic matter can be a problem for UV disinfection systems being used after SSFs, as they may not have been removed. The contact time between the water and UV radiation is important in determining the dose. It is thus important to control both the radiation intensity and the water flow rate past the lamp.

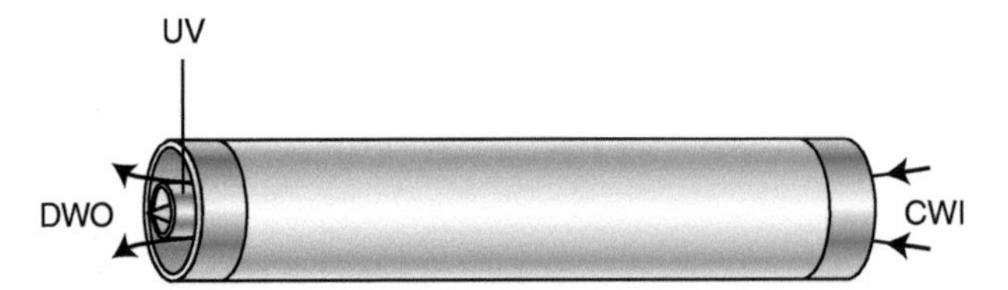

Figure 7.11 An example of a UV lamp. CWI, clean water inflow; DWO, outlet for disinfected water; UV, the UV lamp which is enclosed in a quartz glass tube for protection. This could be installed for a single household or as a multiple array for a large building or group of houses with a centralised supply. Note: There must be an electricity supply available.

The disadvantages of UV disinfection systems are as follows:

- There is no lasting disinfectant residual.
- They need a constant power supply.
- The water to be treated must be of suitable quality.
- They can be subject to fouling.

The advantages are as follows:

- There are no toxic by-products.
- No danger of emissions of products to the atmosphere.
- Produce no taste or smell.
- No storage of hazardous materials.
- Compact and only need a small space.
- Do not affect the mineral composition of the water.
- Minimal environmental impact.

Ozone

Ozone is a powerful oxidising agent capable of destroying microorganisms and breaking down organic molecules in water. It can be used for disinfection of drinking water and de-odourisation of wastewater in wastewater pumping stations. It potentially has greater effectiveness against bacteria and viruses than chlorine and helps remove iron and manganese in water due to its oxidising ability.

Ozone has been used in drinking water treatment since the late nineteenth century and has become quite widely used in Europe and, subsequently, Asia. It was not widely used in the United States of America until the middle of the twentieth century when it was introduced in Los Angeles. Its use grew for larger works in the 1990s, being used in over 90 treatment processes (Langlais et al. 1991). This number is increasing annually with the largest increase among small treatment plants. Some are using it for colour removal, but the majority use is for disinfection.

Ozone is a gas at room temperature and is highly corrosive, toxic and only sparingly soluble in water compared with chlorine. Ozone is an unstable molecule that quickly degrades to oxygen ($3O_2 \leftrightarrow 2O_3$). To disinfect 1 m^3 of reasonably clean water, about 0.5–1.0 g of ozone is required (Casey 1997). The formation of O_3 from O_2 requires a large input of energy. The main method in use is corona discharge. This involves passing air or oxygen past two electrodes paced apart in parallel. A voltage is applied to the electrodes, which produces an electron flow across the gap between the electrodes, the discharge gap. The electrodes are sometimes protected in a glass or ceramic tube. The electrical discharge is between 4000 and 10 000 V (Stevenson 1997). This discharge produces the energy required to cause the O_2 molecules to disassociate and reform into O_3, ozone. In an ozone disinfection system, a supply of air or oxygen is required as well as the generator. The ozone is then fed into a water contacting chamber, which typically has several compartments and then a means of destroying any excess ozone before release to the atmosphere. A typical arrangement is shown in Figure 7.12. If an air feed is used, it is essential that the air should be both clean and dry. It should be free from particulates and droplets (USEPA 1999). Electrical energy is

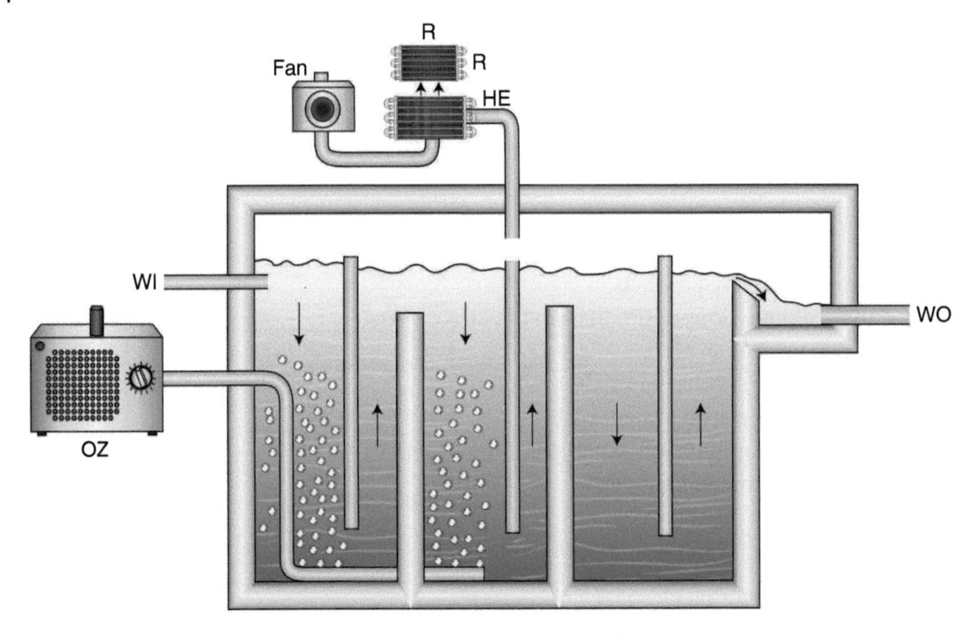

Figure 7.12 Ozone dosing. OZ, ozoniser; WI, water inlet; R, reheater; HE, heat exchanger; WO, water outlet. Note that the heat exchanger, re-heater and fan are for ozone destruction. Sodium bisulphite can also be dosed at the water outlet. The first chamber in the tank provides oxidation and the second disinfection. In the final chamber, the remaining ozone decays.

required to generate ozone, and a large proportion of this is lost as heat. This excess heat must be removed by cooling, usually with water. The contactor may be of different types, but a typical arrangement is that of a bubble diffusor contactor. This consists of several chambers although the number can vary. Ozone is introduced at the bottom of the first and sometimes also the second chamber. The water flows through the chambers, and the ozone is dissolved into the water with an efficiency of up to 90%. The contact time is around four minutes. Any gas not dissolved is passed into an ozone destruction unit. The advantage of a bubble diffuser is that it has no moving parts, an excellent gas transfer and is simple to operate (USEPA 1999). Ozone can also be injected into water, and this method is frequently used in Europe and the United States of America. The ozone destructor changes the O_3 back to O_2 before being released back to the atmosphere. Ozone does not have a residual amount carried into the distribution system, so there is no lasting disinfectant effect. The disadvantages of using ozone as a disinfectant are as follows:

- Higher equipment costs.
- No lasting disinfectant residual for the distribution system.
- There could be unwanted and carcinogenic by-products, such as bromates, aldehydes, ketones and carboxylic acids.
- There can be potential hazards associated with ozone generation.

The above are taken from Water Research Center (2017).

Pre-ozonation is sometimes used to destroy either microorganisms or to oxidise any iron or manganese present before chemical coagulation treatment. Where ozone is used as the primary disinfectant, a secondary disinfectant such as chlorine must be added to minimise and microbial growth in the distribution system.

Ozone does not form THM by-products that could be of concern with chlorination. It may, however, give rise to other organic and inorganic by-products. For example, if bromine is present in the water, brominated DBPs can be formed, which can pose higher health risks than chlorinated ones (USEPA 1999). The Canadian Drinking Water Quality has established a maximum interim acceptable concentration for bromate of 10 μg/l. Other possible by-products of ozone disinfection include aldehydes, ketones and some acids (Earth Tech (Canada 2005)).

Chlorine

Chlorine has been the most widely used disinfectant for drinking water treatment. It was first used in the late nineteenth century and has since been used worldwide. It is a strong oxidant and is effective at destroying microbial pathogens. Chlorine is a brown/orange gas heavier than air. It is extremely toxic. Chlorine reacts with water to form hypochlorous acid (HOCl) and hydrochloric acid (HCl). HOCL is a weak acid, and it quickly dissociates, producing hypochlorite ions that act as disinfectants:

$$Cl_2 + H_2O \rightarrow HOCl + H^{=} + Cl$$

$$HOCl = H_{=}OCl$$

The term available chlorine refers to the sum of the concentrations of the molecular chlorine (Cl_2), HOCl and hypochlorite ion (OCl).

Free available chlorine is a term used to the cumulative sum of the molecular chlorine, HOCL and hypochlorite ion ($Cl_2 + HOCl + OCl$). The proportions of each depend upon the pH. As HOCl is a more effective disinfectant than OCl and HOCl predominates at pH 6, chlorine disinfection is more effective at this lower pH than at more alkaline levels.

Several chlorine compounds, including sodium and calcium hypochlorite (NaOCl and CaOCl), are also used as disinfectants as the former is available in aqueous form and the latter in solid form.

When disinfection with chlorine is considered, the need for plumbosolvency control must also be taken into account. Plumbosolvency control usually involves raising the pH of the water to about pH 8, which causes carbonates to be deposited on the inside of pipes, reducing the possibility of lead dissolving into the water. If such treatment is proposed this must occur after chlorine disinfection as this requires a lower pH and is less effective at the higher one.

There are a number of chlorination technologies that can be used for drinking water disinfection. These include chlorine gas, sodium hypochlorite and calcium hypochlorite. For medium- to large-scale treatment processes, chlorine gas is the most likely choice. Chlorine gas is supplied in drums or cylinders, which must be stored in a secure building. As chlorine is a highly toxic gas, there are strict health and safety regulations regarding its storage and use. Drums and cylinders are heavy, so proper provisions must be made for their handling. All confined spaces where chlorine is being used must be identified by warning signs. To obtain adequate disinfectant protection in the distribution system, a target residual must be set and monitored. The chlorine residual is likely to fluctuate for many reasons, for example raw water quality, pH, turbidity and temperature (WHO 1993, 1996), so there should be an accepted operating range outside which alarms must be activated. As the action of chlorine is not instantaneous in killing pathogens, it must be given time to work. This is called the contact time. To provide this, a contact tank is provided after the chlorine has been added to allow it to act. The length of the contact time depends upon the pathogens to be controlled and the water quality.

For chlorine gas dosing and to minimise any danger, the mechanism is designed to operate under vacuum (Environmental Protection Agency, Ireland 2011). The chlorine gas is fed through a chlorinator that injects it into a monitored flow of water. This leaves the chlorinator as a chlorine solution (HOCl). This chlorine solution is then mixed with the water to be treated and passed into the contact tank. The contact time will vary depending upon the situation but typically can range from 30 min to 4 h. The residual chlorine in the distribution system would typically be targeted at 0.2–0.3 mg/l.

Breakpoint chlorination is the most frequently used technique to ensure that the chlorine residual is sufficient for adequate disinfection within the distribution system. The amount of free residual chlorine varies depending upon the amount of ammonia present. When chlorine is added to water, it will react with many substances already present. This amount is called the chlorine demand. Ammonia reacts with chlorine to form chloramines. Figure 7.13 shows a graphical representation of the amount of chlorine residual relative to the chlorine: ammonia weight ratio. The first chlorine added is used in reactions with compounds in the water, such as iron and manganese, so no residual is left. This is zone 1 in Figure 7.13. When this residual demand is satisfied and depending upon the amount of ammonia in the water, chloramines are produced. As the amount of chlorine increases the chloramines are oxidised by the chlorine (zone 2 in the Figure 7.13). When these reactions

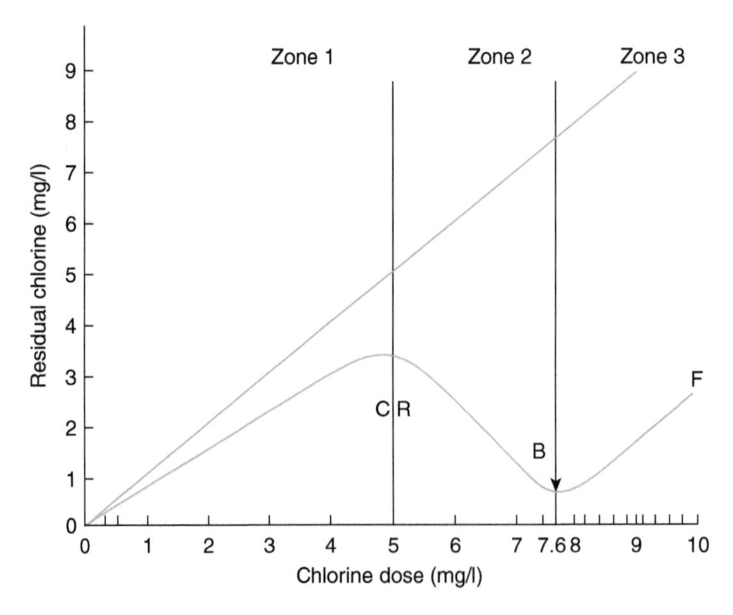

Figure 7.13 A theoretical breakpoint curve for chlorine dosing of water for disinfection.

are complete, the breakpoint occurs, and from then on, added chlorine produces a free chlorine residual (zone 3 in the Figure 7.13). Line A represents the chlorine concentrations that would be expected in pure water, and line B represents those in average treated water, which will have various compounds naturally present.

Disinfection By-products of Chlorination

Although chlorine is an effective disinfectant concern has been raised about possible harmful side effects of certain by-products. Over 300 possible by-products have been reported in the literature and as analytical techniques improve, more are being identified. Chloramines form when aqueous by-products have been reported in the literature (Richardson 2000), and analytical chlorine (HOCl) reacts with ammonia. Chloramines are not generally strong disinfectants but are quite good against bacteria. They are also stable and provide a long-lasting residual disinfectant in the distribution system where they can control biofilms on the inside of pipes and any organisms in water ingress. A number of chloramine compounds can be formed in the chlorination reaction. Monochloramine is useful in the distribution system, but the higher concentrations (dichloramine and trichloramine) can cause taste and odour problems. Chloramines are known to reduce the formation of THMs and other DBPs. Other by-products, such as HAAs, may also be formed that may have health effects (Nikolaou et al. 1999). Chloroform is probably the most common THM and the main by-product of chlorine disinfection. Where bromides are present, brominated THMs are formed preferentially to chloroform. These include bromoform and bromodichloromethane. Chloroform has a WHO drinking water guideline of 300 μg/l. For bromoform, it is 100 μg/l, for dibromochloromethane 100 μg/l and for bromodichloromethane 60 μg/l (WHO 2011).

As most THMs are volatile, it is assumed that most of those present will be lost to the water and transferred to air. Because of this, individuals may receive higher doses when in the shower for any length of time than simply from drinking chlorinated water. To a lesser extent, this also applies to taking a hot bath.

Sodium Hypochlorite and Calcium Hypochlorite Disinfectants

Sodium hypochlorite is manufactured from chlorine and sodium hydroxide and is generally used as an aqueous solution of about 15% w/w Cl_2. Although it can be easier and safer to use than chlorine gas, it is more expensive and bulkier. It is unstable, and with time, it degrades into sodium chlorate. 0.5% caustic soda can be added to improve its stability (Environmental Protection Agency, Ireland 2011). The storage time for bulk hypochlorite solution should preferably be less than 48 h although up to 72 h, should not result in excessive degradation. It is strongly alkaline with a pH of 11–13 and so is corrosive and must be handled with care. Metals should not be used anywhere that sodium hypochlorite is in the system.

Sodium hypochlorite has the advantage of being relatively simple to dose. The most frequently used systems are diaphragm metering pumps.

Calcium hypochlorite is a solid white powder or, for smaller scale use, tablet. The powder form can be dosed using a volumetric feeder that dispensed the required amount into a tank of water accompanied by mixing in order to dissolve it. This solution can then be dosed into the main flow of water. For large volumes of water, sodium hypochlorite is usually favoured as the dosing systems are ore simple to operate. For small systems or individual households, tablets are a convenient method of disinfection.

General Considerations for Disinfection

It is essential before deciding upon the type and amount of disinfection to take into account a number of factors. Each location where disinfection will be applied will have its own characteristics. Some of the considerations for primary disinfection are as follows:

- An understanding of the organic and inorganic contaminants in the water to be treated.
- Regular monitoring of the changes in the above contaminants with time.
- The level of types and abundance of pathogens that are likely to be present.
- The performance of all of the stages of treatment was given to the water before disinfection.
- The expected variations in temperature and pH of the water.
- The performance and efficiency of the dosing system.
- The performance and efficiency of the contact tank.
- Performance monitoring of the entire system for compliance with the relevant national and international standards.

The WHO guidelines for the free chlorine concentration (0.5 mg/l, *C*) and a contact time of 30 min (*t*) is the target value, the *Ct* value. The *Ct* value using the WHO standard would be $0.5 \times 30 = 15$ mg min/l (Environmental Protection Agency, Ireland 2011; WHO 2004). Some general recommendations for Ct values (in mg min/l) exist for bacteria at pH 7 (0.08) and pH 8.5 (3.3), for viruses at pH 7.5 (12) and for Giardia at pH 7.5 and low temperatures (230). These are general guidelines and aim to achieve 99% inactivation. With secondary disinfection where a disinfection boost is given to ensure a target chlorine residual throughout the distribution system, which may occur along trunk mains, service reservoirs, etc., the aim is to maintain at least 0.1 mg/l free residual chlorine even at the extremities of the system (Environmental Protection Agency Ireland 2011). It is important that regular testing for residual concentrations is performed not only at the extremities of the distribution system but also throughout the system.

Sometimes there may be a need to use higher concentrations of chlorine for the purpose of local disinfection, e.g. if there was a need to replace a length of pipes in the system because of a burst or in mains flushing. In such cases, after a suitable contact time the chlorine residual could be too high and give rise to consumer complaints. There are also occasions where chlorinated water may need to be discharged to the environment, and although chlorine is an effective disinfectant and protects human's drinking water, it is toxic to aquatic species in receiving waters. In this case a degree of dechlorination could be required. There are a number of different ways to remove chlorine from water including aeration, filtering through GAC and chemical removal. Aeration is the simplest method although it is slow. It is not effective at removing THMs and is less effective at higher pH values. GAC will remove chlorine by adsorption. It will also remove chloramines, organics and any colour in the water. Sodium thiosulphate is a commonly added chemical for chlorine removal. Other sodium compounds, such as sodium sulphite and bisulphite, can also be used. Sodium thiosulphate is commonly used as it is relatively easy and safe to use. Sodium bisulphite costs less and has a higher capacity for chlorine removal.

References

Ankrah, D.A. and Sogaard, E.G. (2009). A review of biological iron removal. *Thirteenth International Water Technology Conference*. IWTC 13. Hurghada, Egypt, pp. 999–1005.

ATSDR (2000). Toxicological profile for manganese. Atlanta GA, US Dept. of Health and Human Services, Public Health Service. Agency for Toxic Substances and Disease Registry.

Bellinger, E.G. (1968). The removal of algae by microstraining. *Journal of Sociology Water Treatment and Examination* 17: 60–66.

Bellinger, E.G. (1979). Some biological aspects of slow sand filters. *Journal of the Institution of Engineers and Science* 33: 19–29.

Berliner Wasserbetriebe (2016). Water for Berlin, *Neue Jüdenstrasse* 1, D-10179. Berlin.

Betancourt, W. and Rose, J.B. (2004). Drinking water treatment processes for removal of *Giardia* and *Cryptosporidium*. *Veterinary Parasitology* 126: 219–234.

Bundesgesundheitsamt (1991). Umwelt-Survey (Environmental Survey), Vol. 111b. Berlin (WaBoLu-Heft No. 3/1991).

Calvo-Bado, L.A., Pettitt, T.R., Parsons, N. et al. (2003). Spatial and temporal analysis of the microbial community in slow sand filters used for treating horticultural irrigation water. *Applied and Environmental Microbiology* 69 (4): 2116–2125.

Carson, R. (1962). *Silent Spring*. USA: Houghton Mifflin Co.

Casey, T.J. (1997). *Unit Treatment Processes in Water and Wastewater Engineering*, 280pp. Chichester, England: Wiley.

Cech, T.V. (2010). *Principles of Water Resources. History, Development, Management and Policy*, 541pp. Wiley.

Chadwick, E. (1842). The Sanitary Conditions of the Labouring Population and on the means of its improvement. Privately published report available online.

Chorus, I. and Bartram, J. (1999). *Toxic Cyanobacteria in Water*, 461pp. London: WHO, E & F. N. Spon.

Clancy, J.L. and Hunter, P.R. (2008). World Class Parasites.

Corcoran, E., Nellemann, C., Baker, E. (eds.) et al. (2010). Sick Water? He central role of wastewater management in sustainable development. A Rapid Response Assessment. *UNEP/UNHABITAT*.

Culp, R.C. and Culp, G.L. (1971). *Advanced Waste Water Treatment*. New York, USA: VanNorstrand.

Davison, A., Howard, G., Stevens, M. et al. (2005). Water Safety Plans. Managing drinking –water quality from catchment to consumer. *WHO > WHO/SDE/WSH/05.06*. Geneva.

DFID (2008). Water and Sanitation Policy–Water: An Increasingly Precious Resource. www.gov.uk.

Duarte, A.L.S., Cardosa, S.I., and Alcada, A.J. (2009). Emerging and innovative techniques for arsenic removal applied to a small water supply system. *Sustainability* 1: 1288–1304.

Dufou, A., Snozzi, M., Koster, W. et al. (ed.) (2003). *Assessing Microbial Safety of Drinking Water. Improving Approaches and Methods*, 295pp. London: IWA Publishing.

Duncan, A. (1988). The ecology of slow sand filters. In: *Slow Sand Filtration* (ed. N.J.D. Graham), 163–180. Chichester: Ellis Horwood Ltd.

Earth Tech (Canada) Inc. (2005). Chlorine and Alternative Disinfectants. Guidance Manual. Manitoba Water Stewardship, Manitoba, Canada.

Elhadi, S.L.N., Huck, P.M., and Slawson, R.M. (2006). Factors affecting the removal of geosmin and MIB in drinking water biofilters. *Journal American Water Works Association* 98 (8): 108–119.

Environmental Protection Agency (EPA) Ireland (1995). *Water Treatment Manuals; Filtration.* Wexford, Ireland: The Environmental Protection Agency.

Environmental Protection Agency (EPA) Ireland (2011). *Water Treatment Manual: Disinfection.* Wexford, Ireland: EPA, Office of Environmental Enforcement, EPA.

European Commission (b), 1998. Directive 98/83/EC, related with drinking water quality intended for human consumption. Brussels, Belgium

Eurostat (2007). *Europe in Figures; Eurostat Yearbook 2006-7.* Luxembourg: European Communities.

FAO (1988). Requirements of vitamin A, Iron, folate and vitamin B_{12}. Report of the joint FAO/WHO Expert Consultation. Rome.

FAO/WHO (1996). Codex Alimentarius Commission (1996) Report of the Twenty-ninth Session of the Codex Commission on Food Hygiene. (ALINORM 97/13A), 60pp.

Feinstein, L. and Daiess, G. (2019). *Plumbing the Depths: California Without Toilets and Running Water*, 1–17. Oakland, CA, USA: Pacific Institute.

Furukawa, D. (2008). *A Global Perspective of Low Pressure Membranes.* CA, USA: National Water Institute.

Gage, B., O'Dowd, D.H., and Williams, P. (2001). Biological iron and manganese removal. Pilot and full scale applications. *Ontario Water Works Association conference* (3 May 2001).

Gray, N.F. (1999). *Water Technology*, 548pp. Arnold Publishers.

Guchi, E. (2015). Review on slow sand filtration in removing microbial contamination and particles from drinking water. *American Journal of Food and Nutrition* 3 (2): 47–55.

Haarhoff, J. and Cleasby, J.L. (1991). Biological and physical mechanisms in slow sand filtration. In: *Slow Sand Filtration*, 34–98. American Society of Civil Engineers.

Hall, T. and Croll, B. (1997). *Water Treatment Processes and Practices*, 2e, 143–149. UK: WRC Swindom.

Halliday, S. (2009). *The Great Stink of London. Sir Joseph Bazalgette and the Cleansing of the Victorian Metropolis.* Stroud, Gloucestershire, UK: The History Press.

Hashimoto, M., Kunikane, S., and Hirata, T. (2002). Prevalence of *Cryptosporidium* oocysts in the drinking water supply in Japan. *Water Research* 36 (3): 519–526.

Hellier, K. (2000). Hazard analysis and critical control points for water supplies. 63rd Annual Water 2007.

Hendricks, D. (ed.) (1991). *Manual of Design for Slow Sand Filtration.* USA: American Water Works Association Research Foundation.

Hill, R. and Lorch, W. (1987). Water purification. In: *Handbook of Water Purification* (ed. W. Lorch). Chichester, UK: Ellis Horwood.

Ho, L., Hoefel, D., Meyn, T. et al. (2006). Biofiltration of Microcystin Toxins. An Australian Perspective. *Recent Progress in Slow Sand/and alternative Biofiltration Processes.* IWA, London, 162–170.

Huisman, L. and Wood, W.E. (1974). *Slow Sand Filtration.* Geneva: World Health Organization.

Jacangelo, J.G., Adham, S., and Laine, J. (1997). *Membrane Filtration for Microbial Removal.* American Water Works Association Research Foundation and American Water Works Association USA, 187pp.

Jekel, M.R. (1994). Removal of arsenic in drinking water treatment. In: *Arsenic in the Environment: Part 1, Cycling and Characterization* (ed. J.O. Nrigu). New York: Wiley Interscience.

Juuti, P. and Katko, T.S. (2005). Water, time and European cities. History matters for the futures. Tampere University Press.

Karanis, P., Sotiriadou, P., Kaetashev, V. et al. (2006). Occurrence of Giardia and Cryptosporidium in water supplies of Russia and Bulgaria. *Environmental Research* 102 (3): 260–271.

Karsoyiannis, I., Zouboulis, A., Althof, H.W., and Bartel, H. (2002). As (III) removal from groundwater using fixed upward flow bioreactors. *Chemosphere* 47: 325–332.

Katko, T.S., Juuti, P.S., and Pietila, P.E. (2006). Key long-term strategic decisions in water and sanitation services management in Finland, 1860–2003. *Boreal Environment Research* 11: 389–400.

Langlais, B., Reckhow, D.A., and Brink (ed.) (1991). *Ozone in Drinking Water Treatment: Application and Engineering*, 556pp. Boca Raton, FL: AWWARF and Lewis Publishers.

Lewin, J. (1961). *Journal Institution of Water Engineers* 15: Q45tg.

Liessens, J., Germonpre, R., Beernaert, S., and Verstraete, W. (1993). Removing nitrate with methylotrophic fluidized bed technology and operating performance. *Journal of American Water Works Association* 85 (4): 144–154.

Lievremont, D., Bertin, P.N., and Lett, M.C. (2009). Arsenic in contaminated water. Biogeochemical cycle, microbial metabolism and biotreatment processes. *Biochimic* 91: 1229–1237.

Lovley, D.R. (1995). Bioremediation of organic and metal contaminants with dissimilatory reduction. *Journal of Industrial Microbiology* 14: 85–93.

Manz, D. (2011). Arsenic removal using slow sand filtration. *Western Canada Water Conference*, Saskatoon, Saskatchewen (20–23 September 2011), 12pp.

McDowell, B., Royer, J., and Chauvin, F. (2007). Removal of geosmin and 2-methylisoborneol through biologically active sand filters. *International Journal of Environment and Waste Management* 1 (4): 311–320.

Mohseni-Bandpi, A., Elliot, D.J., and Zazouli, M.A. (2013). Integrated resource management, institutional arrangements and land-use planning. *Environment and Planning A* 37: 1335–1352.

Mouchet, P. (1992). From conventional to biological removal of iron and manganese in France. *Journal AWWA* 84 (4): 62–66.

National Drinking Water Clearing House (2000). Ultraviolet Disinfection. Tech. Brief Fact Sheet.

National Research Council (1979). Iron. National Research Council, University Park Press: Baltimore, MD, USA.

National Research Council (2002). *Privatization of Water Services in the United States: An Assessment of Issues and Experience*. Committee on Privatization of Water Services, in the United

Nikolaou, A.D., Kostopolou, M.N., and Lekkas, T.D. (1999). Organic by-products of drinking water chlorination. *Global NEST International Journal* 1 (3): 143–156.

Nissen-Petersen, E. (2006). Water from Dry River beds. In: *Danish International Development Assistance (DANIDA)*. Nairobi, Kenya: ASAL Consultants Ltd.

Nordstrom, D.K. (2002). Worldwide occurrences of arsenic in groundwater. *Science* 296: 243–252.

Northcliff, S., Carr, G, Potter, R.B., and Darmame, K. (2008). Jordan's Water Resources: Challenges for the future. Geographical Paper No 185. University of Reading, UK.

Palmer, M.C. (1962). Algae in water supplies. *U. S Department of Health, Education, and Welfare*. Publication No. 657, 88pp.

Pitkanen, T., Juselius, T., Miettenen, I.T. et al. (2015). *Resources* 4: 637–654.

Pokhrel, D., Virarghavan, T., Asce, F., and Braul, L. (2005). Evaluation of treatment systems for the removal of arsenic from groundwater. *Practice Periodical of Hazardous, Toxic, and Radioactive Waste Management* 9 (3): 152–157.

Putigeni, L. and Menichelle, D. (2010). Global distribution, public health and clinical impact of the protozoan pathogen *Cryptosporidium*. *Interdisciplinary Perspectives on Infectious Diseases* 2010: 753512.

Richardson, S.D. (2000). Drinking water disinfection by-products: What is known. *Int. Workshop on Exposure Assessment for Disinfection By-Products in Epidemiological Studies*, Ottawa, Canada.

Rittmann, B.E., Stilwell, D., Garside, J.C. et al. (2002). Treatment of colored groundwater by ozone-biofiltration pilot studies and modeling interpretation. *Water Research* 36: 3387–3397.

Scholz, M. and Martin, J. (1997). Ecological equilibrium on biological activated carbon. *Water Research* 31 (12): 2959–2968.

Steele, J., Griffith, J., Noble, R., and Schiff, K. (2017). Tracking Human Fecal Sources in an Urban Watershed During Wet Weather. San Diego: Southern California Coastal Water Research Project. *Technical Report 1002*. http://ftp.sccwrp.org/pub/download/DOCUMENTS/TechnicalReports?1002humanMarkerTracking.pdf.

Stevenson, D.G. (1997). *Water Treatment Unit Processes*, 474pp. London: Imperial College Press.

Tavasolifar, A., Bina, B., Amin, M.M. et al. (2012). Implementation of hazard analysis ad critical control points in the drinking water supply system. *International Journal of Environmental Health Engineering* 1: 32.

Tekerlekopoulou, A.G. and Vayenas, D.V. (2007). Ammonia, iron and manganese removal from potable water using trickling filters. *Desalination* 2010 (2007): 225–235.

The Times (1856). Cholera and water supply. *The Times* (26 June 1856), 12 Col B.

UNEP TU Delft (2008). Every Drop Counts. Environmentally Sound Technologies for Urban and Domestic Water Efficiency. *UNEP Division of Technology, Industry and Economics*. International Environmental Technology Centre, Osaka.

UNICEF WHO (2015). 25 years Progress on Sanitation and Drinking Water: 2015 update and MDG Assessment, New York, 81pp.

UNICEF WHO (2019). Progress on household drinking water, sanitation and hygiene, 2000–2017. New York.

United Nations (2012). *The Millennium Development Goals Report 2012*. UN, New York.

United States Environmental Protection Agency (USEPA) (1999). Alternative Disinfectants and Oxidants. Guidance Manual. April 1999.

United States Environmental Protection Agency (USEPA) (2002). Office of groundwater and drinking water. Implementation guidance for the arsenic rule. *EPA Report 816-D-02-005*.

United States Environmental Protection Agency (USEPA) (2006). *UV Manual*. Washington, USA: USEPA.

Vickers, J.C., Braghetta, A., and Hawkins, R.A. (1997). Bench scale evaluation of microfiltration-nanofiltration for removal of particles and natural organic matter. *Proceedings of the AWWA Membrane Technology Conference*.

Wagner, E.G. and Pinheiro, R.G. (2001). *Upgrading Water Treatment Plants*, 216pp. London: Spon Press.

Water Research Center (2017). Ozonation in water treatment, 381pp. http://www.water-research.net/index.php/ozonation.

World Health Organization (1993). *Guidelines for Drinking Water Quality*, Recommendations, 2e, vol. 1. Geneva: WHO.

World Health Organization (1996). *Guidelines for Drinking Water Quality*, Health Criteria and supporting Information, 2e, vol. 2, 541pp. Geneva: WHO.

World Health Organization (WHO) (2004). *Guidelines for Drinking Water Quality*, 3e. Geneva: WHO.

World Health Organization & UNICEF (2006). *Meeting the MGD Drinking Water and Sanitation Target. The Urban and Rural Challenge of the Decade*, 21pp. Geneva: WHO.

World Health Organization (WHO) (2011). *Guidelines for Drinking Water Quality*, 4e, 541pp. Geneva: WHO.

Zouboulis, A.L. and Katsoyiannis, I.A. (2002). Removal of arsenates from contaminated water by coagulation-direct filtration. *Separation Science and Technology* 37 (12): 2859–2873.

Zu, I.X. and Bates, B.J. (2013). Conventional Media Filtration with Biological Activities, pp. 137–166.

8

Wastewater Treatment

Our total water usage is increasing, and thus the amount of wastewater produced is also increasing. If this is not properly treated, it leads to a deterioration in the general water quality and volume available, which is already under some strain in many parts of the world. Over two and a half billion people on this planet still do not have access to proper sanitation (JMP 2010). At the same time, the global human population is steadily increasing, so there is even more pressure to provide appropriate sanitation facilities. The fastest growth rates are in some developing countries where water and sanitation issues are greatest. A lack of proper sanitation is a major cause of the spread of disease in these countries, many of which are in Africa and Asia. An estimated 10% of the global disease burden could be prevented with improvements in sanitation and drinking water provision (OECD 2011). With an increase in the global population more people, mainly from rural areas, are moving to towns and cities for living. Currently, more people live in cities than in the countryside. To solve the problem of both supplying drinking water and the provision of adequate sanitation requires both concerted action and significant financial inputs by both regional authorities and national governments. As the improper disposal of both human wastes and wastewater spreads disease and pollutes surface waters, such as rivers and lakes as well as aquifers, proper waste treatment and disposal are essential. The need to manage the disposal of human waste in a safe manner was realised in ancient times by the Romans, who used running water to remove both urine and faeces from public toilets, private houses and public baths. This was also done in ancient China (Western Han dynasty from 200 BCE).

The early methods of sewage disposal were to land or into the nearest surface water, if available. Although direct discharge to a stream or river might then be acceptable (but discouraged) as long as there is a large dilution factor, there is still the problem of the spread of waterborne disease and adverse impacts on aquatic ecosystems. As human populations grow, the natural breakdown processes in the aquatic ecosystem often cannot cope, and complete degradation and dilution of waste do not occur. Disposal to a designated area of land outside to town or city boundary to areas called sewage farms was attempted, but these areas suffered in the same way as populations grew, and more land on the edge of cities and towns was required for building. In the mid-nineteenth century, scientific investigations into waste disposal were started. In 1868, Sir Edward Frankland carried out an investigation for the UK Royal Commission on the efficacy of filtration for treatment and

Water: Our Sustainable and Unsustainable Use, First Edition. Edward G. Bellinger.

showed that trickling filters could successfully treat sewage waste. In the United States of America, the Massachusetts Board of Health established an experimental filter station at Lawrence in 1887 to assess stream pollution caused by sewage. These studies supported the views of Frankland (1868) on filters, see Buswell and Strickhouse (1928). In the late 1890s, Corbett developed a full-scale filter at Salford (Greater Manchester), UK, with rotating sprinklers to apply settled sewage (Corbett 1903). Filters continued to be developed and gained wider use.

The main aims of treating wastewater are as follows:

1) To render waste materials present in the wastewater, including substances in both solution and as solids, harmless to living organisms. In particular, it is a danger to human health.
2) To allow treated wastewater to be disposed of effectively and without any harm to the environment and ecosystems.
3) To allow these discharges to occur in a controlled manner in both time and volume.
4) To recover and reuse as much of the wastewater as is feasible, including not only the water but also any substances that are in it.
5) To have methods of disposal that are both economic and are properly maintained.
6) The final effluent must comply with any legal standards that will vary with the receiving water and from country to country.

There are many definitions of wastewater, but the following are generally accepted:

- Domestic effluent consisted of excreta, urine and faecal sludge (blackwater) and kitchen and bathing/washing water.
- Water from commercial establishments and institutions, including hospitals.
- Industrial effluent, storm water and other urban run-off.
- Agricultural, horticultural and aquaculture waste either in solution or as solids in suspension.

These definitions are adapted from UN water (2015). Unfortunately, much wastewater is not properly treated before disposal (Corcoran et al. 2010). They estimated that more than 70% of industrial wastewater in developing countries is dumped untreated into surface waters, contaminating them and aquifers often making them unsuitable for other uses. Mining operations have long been a source of water contamination, especially where incorrectly constructed tailings reservoirs are used to collect wastes. There are many cases of tailings dams or bund collapsing. Mining itself can also affect surface water drainage. Below surface mining, there is often water ingress that has to be pumped out. This can lead to problems of acid mine drainage (AMD). AMD can be caused by the action of air on sulphide-bearing rocks being converted, in the presence of water, to sulphuric acid. When acidic water is pumped out, it can have a devastating effect on ecosystems. It is difficult and expensive to treat this, and the problem can last for many years. Examples of AMD problems have occurred in many countries, including the West Rand Goldfield (Africa), Tinto River (Spain), Iron Mountain Mine (California), Pronto mine tailings (Canada) and Ok Tedi Mine (Papua New Guinea).

Different types of wastewater are often treated differently depending upon requirements.

Natural water systems have populations of microorganisms that metabolise many organic and inorganic compounds that occur both naturally and through pollution.

These metabolic processes help to purify the water, enhancing the ecosystems present. Many of these processes, depending upon local conditions, operate relatively slowly, so if they are to be used for breaking down human and some industrial wastes, their actions need to be speeded up.

We call the wastewater produced by a human community sewage that can include domestic waste, industrial waste, surface and flood water and, in some cases, infiltration from groundwater. The average composition of domestic sewage is provided in Table 8.1.

For larger conurbations, it is more efficient to bring together all sewage waste to a central work for treatment. This is achieved by connecting each dwelling to a piped network. This was proposed by Bazalgette after the 1858 'Great Stink' of London (Halliday 1999). The use of these sewage collection systems greatly improved public health. Earlier systems collected sewage and surface water run-off, which included not only storm water but also road drainage in the same pipes. These were called combined systems. In later systems, sewage and surface run-off were collected in separate pipes that gave the potential to treat each separately. These are termed dual systems.

Many towns and most cities will have some industrial or commercial discharges to the sewerage system in addition to domestic discharges. These non-domestic discharges will vary greatly in their strength, organic content and chemicals present and may need on-site treatment before discharge and treatment at a centralised treatment work. Some examples of different strengths of effluents from different works compared with a typical activated sludge treatment work effluent are provided in Table 8.2. Some effluents will require either considerable dilution or special pre-treatment before discharge.

Table 8.1 The average composition of domestic sewage in g/person/day.

Component	Weight	BOD_5
Suspended solids	15	10
Settleable solids	20	20
Dissolved solids	125	30
Total	60	60

Source: Data from Imhoff (1983).

Table 8.2 A comparison of different industrial effluent strengths compared with the final effluent from an average activated sludge plant (ASP).

Industry	Effluent BOD
Brewery	850
Dairy	600–1000
Slaughterhouse	1500–2500
Crude sewage	250–350
ASP effluent	10–20

Proper wastewater treatment is essential to prevent the pollution of watercourses into which the wastewater effluent is discharged in order to minimise the risk to both public health and ecosystems. In standard municipal waste treatment systems, this protection is achieved by breaking down or removing the waste materials, converting them into less harmful ones. Any highly toxic or persistent substances should be treated before discharge to sewers. To ensure that the municipal waste treatment system carries out this treatment, it must be properly constructed, managed and reliable as well as should be able to function continuously. Each stage of the process should be regularly monitored. It is also important that whichever treatment process is being used, it is carried out at an acceptable cost to the community, and to this end, recovery of any valuable materials is sometimes practiced. It is also essential that the final effluent complies with the national and international standards. The waste breakdown mechanisms most commonly involved are biological, and biological wastewater treatment is the most widely used secondary treatment globally. Biological wastewater treatment relies upon bacteria and other microorganisms to change potentially harmful compounds into less harmful compounds that can be safely discharged into the environment. The range of microorganisms found in biological wastewater treatment is all found in nature as are their breakdown processes of waste, but in nature, these processes take place relatively slowly. In order to speed them up in a controlled way, the biological treatment is carried out in either a fixed film or a fluidised system. In both of these systems, the organisms are encouraged to grow under controlled conditions that allow the waste to be purified as quickly as possible.

The breakdown processes carried out by organisms are enzyme-based reactions performed both inside and outside cells. Higher molecular weight organic compounds in wastewater are broken down by enzymes produced by organisms to produce lower molecular weight compounds, releasing energy and compounds that can be used in cell growth. To optimise these reactions, a suitable environment must be provided, such as the correct pH range and the absence of toxins or other enzyme inhibitors. The rate of enzyme-based reactions is also temperature dependent. As an approximation, the rate of reaction doubles for every 10 °C rise in temperature within its operating range. Similarly, it can halve for a similar drop in temperature. Although municipal sewage, depending upon the geographical location, will not usually be below 5 °C and mostly not above 20–25 °C, the speed of breakdown reactions will vary with season, and thus, the demand of the microorganisms for resources (e.g. food and oxygen) will also vary. Chemicals that inhibit enzyme action include cyanides and heavy metals, both of which can be present in industrial effluents. Some chemicals can stimulate enzyme activity. Examples include certain inorganic salts, such as chloride ions, which stimulate amylase, and certain dehydrogenases require manganese, zinc or magnesium (Mudrack and Kunst 1986). The biological breakdown process can be aerobic (requiring oxygen) with ultimate end products of carbon dioxide and water or they may be anaerobic (carried out in the absence of oxygen) in which the end products are methane and carbon dioxide. Aerobic processes include activated sludge and trickling filters, and anaerobic processes are used in some waste stabilisation ponds (WSPs) and in gas generation. As breakdown reactions take a certain amount of time, which varies depending upon the chemical compounds present, not all waste may be fully broken down in the time available. Many simple molecules will be completely decomposed, but other molecules, often more complex ones, are poorly or even largely non-degradable in the time available. Compounds such as cellulose are slow to decompose, and many man-made substances such as DDT (dichloro-diphenyl-trichloroethane) may not be broken down at all. Substances that are slow to degrade within normal retention times for the biological stage of treatment can often be recycled to the start of the process, which gives a longer time for breakdown to occur. Two main types of microbiologically induced biochemical reactions are important in the breakdown of organic waste. The majority of these processes in activated sludge and biological filters are aerobic although some anaerobic processes will also occur. The main anaerobic processes occur in sludge digesters that are designed to produce methane as well as to decompose sludge.

Not only do microorganisms have to deal with a wide range of compounds, they also have to a range of varying flows both in the short term (hourly and weekly) but also in the longer term (seasonally). Figure 8.1 shows typical hourly variations in flow throughout a day. Seasonal flows occur in both hot weather in summer and cold weather in winter when reaction times are slower. The overall treatment of waste consists of several stages depending upon the strength of the sewage and the compounds present. The main purpose of the treatment is to oxidise carbonaceous matter and nitrogenous compounds. It can also be important to remove some phosphorous as well in order to protect aquatic ecosystems. The impact on the ecosystem of the receiving water does, to an extent, depend upon the amount of dilution that occurs. The original standard adopted in the United Kingdom derived from the Royal Commission required at least an eightfold dilution and an effluent standard from the treatment works of 20 biological oxygen demand (BOD) and 30 suspended solids (SS). If the dilution factor was, for example, 300-fold, then a higher concentration of BOD and SS (100 and 60) could be acceptable. This dilution-dependent standard has now been superseded and follows the European Union Waste Water Directive (91/271/EEC). The composition of sewage depends upon its origin, but average values are provided in Table 8.3.

The required quality of the final discharge can have further restrictions placed upon it if the receiving water is of a very high quality and a vulnerable type. For example, a pristine

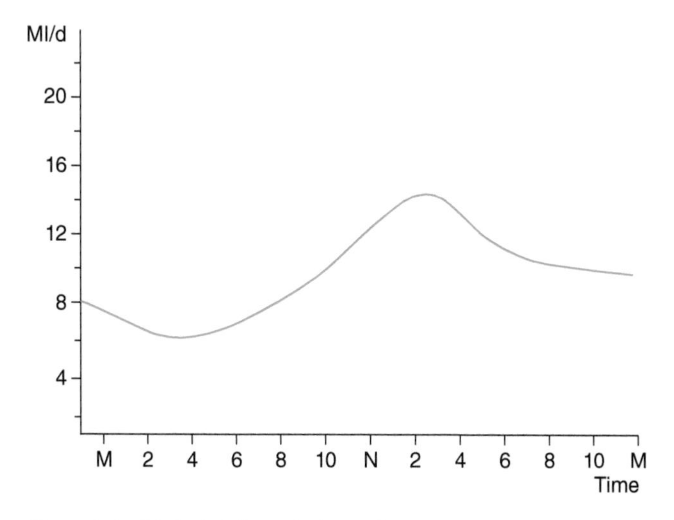

Figure 8.1 Typical daily variations of domestic sewage flows.

Table 8.3 Average composition of untreated domestic sewage.

Component	Concentration (mg/l)
Five-day biochemical oxygen demand (BOD_5)	100–300
Chemical oxygen demand (COD)	250–1000
Total dissolved solids (TDS)	200–1000
Suspended solids (SS)	100–350
Kjeldahl nitrogen (KN)[a]	20–80
Total phosphorus (TP)	5–20

[a] This is the total amount of nitrogen measured by the Kjeldahl test.

salmon or trout river or a pristine chalk stream's ecosystems would be particularly susceptible to pollution, so standards would be more stringent.

If any industrial discharges are present, the above concentrations can be altered considerably, and additional compounds, such as fats, oils and grease (FOG), may be added together with inorganic chemicals, such as metals, detergents, solvents, etc.

As with drinking water treatment, there are multiple stages and barriers created to prevent the waste from contaminating the environment. Examples of typical operation sequences are given below. There may be additional process stages depending upon the nature of the waste being treated and any restrictions/requirements on the quality of the effluent discharge required. Where there are significant industrial discharges into the sewer, there will probably be a requirement on the industry concerned to pre-treat their waste before discharge to sewer.

An outline of treatment stages for a conventional waste treatment plant could be as follows:

- Crude sewage (may include surface run-off and industrial waste)
- Screening to remove large objects*
- Grit separation*
- Primary sedimentation**
- Biological treatment***
- Final sedimentation/clarification
- Effluent discharge

*Preliminary treatment. **Primary Treatment. ***Secondary Treatment.

Preliminary Treatment: Screening and Grit Removal

Preliminary treatment is designed to remove larger solids and grit that have been washed into sewers through surface drainage. If oil and grease are present in significant amounts, they may also be removed at this stage.

Any large objects such as twigs, pieces of wood, plastic bags, rags, etc. and large particles of coarse grit need to be removed to prevent damage to later treatment process equipment.

These screenings are fed into a comminuter that macerates the large particles into small ones. These screenings are mixed with faecal material and are collected together and disposed of to landfills or incinerated. In larger works, the screenings may again be fed into a comminuter prior to disposal. The velocity of flow through the screens should not be too great, less than 1 m/s, so as not to dislodge the materials trapped on the screen. Screens usually consist of mild steel bars usually set at about 15–30 mm apart (Figure 7.1). They are usually set at a sloping angle of about 60° to aid both manual and mechanical cleaning. If manual cleaning is being used, the screens usually require raking at least twice a day. In some cases, finer screens with an aperture size of 3–15 mm are occasionally used forming a rotating drum. These are often used when certain industrial effluents are being treated. These are more expensive to fit and operate, so they are not usually used in small works or developing countries. Grit, sand and fine mineral particles are not significantly removed by screening. Grit has an average density of 2.5 and is thus heavier than organic particles and settles faster. This faster settling rate can be used to advantage by greatly reducing the velocity of flow in controlled conditions, allowing the grit to settle out. Constant velocity grit channels provide the required environment for settling. These may be in the form of a properly constructed channel (Mara 2013) or cyclone separators and detritors (Gray 1999). Grit channels usually have a parabolic cross section with a cross-sectional area proportional to the flow. A number of devices are available for removing grit, including cyclones and vortex generators. These are hydrodynamic devices that produce a vortex that spins the solid grit particles out of suspension and collects them together for disposal. The principle of their use was first observed by B. Smisson, an engineer from Bristol in the south-west of England (Smisson 1967). Vortex grit separators consist of a circular tank either with the fluid being introduced at an enhanced velocity tangentially or with a turbine rotor to control velocities. The grit particles are thrown to the outside and fall to the tank bottom of the tank to a sloping conical floor. They can then be drawn off for disposal. The grit-free fluid can overflow at a channel at the top of the tank. An idealised vortex separator is shown in Figure 8.2. These grit separators have developed since that time and have the advantage of not only efficient grit removal (depending upon the density and size of the particles) but also taking up less space (Faram et al. 2004). Different variations in the basic design have been developed including Sullivan et al. (1982) and Brombach (2004). Others are described in Gray (1999) and Field et al. (1997). Some treatment works also have the additional problem of having large amounts of FOG, which has to be treated. FOG from industrial sources and cooking wastes simply poured down domestic sinks will cool and congeal and stick to the sides of the pipes, reducing their diameter and eventually blocking them.

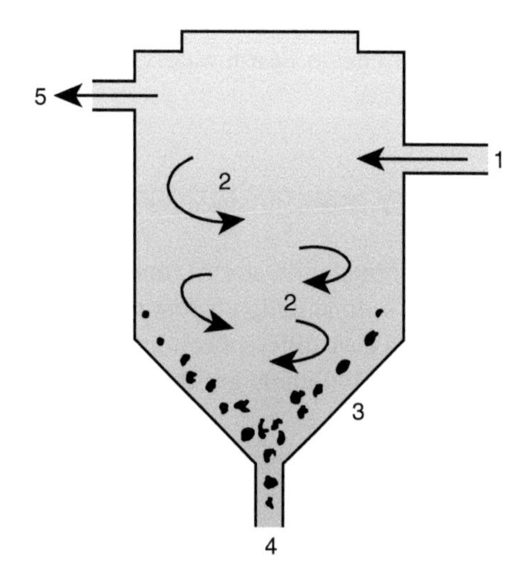

Figure 8.2 Vortex separator for grit and particle removal. (1) inflow; (2) vortex flow; (3) collected grit and particles thrown to side; (4) outlet for collected grit, etc.

FOG problems can be reduced by installing traps or flotation devices in all restaurants or, in the case of cooking oils recycled. Fats, grease and oils are less dense than water so will float. This floating matter can be removed in unmixed tanks, but to speed up their removal, dissolved air flotation (DAF) is often used. The fine air bubbles attach themselves to the FOG particles, enhancing their flotation where they form a surface scum that can be removed by scrapers and taken for disposal.

Storm Water

Surface storm water draining from roads and other paved areas can discharge to sewers. If the storm is intense, it can cause a rapid rise in the flows to the sewage works. Some storms are not so intense but are heavy and more prolonged, so their overall effect on sewage works is the same. This is because the flows can exceed the capacity of the works, and quite often, they will contain more debris, grit and chemicals washed off the road. It is therefore essential that the works be provided with extra capacity to contain the extra volumes. If the sewer system cannot cope with the flow, water cannot get into the drains and will back up and overflow causing local flooding. The first flush of storm flow usually has the highest concentrations of contaminants, including stale sewage scoured from pipes. After this first foul flush, the amount of contamination should significantly decrease. Traditionally, sewage works are designed to give full treatment to flows up to as much as three times the dry weather flow (DWF), and the treatment stages need to be protected against excessive flows. If the flow is between three and six times the DWF, the excess should be diverted and contained in storm water holding tanks to allow settlement before treatment or discharge to the receiving water. Above 6× DWF, flows will be screened and then discharged, although depending upon the size and duration of the storm flow the water could be diverted through storm tanks. The first hours of storm flows being the most polluted should be fully treated before discharge. Storm waters are usually separated from normal flows by overflow weirs (Figure 8.3).

Primary Sedimentation

Under normal (non-storm) conditions, sewage flows are passed through pre-treatment and primary sedimentation before undergoing secondary biological treatment. Sedimentation tanks can have different configurations, e.g. radial flow, upwards flow and horizontal flow (Gray 1999). Radials are the most frequently used in municipal wastewater treatment

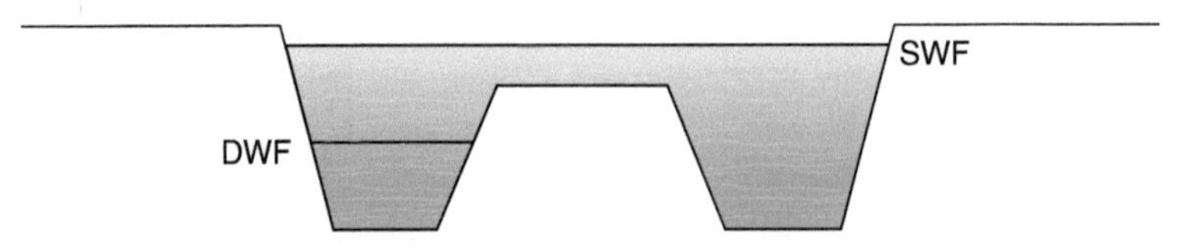

Figure 8.3 Overflow channels for inflowing water. DWF, channel for dry weather flow; SWF, overflow channel for storm water with weir separating it from DWF and taking it to a storm water holding tank.

plants although horizontal flow tanks are sometimes used. Upwards flow tanks are more frequently used in smaller works. The flow through these tanks is slow enough to allow fine particles to settle out under gravity, and some flocculation may also occur. These settled solids can then be removed by scrapers at the bottom as primary sludge and taken away for treatment. Any solids that are less dense will then accumulate at the surface and can then be removed with skimmer blades. The settled clarified liquor overflows at the surface via castellated notched weirs. The tanks usually have a retention time of slightly over two hours. This stage of treatment would be expected to reduce the BOD by up to 40% and the suspended solids by up to 70%. A range of processes take place within the tank, including sedimentation, flocculation and adsorption, all of which aid the removal of suspended solids (Casey 1997). Although primary sedimentation tanks can be rectangular or round, the latter are the most common.

Activated Sludge (AS) Process

After primary sedimentation, the liquor passes on to a secondary treatment stage. This may be based upon a fluidised or suspended micro-biota or one living as a solid-based fixed film system. The activated sludge process is a suspended fluidised system. The AS process is the most widely used system of wastewater treatment worldwide. It has been developed progressively over the last 100+ years. In this fluidised system, the microorganisms are kept in suspension and brought into contact with the waste organic matter. The system is aerobic, so an adequate supply of oxygen to the organisms is required. The reactions carried out by aerobic organisms are usually faster than those carried out by anaerobic organisms.

The process was first described by Arden and Lockett in 1913 at the Davyhulme Sewage Works in Manchester, UK. The results of their work were published in 1914 (for a general review of their work and contribution of others, see Coombs (1992)). Arden and Lockett called the process they studied the activated sludge process. This relies upon aerobic bacteria and other microorganisms to break down unwanted organic matter in the settled sludge and produce a floc or culture that can then be removed, leaving an effluent that can be safely discharged into receiving water. Although pure oxygen can be injected into liquor to meet the needs of aerobic organisms, it is less expensive to use air. It is also more economical to regulate the final oxygen concentration to about 2 mg/l as higher concentrations do not produce significantly greater efficiency but involve a much greater energy use and hence higher operating costs. Currently, there are various ways of introducing air, allowing oxygen to dissolve in water. The main ones are diffused air and mechanical aeration. Both approaches mix the water to ensure that the system is uniform throughout, keep the biomass in suspension and mix the incoming settled sewage with the existing biomass.

Diffused Air

In a diffused air system, the air is introduced through a blower or compressor, after passing through a filter to remove unwanted and possibly damaging particles, into a system of pipes along which are mounted diffuser domes that release a stream of fine bubbles that rise to the surface. As these rise through the water, oxygen dissolves, and circulation currents are generated. The diffusers are generally designed to produce small bubbles, about 2 mm in size (CIWEM 1997), to provide a large surface area for O_2 to dissolve (Figure 8.4a–c). A pressure measuring device is generally provided in order to balance the flow between different banks of diffuser as and when this is required. This may be needed as after some time biological growths may occur on the diffusers partially blocking them on both the inside and outside impeding flows. Large works have thousands of these diffuser domes, and they should be checked periodically for leaks and blockage, especially for biological growths that could cause uneven distribution of both oxygenation and circulation.

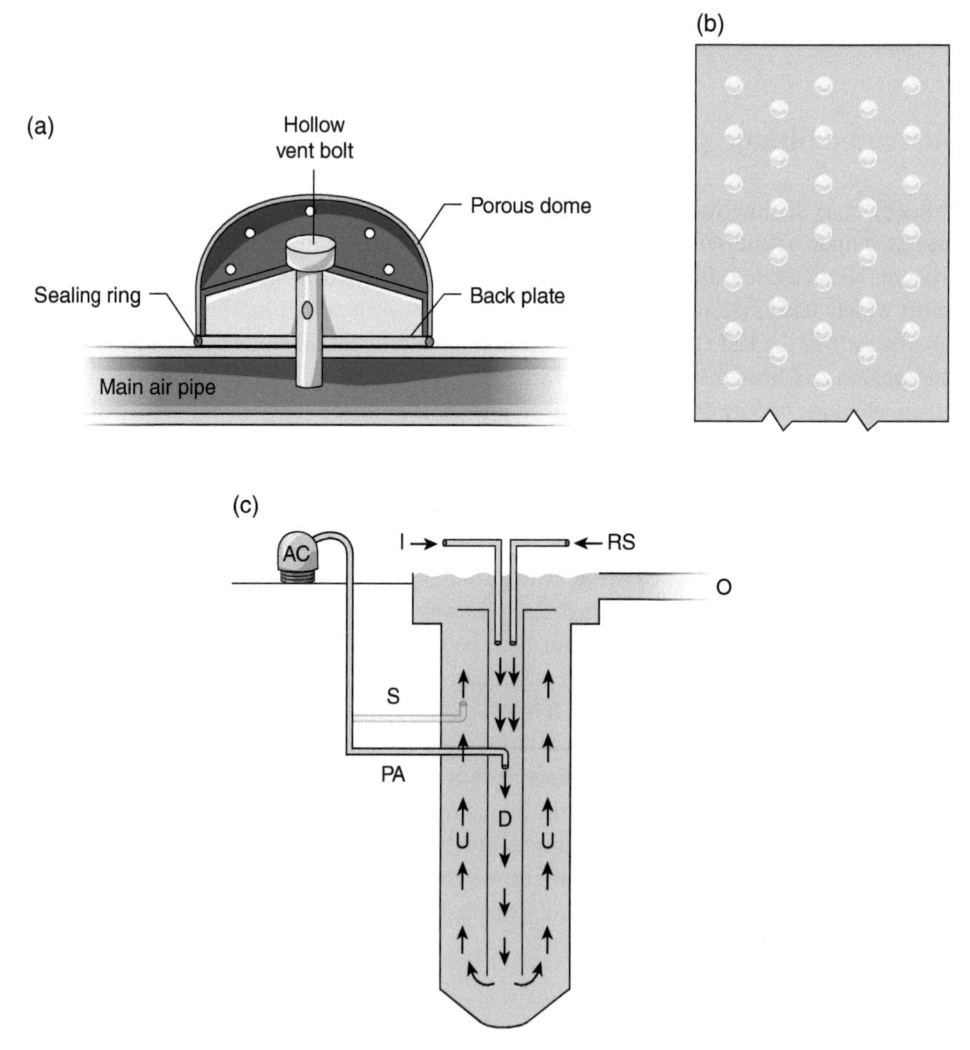

Figure 8.4 (a, b) A diffused aeration system and layout. (a) Individual diffuser dome structure. (b) Array of diffuser domes in treatment tank. (c) Vertical or deep shaft aerator. AC, Air compressor; S, start up air; PA, main process air flowing downwards; I, inflow of settled sewage; RS, sludge recycle; D, downward flow; U, upward flow; O, outlet.

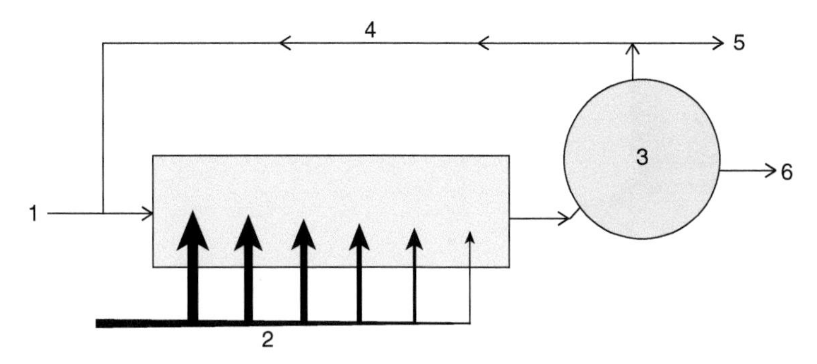

Figure 8.5 Tapered aeration. (1) Feed water, (2) arrangement of decreasing amounts of aeration along the tank, (3) sedimentation tank, (4) return sludge, (5) excess sludge going to waste and (6) effluent.

This may require the aeration tank to be drained and cleaned out regularly. Biological fouling of domes can be monitored by suspending a test dome on the end of a flexible air pipe, which is then suspended in the aeration tank. This can then be raised and inspected at regular intervals for the degree of fouling. The surface pattern of bubbles over the whole tank should be even so regular visual observations should pick up any irregularities in the surface bubble patterns indicating a possible malfunction of one or more domes.

In conventional diffused air plants, the diffusers are arranged in a uniform grid across the floor of the tank. In some situations, a modified arrangement has been shown to be more effective. These variants are tapered aeration and step feed aeration. The designs for these are shown in Figure 8.5 for step and tapered aeration. In the tapered system, most oxygen is supplied where the demand is greatest, so there is a greater concentration of diffusers towards one end. In step feed aeration, instead of the sewage feed being all at one end, it is introduced at different points along the tank. This has the effect of ensuring that all of the biota have enough oxygen for them to act upon the lower concentrations of sewage along the tank.

Surface Aeration

Surface aerators work by vigorously mixing the surface water through the action of blades of various designs. This results in turbulence and spray, which allows oxygen to dissolve as well as promote circulation. The depth of immersion of the aerator is critical. Its speed of rotation can be altered to change the amount of aeration.

Vertical Shaft Rotors

These consist of a motor-driven vertical shaft at the bottom of which is attached to a series of blades that cause turbulence, producing both aeration and circulation. There are various designs of aerator. An example is given in Figure 8.6. The rotors are usually run at speeds of 30–60 rpm and at depths between 1 and 4 m. Often a vertical tube is fixed below the rotor

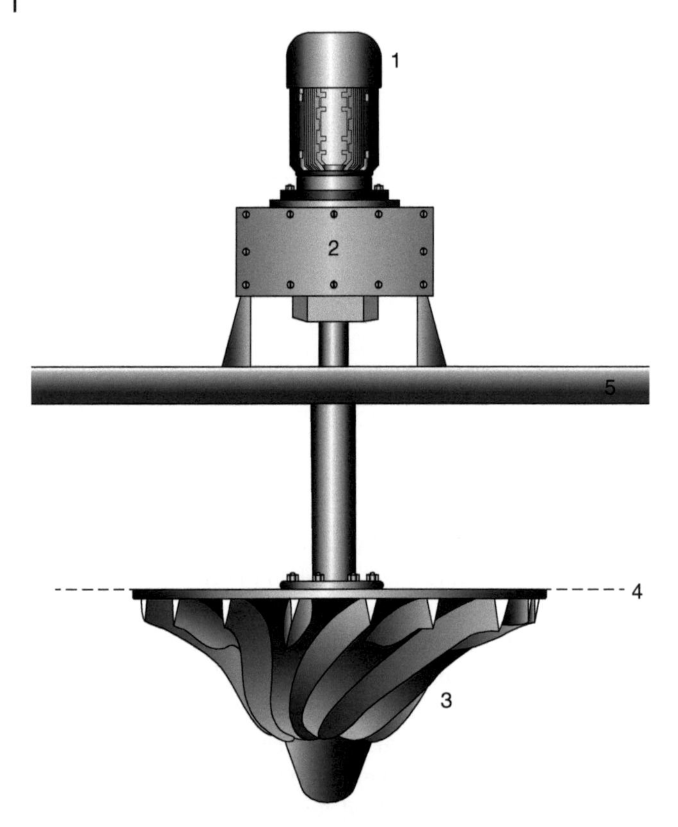

Figure 8.6 Example of surface aerator. (1) Motor drive; (2) reducing gears allowing rotor speed to be selected; (3) impeller head; (4) normal liquid level; and (5) solid surface to mount motor.

for improving the water circulation within the tank. Each aerator is driven by an electric motor. This form of aeration can be noisy which is unacceptable in some locations. These types of aerators can also be used on floating pontoons, especially when they are used in a lagoon or small lake.

Horizontal Shaft Aerators

These consist of a horizontal shaft onto which blades are fixed. As the shaft rotates, the blades create a wave motion as well as turbulence at the surface. They also entrain air that aids the dissolution of oxygen. The paddles are immersed in the liquid to a depth of about 40 cm and rotate at about 30 rpm. This paddle design is less frequently used now in its simplest form, but recent developments like the Mammoth horizontal rotor are more efficient and are still used (Gray 1999). The Kessner brush is widely used and has proven quite efficient. This was developed in the Netherlands and consists of rows of steel brushes attached to a horizontal shaft. The whole is usually 0.4–0.8 m in diameter and operates at about 50–70 rpm. These are frequently used in oxidation ditches to provide both mixing and aeration (Figure 8.7).

There are different ways in which an ASP can operate. These factors determine the nature of the treatment being given. These are termed high rate, normal rate and low rate.

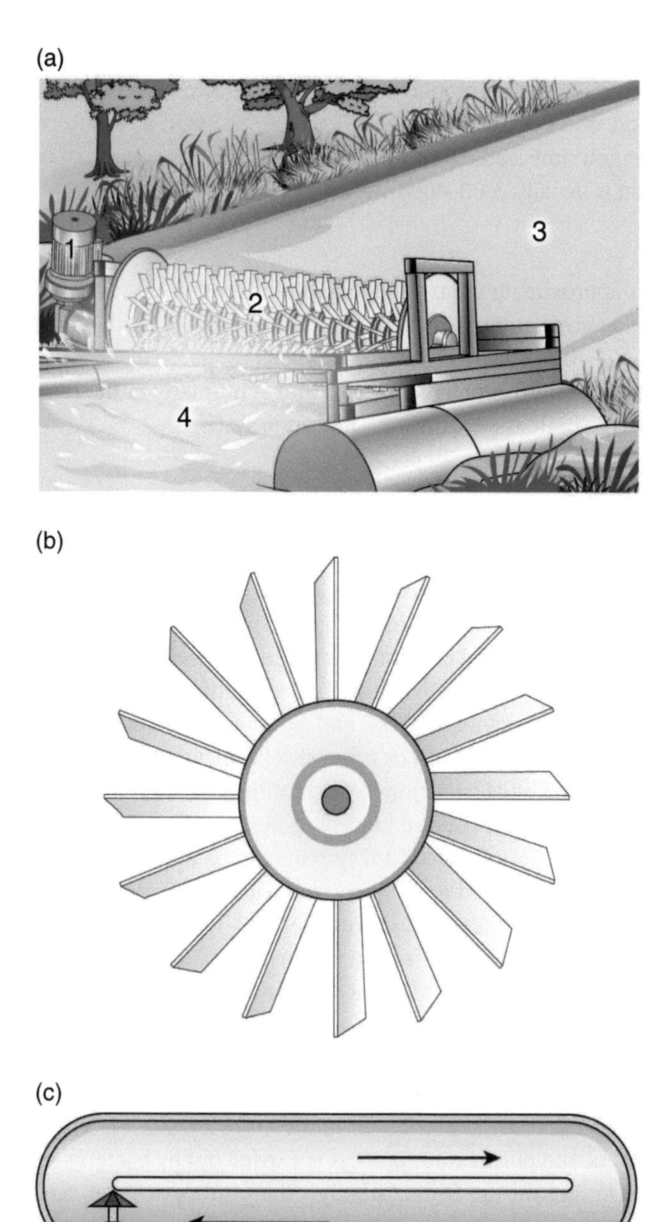

(c)

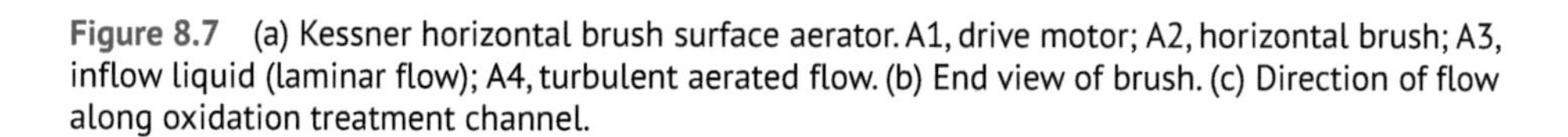

Figure 8.7 (a) Kessner horizontal brush surface aerator. A1, drive motor; A2, horizontal brush; A3, inflow liquid (laminar flow); A4, turbulent aerated flow. (b) End view of brush. (c) Direction of flow along oxidation treatment channel.

High-rate plants operate with a shorter retention time in the aeration tank and a low level of mixed liquor-suspended solids (MLSS) at about 1000 mg/l. These tend to be used as a pre-treatment for some industrial wastes. In some cases, they precede treatment by biological filtration. Some BOD is removed, but there is no reduction in ammonia. A normal rate plant has longer aeration (6–12 hours) and MLSS levels of 2000–4000 mg/l. A large proportion of the BOD is removed as well as some ammonia. Low-rate plants (extended aeration plants) have very long retention times of up to 48 hours and are designed to treat low-strength sewage. This long retention time leads to an overall reduction in the volume of the sludge. The MLSS concentration is usually 2000–4000 mg/l.

Mixed Liquor Suspended Solids

The MLSS concentration is an approximate measure of the concentration of suspended solids in the aeration tank. It also gives an approximation of the amount of biomass present. This will also give an idea of the amount of microorganisms available for breaking down the organic matter present. It is also expressed as mg/l and is normally in the range 1500–3500 mg/l but this could be as high as 8000 mg/l for high-rate systems.

Biological Filtration

Unlike in the activated sludge process, where organisms are kept in suspension in biological filtration the biomass is attached to an inert solid medium. The settled sludge to be treated is passed over the medium and attached biofilm, where the organisms break down the organic matter. There are three main types of fixed film biological filtration. These are trickling filters, rotating biological contactors (RBCs) and submerged aerated filters. Although the term filter is used, these do not strain or remove solid material in the same way as sand filters but rely upon microbiological action in the biofilm to remove unwanted materials. In traditional units, the medium is fixed in place, but in some modern developments, the medium moves through the liquor. Fixed film systems can be used as partial or complete treatment. The filter beds are designed to allow contact with air in order to maintain high enough oxygen concentrations to keep the process aerobic. In some specialised filters, anoxic conditions are created so that denitrification can take place. These filters are permanently submerged under the wastewater. The most common configuration for bio filters is aerobic, and the wastewater is allowed to trickle over the biofilm to be collected in the bottom, hence the name trickling or percolating filters. These filters are usually preceded in the treatment stream by primary sedimentation to remove the majority of suspended solids.

As with activated sludge systems, the purification process is carried out by bacteria and other microorganisms that live in and on the film. Although these processes can occur naturally in streams, lakes and ponds, they are concentrated, and the reactions are accelerated in wastewater treatment. Biological filtration is probably the oldest form of sewage treatment. In modern times, probably the first full-scale treatment plant was built by J. Corbett in 1897 (Corbett 1903). Biological filters are generally circular or rectangular in shape partly depending upon available land. Whatever the shape, the basic removal mechanisms are biological and occur in the biological slime layer that consists of bacteria and certain other organisms that develop over the surface of the inert medium. In mature filters, other more complex organisms may develop (Figure 8.8a). Examples of some of the protozoans that may be found are illustrated in Figure 8.8b. The film increases in thickness as the film develops as does, to an extent, the range of species. The thickness needs to be

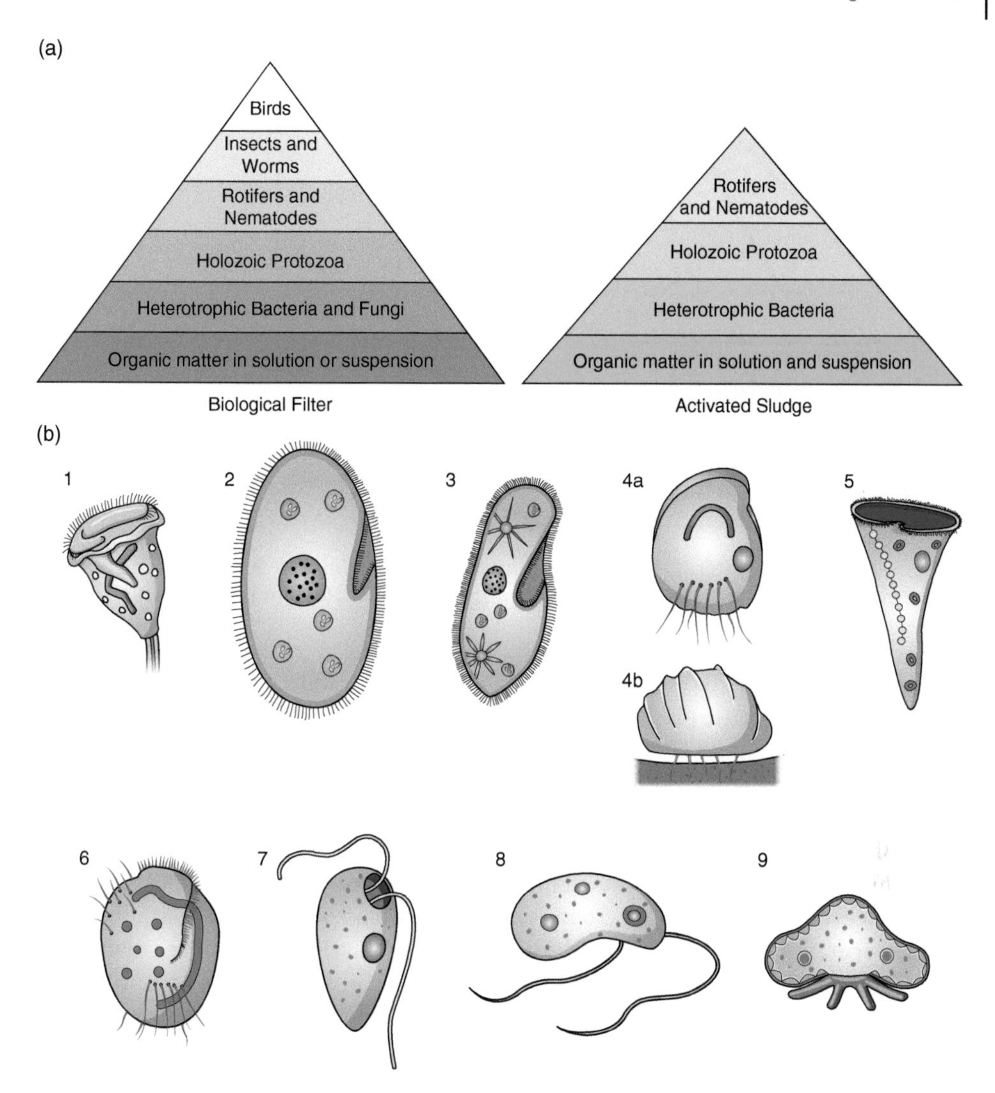

Figure 8.8 (a) The relative abundance of organisms in biological filters compared with an activated sludge plant. (b) Example of the common protozoa found in an activated sludge plant. (1) Vorticella, (2) Colpidium, (3) Paramecium, (4a and 4b) Aspidisca, (5) Stentor, (6) Euplotes, (7) Bodo, (8) Pleuromonas and (9) Arcella.

controlled, as if it is allowed to thicken too much, it could block the spaces between the pieces of media, stopping both the flow of liquid and the passage of air. Excess film thickness can be removed by erosion as the liquid flows over the film, or in some cases, by jet washing with clean water. An increase in the flow velocity will increase erosion so that some parts of the film will slough off. This is often because parts of the film become unstable and break away. The media chosen for the film to grow needs to be inert, hard to wear and low cost. A rough faced mineral is often used with a large surface area for film growth. Although fairly uniform in size (up to 200 mm), it needs to be randomly angular in shape with a large surface area and good void spaces between each piece so that liquid and air can

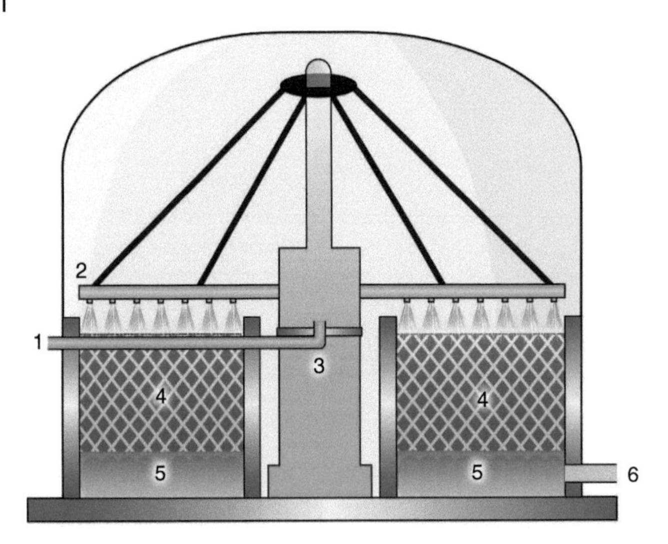

Figure 8.9 Biological filter. (1) Inlet pipe to central distributer with inlet at 3, (2) rotating distributer arms, (3) central seal, (4) filter medium, (5) under drains, (6) filtered liquid outlet.

flow freely over them. If a mineral medium is used, it is usually to a depth of 1.5–2.0 m. Many modern plants use plastics of various shapes, e.g. short tubes and corrugated sheets. These have the advantage of being light, meaning that a treatment tower above ground can be built up to a height of 10 m, thus requiring much less land. The construction of a typical biological filter is shown in Figure 8.9. The settled sewage liquor must be evenly distributed across the surface. It is usually applied using a rotary distributor consisting of a central pipe into which the sewage flows and then into distributor arms and a jet sprinkler that sprays the liquor evenly over the surface. These jets are usually placed further apart near the centre and closer at the periphery to allow a greater distance of travel in a circular bed towards the edge. This keeps the distribution of liquor even from the centre to the edge. Although the distributor arms can be mechanically driven, there is usually enough pressure from the jets to drive them around. As the liquid flows downwards over the medium, it is only in contact with the film for less than a minute but this is enough time for material to be absorbed into the film and acted upon by the bacteria. The main organic breakdown processes are aerobic, so it is essential to maintain a good and continuous supply of air/oxygen throughout the bed. Although air can be blown through the bed mechanically, this can be expensive, and usually, as long as there are enough void spaces between the media, the natural circulation of air, be it upwards or downwards, is enough. Modern plastic media are usually designed to have an optimal flow of air through them. For a cleaning process to work, there must be a transfer of organic matter and oxygen across the film to the aerobic bacteria. As the bacterial populations grow, the film grows thicker and must be controlled to allow air and sewage flow to continue. If not anaerobic layers of film next to the media can develop and this can become unstable and be sloughed off. At the base of the filter, which is usually constructed with a slope, the filtered liquor is collected into underdrains. This filtered liquor should contain little dissolved organic matter but may contain a high amount of suspended matter derived from dead microorganisms and pieces of biofilm that have broken away. It may also contain larger grazing invertebrates, such as fly larvae and worms. Because of this, further humus settlement is required, which is carried out in secondary sedimentation tanks of similar design to the primary ones. After this, the effluent can then be discharged to a suitable water course. The filter's performance is affected by low temperatures. In cold weather, the metabolism of organisms, especially bacteria, slows down. The larger microfauna stop breeding and migrate lower into the bed where it is warmer. Reactions such as nitrification decrease significantly below 10 °C, so ammonia concentrations in the effluent may increase (EPA 1997).

There are a number of ways that filters can be operated, which are as follows (see Figure 8.10):

- Conventional single pass
- Recirculation
- Double filtration
- High-rate filtration
- Alternating double filtration (ADF)

Conventional single pass operations are the most frequently used. In this case, the settled sewage is passed through a filter once. The effluent is then passed into a humus tank (secondary sedimentation) and discharged to a water course. In the recirculation mode,

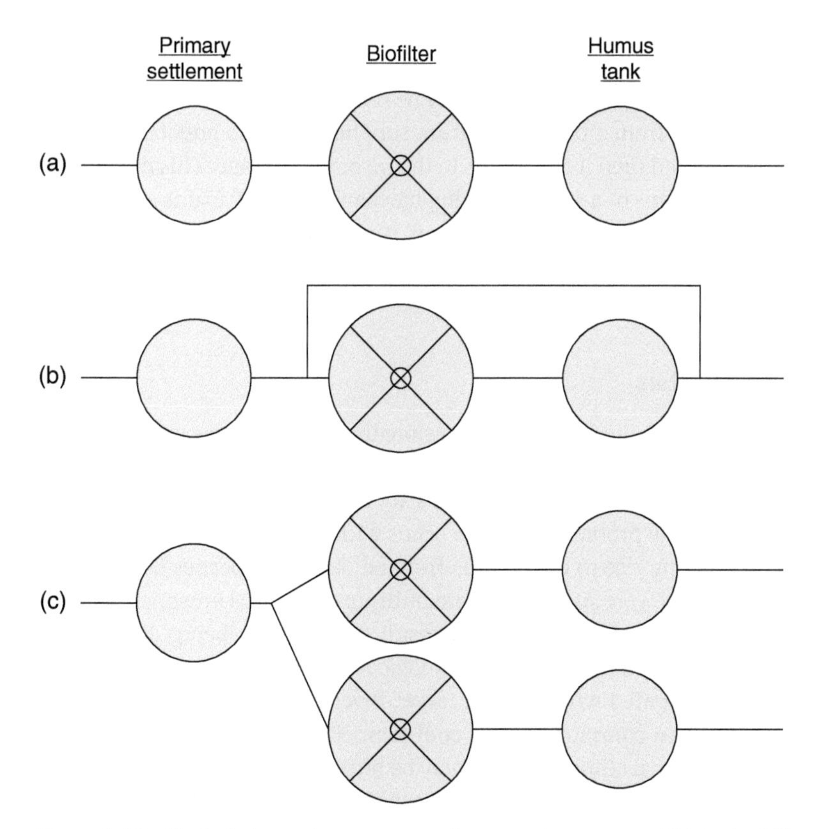

Figure 8.10 Typical alternative sequences for operating groups of biological filters. (a) Single filtration. (b) With recirculation of filtrate. (c) Alternating double filtration.

a proportion of the humus tank effluent is returned to the settled sewage stream at a ratio of about 1:10, where it is then filtered again. This gives more prolonged biological treatment, so more difficult compounds are potentially broken down. Recirculation can also be used in smaller works, and this keeps the bed from drying out intermittently and provides a continuous daily flow through the bed. In double filtration, the settled sewage is passed through one filter and a humus tank and then passed through a second filter, giving additional treatment. At lower temperatures, the first filter can have a tendency to pond causing problems. If plastic media are used in the first filter, this can reduce problems as well as have a smaller land requirement. High-rate filtration is a form of double filtration that uses larger beds with larger sized media or plastic media. Plastic media have the advantage of having very good aeration and oxygen transfer to the biofilm. This means that it is able to take a much higher organic loading and/or work at higher hydraulic loads. ADF is a form of double filtration in which the order of flow between the leading and second beds can be reversed at intervals. A higher load can cause ponding to the first bed, so if the flow is changed so that the first becomes the second, it will then receive weaker sewage and allow the biofilm thickness to be controlled. The reversal time will depend upon local conditions.

Some works receive raw sewage that does not produce good bacterial populations. This could be due to several factors, including (i) an incorrect nutrient balance for the bacterial populations, (ii) toxic or inhibitory substances in the sewage and (iii) incorrect hydraulic loading that can cause excessive biofilm sloughing and removal.

In the first case, it may be possible to add a suitable bacterial culture capable of living on the balance of nutrients present. This could possibly be obtained from other works having a similar raw sewage feed or from a bacterial culture supplier. It is also possible knowing which nutrient is deficient and then it to add this to the incoming sewage. This might need to be added constantly by means of a suitable dosing mechanism. If any toxins are present, their source needs to be identified and controlled. If it is from an accidental discharge, industry could be required to pre-treat the waste to remove the toxin prior to discharge to the sewer.

Possible Problems with Filters

Biological filters are generally quite robust, but occasionally some problems may arise. The most common problems are ponding and excessive sloughing of the biofilm in the spring. Some of the causes of ponding and possible solutions are listed in Table 8.4.

Some of the other additional problems that can occur with filters are mainly biological. The most common one giving rise to complaints from adjoining properties is with flies. Although most sewage works were sited away from built-up residential areas, subsequent pressure on land by increasing populations can result in dwellings being constructed right up the sewage work perimeter. Because biological filters have an extensive and diverse food web when compared with an ASP, larger invertebrates such as worms and fly larvae colonise them. The common flies in cool temperate climates are not usually harmful, but in warmer tropical climates, some may be biting species. In cooler climates, the species have only weak powers of flight so that when adults emerge from the larvae that live in the filter, they can only fly a small distance from the filter unless carried by air currents. The adults lay their eggs in moist sheltered locations as are found in the

Table 8.4 Potential problems and causes and some possible solutions in biological filters.

Cause	Effect	Solution
Too high organic loading	Rapid growth of biofilm clogs voids	In short term pressure wash to remove film from voids. In longer term consider increasing hydraulic loading
Medium too small	Voids not big enough	Dig or rake medium to remove excess Biofilm for short term relief but in long term replace the media
Initial screening of incoming sewage poor	Debris not removed and clogs filter	New efficient screens needed

Source: Partly adapted from EPA (1997).

upper layers of a biological filter, and the larvae emerge and become part of the grazing fauna within the biofilm. The emergence of adult flies tends to be synchronised and corresponds to increasing temperatures. In Europe, two of the most frequently occurring species are *Psychoda* and *Scatella*. Methods of control include recirculation of the effluent, reducing the accumulation of film, cutting any grass areas around the filters so there are no suitable places for the flies to land, planting trees around the perimeter of the works and as the flies flight is weak they cannot fly over them. It is possible to apply insecticide to the filters, although using these filters may require the permission of the local health authority and environmental regulator as ecosystems in the water receiving the final effluent could be affected. Some works have applied plastic netting to cover the filters, preventing adults from flying away. Another problem with filters is that moss grows over their surface. Moss tends to grow in drier but damp areas and can be controlled by increasing the flow. Moss growth can also form a substrate for other weeds to grow. All of these plants will only grow if they have access to light and can be controlled by forking or raking over the top few centimetres.

Humus Tanks (Secondary Sedimentation Tanks)

Humus tanks remove any solids from the filtrate and are of similar design to primary sedimentation tanks. They may be circular or rectangular and have a radial, horizontal or upwards flow. Circular tanks with a radial flow are the most common for larger works. They are required as the filter effluent will contain pieces of detached biofilm, dead organisms, fly larvae, etc. which have to be removed. It is important to ensure that the exit weirs are all at the same height so that the outflow is even over the whole tank. Poor effluent quality can occur if the collected sludge is not regularly removed as gas bubbles can be produced within the settled sludge, causing it to float and possibly float over the weirs. Scum boards may be provided to hold any floating matter. Humus tanks generally have a retention time of about two to three hours. Settled sludge is removed from the sloping tank floor by scrapers. Desludging at regular intervals is important and can be carried out automatically or manually although the latter is usually only used in small works.

Rotating Biological Contactors (RBC)

For smaller works and even for only one or two houses, a number of package units are available. Package plants are units that are manufactured in various sizes to suit various population sizes. They are self-contained units requiring only a small amount of pipework and installation. Their aim is to provide acceptable treatment for waste for small communities. The actual treatment processes are similar to those used in conventional municipal sewage works that have been downsized. It should be noted that some have cost implications. One such unit is an RBC. The principles behind RBCs were first outlined at the start of the twentieth century, but it was not until nearly 70 years later that the first units were installed. Sufficient treatment is provided so that the effluent, after final settlement, can be discharged to the environment. Sludge is produced and collected for separate disposal. The RBC unit, illustrated in Figure 8.11, is made up of four zones: the primary settlement zone,

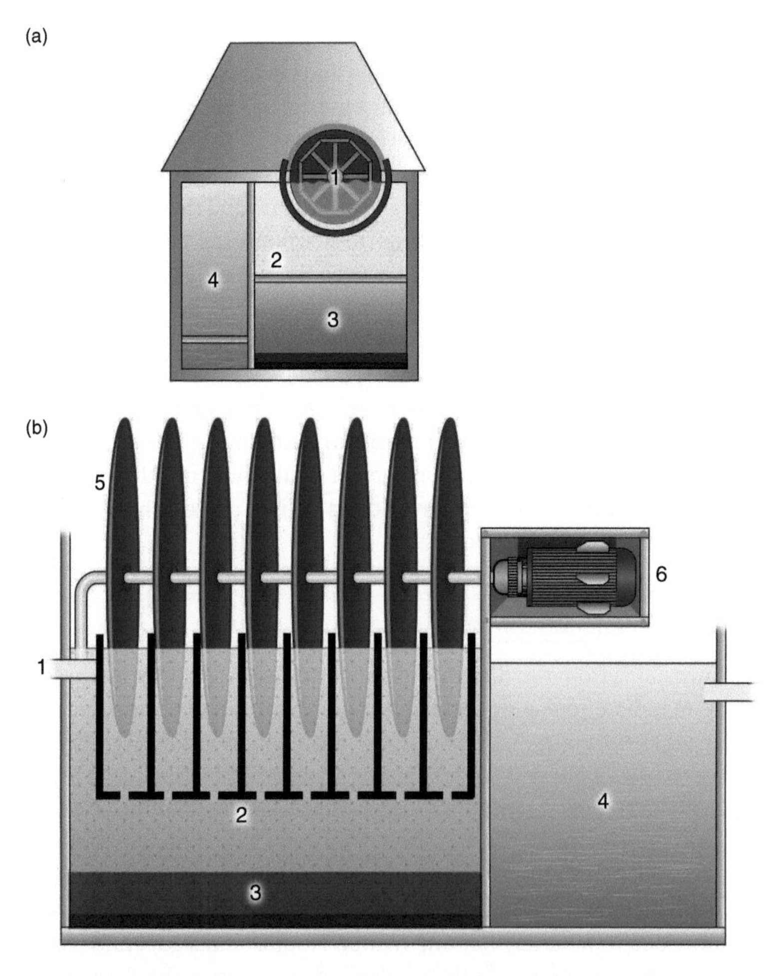

Figure 8.11 Rotating biological contactor. (a) End view. (1) Biological disc, (2) primary settlement zone, (3) sludge settlement zone, (4) final settlement zone. (b) (1) inflow, (2) primary settlement zone, (3) sludge settlement zone, (4) final settlement zone, (5) biodiscs, (6) motor drive to rotate discs.

the main biologically active zone, an area for sludge storage and the final settling zone. The rotating discs, which can be up to 3 m in diameter, are generally made of plastic. They are mounted on a drive shaft that allows them to rotate so that only 40% of the disc area is immersed in the liquid at any time. The multiple discs are mounted about 2–3 cm apart and are rotated at just under one revolution per minute although the speed can be varied. The different zones in the unit are usually separated by baffles which may be stepped down left to right to cause the sewage liquor to flow along a zigzag path from one compartment to the next. As the discs rotate, they are alternately immersed in liquor and then air. The sewage enters zone 1 of the unit with a velocity of about 0.35 m/s. Two fast a rotation of the discs and too great a velocity of flow can cause sloughing of the film. Too slow a rotation can result in too little aeration. The biofilm builds up, which is similar to that in conventional biofilters, on the discs that absorb organic matter from the sewage as it dips into the liquid and absorbs oxygen as it moves through the air. The discs are normally rotated by an electric motor. The flow of liquid and the rotation of the discs keeps the sewage liquor mixed and provides a constant movement over the film. The pollution load decreases as the sewage passes along the different zones, and the change can be observed in the film colour that tends to be grey at the start and becomes brown and smooth by the end. Solids from both the incoming sewage and detached biofilm settle to the bottom sludge collection zone, after which the effluent can be discharged. The film growth is greater near the inlet end and if the thickness needs to be controlled to prevent clogging between adjacent discs. Either the speed of rotation or the velocity of flow can be increased to slough off some excess film. Accumulated sludge should be periodically removed. As the whole unit is covered, it is protected from cold air, and heat is retained within the tank, aiding oxidation of the wastes. RBCs can be manufactured with green plastic covers, so they are inconspicuous.

Sludge Treatment and Disposal

The sludge produced from both biological filters and settlement tanks must be treated before disposal. Typically, the sludge will consist of about 90% water. For disposal, the sludge is thickened to remove most of the water. Although the composition of the sludge will vary depending upon the type of wastewater treatment system and the composition of the raw sewage, dry sludge would have a composition as shown below:

- 45–50% organic matter
- 25–55% inert mineral material
- 2–4% nitrogen
- 0.5–1.0% phosphorus
- 0.2–0.4% sulphur, chlorine
- 0.1–0.2% heavy metals
- Other organic contaminants

Source: Delft University of Technology (undated).

Sludge from the primary settlement tanks and the humus tanks can initially be thickened by further gravity settlement, which can remove up to a further 70% of the water. Other less frequently used methods include flotation and centrifugation. Gravity settlement is usually carried out in circular tanks similar to primary settlement tanks. They need several days of gentle stirring to enhance the concentration and allow solids to settle to the bottom when they are drawn off and either further dewatered or pumped into digesters. Water from the sludge will be rich in organic matter, more than would be present in normal raw incoming sewage, so it needs to be recycled into the treatment process before discharge. Before dewatering, the sludge may need to be stabilised. This involves either thermal, chemical or further biological treatment to prevent anaerobic breakdown and the production of unwanted odours. Of these options biological processes are the most widely used. In larger treatment works, the sludge is gently heated in anaerobic digesters. The resulting microbiological digestion breaks down the larger organic molecules by hydrolysis and then acidifies the resulting products before being acted upon by methanogenic bacteria to produce methane gas and carbon dioxide. The methane is collected and used as a source of energy for the works (Delft University of Technology Undated). The energy generated is a valuable by-product for the treatment works as large amounts of energy are consumed in various stages of sewage waste treatment. The sludge can, if no toxins are present, be composted. Heat treatment can be used on dewatered sludge to produce a dry granular powder that can be sold to the public or for use as an agricultural fertilizer. This heat treatment can be expensive; if this treatment pathway is not used, further dewatering may be carried out by squeezing the sludge in filter presses, centrifugation, drying beds or even reed beds. The aim is to produce a cake containing about 15% water at most (Gray 1999). In some countries, much of the sludge is incinerated for disposal. Disposal to landfills has certain risks as any leachate may contain metals and pathogens, and additional methane can be produced during decomposition. Landfill disposal has been widely used in European countries, for example in Denmark where about 45% goes to landfill, the United Kingdom 26%, Ireland, 51% and France 50%.The EC Directive on landfill requires that any sludge to be disposed of must have a minimum solids content of 35%. Sewage sludge has a high content of organic matter, nitrogen and phosphorus, so it could be a useful fertilizer and soil conditioner. Potassium is usually only present in low concentrations, so it may need to be added. The actual composition of a particular sludge will depend both on its origin and the type of treatment given.

Irrespective of whether the sewage originated from purely domestic sources or partly from industry, it will contain some metals. As many metals occur in natural food chains and can be biomagnified through the natural food chain and they are long lasting in the soil care must be taken as to the amount of sludge used on agricultural or other land. The accumulated metals can, depending upon the crop, be consumed directly by humans or be accumulated by grazing animals. The most common metals found in sludge are copper, zinc, lead, chromium and nickel. Examples of metal concentrations found in sludge are given in Table 8.5.

Waste Stabilisation Pond Treatment

Waste stabilisation ponds are constructed shallow basins that rely on natural processes to treat sewage. They do not rely on mechanical aids or consume large amounts of energy, but their action is slower than conventional mechanical methods such as activated sludge and biofilters and have retention times measured in days rather than hours. Because of their

Table 8.5 Average concentrations of selected metals found in sludge.

Metal	Average concentration (mg/kg of dry sludge)
Cadmium	29
Chromium	744
Copper	613
Lead	550
Nickel	188
Zinc	1820

Source: Abstracted from Gray (1999) and National Water Council (1981).

Table 8.6 Advantages and disadvantages of different sewage treatment systems.

Parameter	AS	BF	RBC	WSP
BOD_5 removal	**	**	**	***
SS removal	***	***	***	**
FC removal	*	*	*	***
Virus removal	**	*	*	***
Construction costs	High	High	High	Low
Operator skills	High	Medium	Low	Low
Maintenance cost	High	Medium	Medium	Low
Energy use	High	Low	Medium	Low

Key: *, low; **, medium; ***, good. The terms high, medium and low are relative to each of the techniques.
Source: Adapted from Arthur (1983).

simplicity, their operation and maintenance do not require high skill levels and can be operated by local people with simple agricultural skills and a small amount of training. Construction of the ponds is by simple earth walls and base either with a natural or artificial impermeable lining. A brief comparison of the advantages and disadvantages of different types of sewage treatment is provided in Table 8.6.

In summary, the three main advantages of WSPs are that they are simple to construct and operate. They cost less to build and run and they are efficient and can withstand shock loads. Some comparison overall costs with percolating biofilters show them to be over 35% less expensive (Mara 2013). The main disadvantages are that they take up more land and purify the waste more slowly. If not properly designed or are overloaded, they are likely to have odour problems. Because micro algae abound in the final two ponds of a sequence, it is quite likely that they will be present in the effluent. This can raise the BOD of the effluent, largely because the BOD test is carried out in the dark. However, because algae are photosynthetic, they will actually contribute oxygen to the receiving water, which is in the light.

There are three main types of WSPs: anaerobic, facultative and maturation ponds. Each of these has a different function. Aerobic and facultative ponds are the main BOD removers, while maturation ponds help remove the majority of faecal bacteria. Microscopic photosynthetic algae play an important role in facultative and maturation ponds by helping supply the oxygen required by the bacteria, which in turn supplies CO_2 for the algae. The three pond types function in the following manner.

Anaerobic Ponds

These ponds receive raw sewage and are thus the most heavily loaded. Their appearance is usually very dark brown, and they may be malodorous. They may be up to 5 m deep and. Because of the intense bacterial activity, they contain little or no dissolved oxygen, and apart from an occasional thin film on the surface, no algae. The purpose of this pond is to remove organic matter and BOD (as much as 60% of the BOD can be removed at 20 °C). The retention time is often less than 1.5 days. In addition to bacterial action, large particles are also removed by sedimentation. Anaerobic breakdown of the waste occurs in the sludge layer and can be quite fast at temperatures above 15 °C. If there are industrial wastes mixed with domestic wastes, there could be toxins present that could affect the action of the microorganisms as well as the two later ponds. In the anaerobic pond, these may be precipitated and removed into the sludge, thus protecting later stages. Some floating matter may be present, and this can be skimmed off for disposal. Over time, usually every two to three years, the sludge solids accumulate, and when the pond is about one-third full, the pond must be de-sludged. Excess sludge can be spread over drying beds or a suitable land disposal site. The main odour problem with anaerobic ponds is hydrogen sulphide generation. This is not usually a problem if suitable design loadings are adhered to (Mara 2013). Methane may also be generated, and if the pond is covered, it may be collected and used as an energy source. Although for domestic wastes, a single anaerobic pond is usually sufficient, and if there are strong industrial organic wastes, for example from agriculture or food processing, a series of two or more ponds is often advantageous.

Facultative Ponds

Facultative ponds usually receive settled wastewater from an anaerobic pond or, in some circumstances, raw wastewater after preliminary treatment. They are designed to receive low BOD loadings. The water colour in these ponds is bright green because of the abundant algal growth. These algae produce large amounts of oxygen through their photosynthetic activities, which are used by the bacteria during their decomposing activities. There are often many species of algae, although at any time, one might dominate, ranging from motile *Chlamydomonas* and *Euglena* to species of *Cyanobacteria* (blue–green algae). Because algae are photosynthetic, they only produce oxygen during daylight. At night, they respire and together with the bacteria consume oxygen. Bacteria consume oxygen both day and night, but algal oxygen production exceeds bacterial consumption during the day. This results in marked diurnal oxygen fluctuations. In warm climates, there is often a more windy period that results in complete mixing of the water, producing an even temperature and algal populations throughout the pond. As the day progresses and the winds drop, solar radiation increases and thermal stratification develops, including the formation of a thermocline. Above the thermocline, the water becomes warmer, but below it, the temperature changes very slowly or not at all. In the late afternoon or evening, especially if an evening breeze develops, a period of mixing can occur initially in the upper layers but then throughout (Marais 1966). This pattern of stratification and mixing occurs in many shallow warm climate lakes and usually leads to layering of the algal populations as well. Active photosynthesis by the algae not only results in higher oxygen concentrations but also a significant rise in the pH to as high as 12, which causes a rapid die-off of faecal bacteria. The algae, during stratification, concentrate in the upper layers and can be passed out in the pond effluent (depending upon the draw off level). This, together with diurnal mixing, can have a marked effect on the outlet water quality. Some ponds have assisted mixing, either wind-assisted aerators or, rarely, electric motor-driven aerators. The action of the mixers generally improves the reduction in the BOD of the pond effluent, probably due to an improved oxygen transfer throughout the pond water column.

Maturation Ponds

Maturation ponds are the last in the series of WSP. Their function is to remove any remaining pathogens, such as viruses and faecal bacteria, as well as reduce the concentrations of nutrients such as nitrogen and phosphorus. The processes in these ponds tend to be slower and do any further removal of suspended solids and BOD. This final removal is especially important when the effluent is going to be used in agriculture for irrigation. Maturation ponds are aerobic at all depths, and their algal flora, although diverse, have smaller biomass than facultative ponds. Maturation ponds are often used as a series of two or more ponds. They are shallow, 1 m or less deep, allowing greater light penetration, which helps destroy viruses. It is generally advisable to line these ponds with a good impermeable membrane to stop the growth of rooted aquatic macrophytes, which could encourage insects such as mosquitos. Maturation ponds usually have a retention time of about 10–15 days. Nutrient removal, in particular phosphorus removal, is important for the protection of the receiving water and preventing eutrophication. Mara (2013) suggests that two types of processes are involved in bacterial removal. These are 'dark mediated' and 'light mediated'. The dark-mediated ones include sedimentation, predation by free-living protozoans and invertebrates and senescence. Light-mediated processes include the length of time exposed to sunlight, temperature (although bacterial growth rates increase with temperature within their tolerance range, so do their death rates as the temperature nears their upper limit or exceeds it), and water pH. The pH increases with active photosynthesis and then due to the algal and bacterial die back can rise above 9.4. More than one maturation pond may be used. For example, three could be used in sequence, with each having between three and four days of retention. This sequence will give a better quality effluent capable of being used for unrestricted irrigation or discharge to more sensitive surface waters.

Water Storage and Treatment Reservoirs

This type of reservoir, sometimes called purification lakes or effluent storage reservoirs, was developed in Israel to hold whole-year treated wastewater for future use as irrigation water. A variant of these has been used in Germany and England to store water suffering from a degree of pollution. In both cases, particles are removed by sedimentation, and bacterial numbers also decrease. In Israel, the reservoirs usually have a depth of about 25 m and are used after anaerobic ponds. During the irrigation season they are drained down gradually and used as the crop requires, but as anaerobic pond effluent is continually being added and as the volume of water in the reservoir decreases, the proportion of anaerobic effluent relative to the volume of water in the reservoir increases, so at the end of the growing season, the water quality becomes progressively lower. As a result, this late water can only be used for restricted irrigation, i.e. crops not for human consumption. In the case of the German and English use of these reservoirs, the water is discharged to a watercourse but with a reduction in both BOD and suspended solids (Woods et al. 1984).

Constructed Wetlands (CWs)

Natural methods of waste purification have been used for many years. These methods, termed phytoremediation, can use rooted emergent plants, submerged rooted plants or floating plants. Here, we will only consider the use of emergent plants and subsurface flow systems.

Natural wetlands have long been recognised as a natural cleansing interface between land-based and water-based ecosystems. They help trap and remove sediments and nutrients leaching from the land, thus helping to prevent enrichment of the aquatic system. They have also been used for many years to treat domestic wastewater. CWs are relatively simple to construct, operate and maintain, so they are ideally suited for small communities. Constructed wetlands, also known as reed beds, are designed to enhance and control these natural properties and functions. They consist of rooted aquatic plants (macrophytes as opposed to microscopic algae found in WSPs) grown in sand, gravel or well-drained soil into which domestic sewage, after or without primary treatment, is passed. Typical aquatic macrophytes used in constructed wetlands are the common reed *Phragmites australis,* cat tail, *Typha latifolia,* and the soft rush, *Juncus effuses*, although there are many other suitable local aquatic plants that may be used (see Table 8.7 for examples of wetland plants). Wetlands can be used for secondary or tertiary treatment. They may be surface or

Table 8.7 Examples of wetland plants.

Species	Common name	Water depth (m)
Emergent species		
Scripus spp.	Bulrush	0.1–1.5
Iris sp.	Iris	<0.05–0.2
Juncus spp.	Rushes	<0.05–0.25
Typha latifolia	Bulrush	0.1–0.75
Typha angustifolia	Narrow leafed reedmace	0.1–0.75
Carex spp.	Sedges	<0.05–0.25
Decodon sp.	Water Willow	0.1–0.5
Salix spp.	Willows	0.1–0.5
Phragmites spp.	Common reed	0.1–1.0

Source: Adapted from Kadlec and Knight (1996) and other sources.

subsurface flow, either horizontal or vertical flow (VF) or any combination. Wetlands are less expensive to build and can be made using local materials. They are simple to operate. They do, however, require more land, and care is required in their design based upon local conditions and climate.

Construction of horizontal flow subsurface wetlands.

Constructed wetlands are usually rectangular in shape but longer than wide and relatively shallow. They are usually constructed according to the formula of Kickuth (1984), which is

$$A_h = \frac{Q_d\left(\text{In}\,C_1 - C_e\right)}{K_{BOD}}$$

where

A_h = surface area of bed
O_d = average daily flow rate of sewage
C_1 = influent BOD_5 in mg/l
C_e = influent BOD in mg/l
K_{BOD} = rate constant (m/d)

The bed can also be sized based upon the population equivalent per day, which is about 40 g BOD and an area per population equivalent (pe/d) of 1–2 m^2. The depth of the bed varies but is about 60 cm in Europe but 30–40 cm in the United States of America, although some studies in Spain have concluded that even shallower beds (27 cm) give better results (Garcia et al. 2004). To avoid excessive surface flow and depending upon local climate conditions a depth of 40 cm is usually required. The width of the CW can vary depending upon the local topography, but regardless of the width chosen, an equal distribution and flow of settled sewage is essential. To achieve this for wider beds (>15 m), the bed may need to be divided into two or more lengthwise channels to avoid short circuiting. The medium used should provide a suitable substrate for rooting the macrophytes and provide a suitable surface area on which the microbial film can develop. The medium should also act as a filter. The grain size can be between 0.2 and 30 mm (USEPA 2000). The media at the inlet and outlet zones are of considerably larger size, 40–80 mm, to avoid clogging. The inlet should give an even spread of waste across the bed width, and the outlet should gather all of the effluent into a single collecting pipe.

The flow of settled sewage is mainly sub-surface through the root zone of the plants. The beds are constructed with an impermeable liner and a small slope across its base to encourage flow from entry to exit. The surface of the sand and gravel provides a surface for microorganisms to colonise and grow as does the surface of the plant roots. All of the macrophyte roots do not grow vertically downwards, and many horizontal rhizomes are also produced. The stems and roots of many wetland plants have air spaces (aerenchyma) within them so that gases may pass down the stems to the root zone, providing oxygen to the cells and preventing anoxic conditions. This results in an aerobic zone surrounding the roots. BOD and nutrients are removed in the root zone by a combination of microbial activity and absorption by plants. The sewage waste entering the bed is normally given preliminary/primary treatment first. In the bed, there are several removal mechanisms in action, including sedimentation, filtration and adsorption, bacterial degradation, absorption and volatilisation (UN Habitat 2009). The substrate on which the plants grow is also able to absorb and adsorb metals further helping to clean the wastewater. During the non-growing season, the plants do die back, and some species die back more often than others. Some sub-surface rhizomes die and disintegrate, providing passages that can help the flow of liquid. After a while the bed becomes clogged and may need desludging or even digging out and re-planting. In addition, organic matter breakdown by bacterial compounds of nitrogen and phosphorus can also be removed. Phosphorus removal can occur by adsorption, complexation, precipitation, plant uptake and assimilation (Watson et al. 1989). Nitrogen can be removed by a number of processes as in other forms of biological treatment. These include ammonification, nitrification, denitrification, volatilisation, adsorption and plant uptake. Microbial nitrification/denitrification are the main routes. Removal of metals can also occur in CWs. Indeed, wetlands have been used in metal refining plants for effluent treatment. The mechanisms involved include precipitation, ion exchange, sedimentation, filtration and plant uptake. Many of these are mediated by microbial activity (Watson et al. 1989). Pathogens are removed by similar processes that occur in biological filters, including natural dieback and predation (Cooper et al. 1997). As constructed wetlands are most often used for smaller community waste treatment as well as industry, certain agricultural units and environmental cleansing (this includes road run-off and diffuse land pollution), it is best to keep designs simple, requiring minimal maintenance, needing as little external energy input by using gravity flows wherever possible and making sure that the design can deal with changing flows and changing climate (USDA 2015). The design should also fit into the local landscape, and it should also be noted that a newly planted bed requires a reasonable time to settle down and function properly. In most, if not all wetlands, the vegetation could need harvesting to prevent the annual dieback of leaves, rotting down and clogging the surface. In some countries, burning dead vegetation is used to clear the bed of debris. In any natural system, seeds of unwanted species may be brought in from the surrounding land. Regular inspection of the bed is needed, and if any unwanted species are present, they should be removed.

Vertical Flow Wetlands

An alternative to horizontal flow sub-surface wetlands is a VF system. These operate on similar principles to horizontal flow systems except that the settled sewage is introduced evenly across the surface and allowed to percolate down through the bed to be collected at the bottom. They consist of a lined rectangular flat bed with usually a gravel layer, about 25 cm deep, over which is a layer of fine gravel or coarse sand, 60–80 cm deep. The vegetation is planted into the top coarse layer (Figure 8.12). It is important to ensure that the distribution of sewage liquor is even across the surface, and this requires a series of distribution pipes and a pumped delivery system that should have the facility for intermittent dosing as the water should be fed in batches to prevent flooding and anaerobic conditions developing (Vymazal 2010). Successive doses are only added when the previous dose has percolated through the bad, thus allowing air/oxygen to penetrate the upper bed layers. This mode of operation means that the bed is largely aerobic, unlike horizontal flow systems, where parts of the bed away from the plant roots can become anaerobic. The aerobic

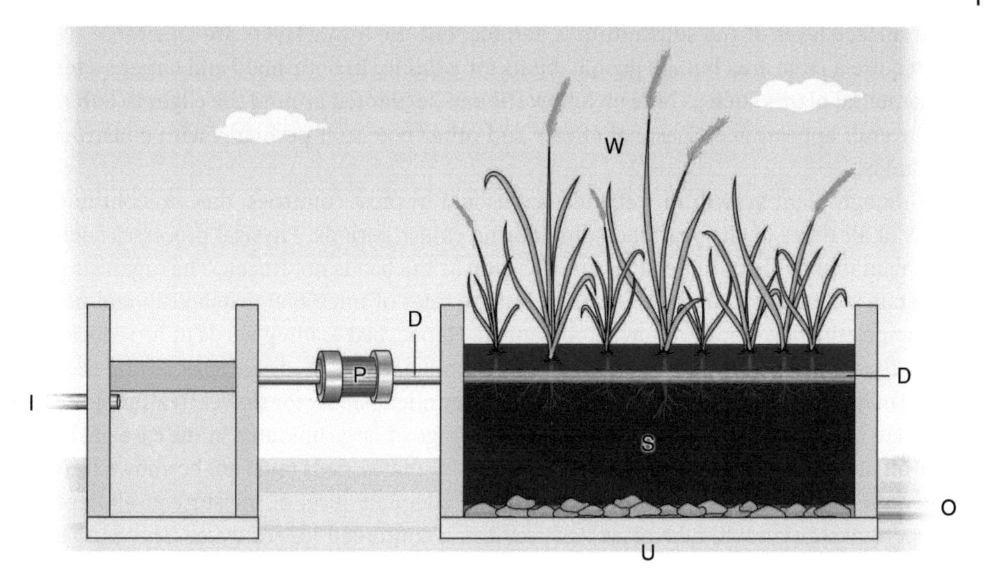

Figure 8.12 A vertical flow wetland. (I) Inlet and sedimentation tank, (P) pump, (D) distribution pipes, (U) underdrains, (S) sand or sandy soil, (W) wetland plants, e.g. *Phragmites australis*, and (O) outlet pipe.

conditions favour nitrification but not denitrification. VF systems are good at removing suspended solids and reducing BOD. They also require less land but do require pumping equipment, use more energy and require more maintenance. As with HF wetlands, VF can be used to treat not only domestic wastes but also surface run-off, food production waste, agricultural wastes and some industrial wastes. In some cases with VF systems, upwards flow rather than downwards flow has been used. In these cases, the filtrate can be collected at the top of the bed. All wetlands are efficient at removing BOD (between 75% and 90%) and suspended solids (75–89%) (Vymazal 2010).

Whatever the type of wetland, a start-up period is required to allow the plants and microbial populations time to develop. During the start-up period, the bed should be kept moist but not flooded; otherwise, it could become anaerobic. Odour problems are uncommon with wetlands if they are not overloaded. Constructed wetlands, as with all treatment systems, should be regularly monitored regarding their performance. This should include monitoring total suspended solids, BOD, ammonia, nitrate, phosphorus and faecal coliform removal. Depending upon the receiving water source protozoan and helminth parasites should also be monitored.

Constructed wetlands can also be used for sludge drying. They have the advantage of reducing both the water content and the organic solids in the sludge. They are simple to operate but can require a large area of land. Beds for sludge drying usually have a similar construction to VF wetlands. The sand/gravel usually has a depth of up to 0.75 m.

Wetland systems can be scaled down to suit small communities or even individual dwellings. They may also be used in surface flow mode where there is free water on the surface at all times. The water is maintained at a shallow depth to allow plants to grow.

The surface layer of the substratum is aerobic, but the lower layers are anaerobic. They do require a large area but are good habitats for wildlife. In both small and larger systems, ornamental plants such as blue or yellow iris can be planted around the edges to enhance the overall appearance. Seasonal effects and other potential problems with constructed wetlands.

Although plant growth in wetlands is seasonal in most countries, they do continue to work at all times of the year, including during colder periods. Physical processes such as sedimentation will continue at all times as long as the bed is not frozen. The organisms do generate some heat from their activities, but the rates of microbial metabolism and hence decomposition are slower at low temperatures. Hence, bed loading needs to be reduced at these times. If there are high organic loadings, as might occur in certain agricultural wastes, some may need to be stored and applied at a lower rate to allow for slower treatment. If the beds are used for surface run-off again, some storage of large amounts in the case of storm run-off and snow melt may be required. In warm climates, beds must not be allowed to dry out, so flows must be regulated to maintain an adequate level of moisture at all depths. Evapotranspiration will take place and result in a continued loss of water. This can be as high as 80% from the surface of shallow water bodies in warm or hot months, and this must be compensated for. Beds used for industrial wastes that may contain dangerous substances, such as metals and pesticides, may more often need the sludge to be removed and safely disposed. Persistent organic compounds can also pose a problem of disposal. In systems designed specifically for wastewater treatment, weeding may be needed to maintain the population of selected plants most efficient for treatment, but in free surface water wetlands, it can be more difficult to control invasion by unwanted species. In addition, a natural ecological succession may develop as areas silt up, allowing for the colonisation of unwanted woody species. Adequate reductions in BOD may still occur, but flow patterns may change adversely. Care must be taken with all constructed wetlands to maintain an evenly distributed flow and not allow short circuiting. Routine maintenance should include repairs to surrounding walls, especially if they are earth walls, as, depending upon local conditions, burrowing animals can cause considerable damage both to the walls and liners. Sometimes wire screens may need to be installed to prevent this. Flying insects, including mosquitoes, can be a problem if there is stagnant open water. Normally, surface ponding should be avoided, and any surface waters should be flowing to discourage adults from laying eggs and their larvae developing.

Constructed wetlands can be used for treating grey water discharges from domestic dwellings. They are usually provided with a primary settlement tank before wetland treatment. Whatever system is used it needs to be deep enough to support the growth of suitable plants. About 0.6 m is enough for *Phragmites* to grow. If necessary, there may be a need for multiple inlet pipes to ensure an even distribution of liquor across the surface.

A different approach to wetland configuration is popular in France. This is a two-stage vertical flow constructed wetland (VFCW). One advantage of this system is that raw sewage can be fed onto the bed, making management of the sludge easier (Boutin et al. 1997; Molle et al. 2005). They recommended two-stage filters, the first of which is divided into three filters and the second into two filters. Each primary unit is fed raw sewage for a period of up to four days. They are then rested for 7–10 days. This alternating approach controls the biofilm growth. The effluent then passes onto the second-stage filters that complete the treatment. This approach is beneficial as the amount of sludge produced is less, and through its mineralisation, the system only needs desludging every 10–15 years.

Septic Tanks

Septic tanks are containers designed to receive domestic waste from individual or small groups of dwellings. Unlike cesspools, septic tanks do provide some treatment rather than being simple storage units. The units can be made of concrete, plastic or PVC. They are usually divided into two or more chambers through which the waste passes in sequence. Each section provides a different stage of treatment before, and at the end, treated wastewater can be discharged. A typical design is shown in Figure 8.13. Initially, the waste will pass into the tank, and the solids settle to the bottom. The liquid flows out of the top and can be dispersed by percolation through pipes over a suitable area of land. A certain amount of anaerobic digestion will occur in the settled sludge, but this is not enhanced and depends upon local conditions. Up to 50% of the BOD and 80% of the solids are removed. Many more modern tanks act as mini treatment plants. These can have three chambers per tank. The first allows the settlement of solids. This may have aeration included to break down larger solid particles. The second chamber is an aeration stage allowing aerobic bacteria to break down the organic matter as in an activated sludge process. In some cases, pieces of freely suspended plastic media are included to provide a surface for a microbial film to develop. Continuous aeration is provided by a submerged bubble diffusor. The third chamber receives the aerated and treated sewage liquor and acts as a clarifier collecting any remaining solids and allowing a clarified effluent to pass to the exit via a scum baffle. The construction and operation of septic tanks and mini-treatment plants are usually regulated by national and local rules. The European Union, for example, gives standards for septic

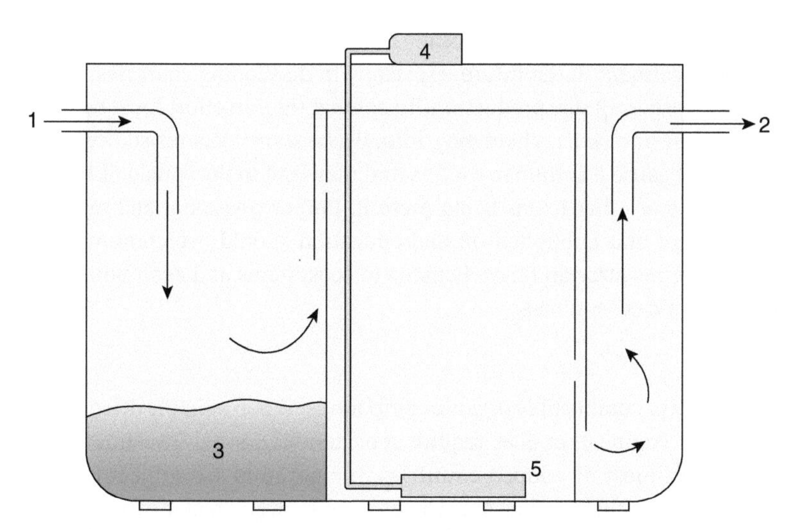

Figure 8.13 A schematic diagram of a typical domestic sewage plant/septic tank design. (1) Inflow, (2) outflow, (3) settled sewage sludge, (4) air pump and (5) aerator.

tanks. Only certified tanks should be used. In the United Kingdom, a permit must be obtained registering the system. Account must also be taken of the ground to which the effluent discharges. If this is a sensitive area, a special permit must be obtained. Direct discharges to surface waters are not usually allowed. In the United States of America, these systems are classified as on-site or decentralised systems. It is the owner's responsibility to properly maintain the system (USEPA 2015) and keep them in a state providing suitable treatment. Discharges to sensitive waters are separately regulated often on a local basis.

Disposal of Sewage Sludge

Sewage sludge, if properly treated, can be a useful fertilizer and soil conditioner in agriculture. If, however, the waste treatment works producing the waste also has part of its raw inflow from industry, then it may not be suitable for agricultural use and must be disposed of differently. Many countries have regulations governing the disposal of sewage sludge. For example, the European Union (EC Urban Waste Water Treatment Directive (91/271/EEC)) and the United States (USEPA Biosolids Rule (Federal Register 5FR9248 to 0404)) are known as the 2 Part 503 rule. Because the amount of sludge produced by treatment works has increased worldwide, there is increasing regulation regarding its disposal to protect the environment and any potential users of the sludge. Sludge in both the United States of America and Europe is called 'biosolids', although this term is sometimes used to denote treated sludge only. Sludge can be disposed of in the following ways: to agriculture, to land other than agriculture and incineration. Disposal to sea has been used in some countries and, although banned in many countries, is still used by some countries. It is not favoured because of its environmental impact. It can also be used in forestry, landfills and building materials, but these form only a small percentage of the total. The disposal method should distinguish between whether the public will have access to the site or not. If there is access, care must be taken to ensure that there is no danger to public health. For all types of disposal, there should be a comprehensive analysis of the sludge available and an environmental risk assessment concerning its disposal. It is inevitable that there will be improvements in waste treatment in the future, especially in developing countries, which will also result in an increase in sludge production. To combat this, an effort must be made to improve local knowledge and skills. There may, initially, be some local resistance to the use of sludge on land (i) because it is human excreta and may lead to the spread of human pathogens and (ii) because of other toxins being present. Proper treatment and management of both the treatment and public liaison and education should overcome worries. Sludge could be a valuable resource and have benefits to ecosystems and dealt with properly should not harm aquatic ecosystems.

Incineration

Sludge is composed mostly of combustible organic compounds. When suitably dried, it will readily burn. To carry out combustion does require considerable capital investment, and running costs are high. In most developed countries, incinerators are subject to strict air pollution control standards that, as they are stricter, require even greater expenditures. Incineration is, however, becoming more popular in densely populated areas and cities. Japan, for example, incinerates more than 70% of its wastewater sludge, and Germany and the Netherlands incinerate 34% and 58%, respectively (UN Habitat 2009). Sewage sludge is also being investigated as an alternative to fossil fuels for energy production because of its lower cost. The material from incineration is sterile ash containing some metals. Some of these metals may be toxic. Even purely domestic sewage contains some metals as these may be present in many household products. Incineration greatly reduces the weight and volume of the sludge that can then be disposed of at a regulated site as hazardous waste. If the amounts of hazardous materials can be reduced, ash can be used as a component of infill in construction projects. The sludge may have <30% solids, so additional fuel may be needed to obtain full combustion, and thus it is used in joint incineration with other wastes/fuels. Although incineration is an expensive option in countries where there has been public concern over agricultural use, incineration has still increased.

Disposal to Land

Application to land includes the spreading, spraying or injection of sewage sludge or anything derived from sewage sludge into or below the surface of the land to provide the soil enhancing properties of the sewage sludge. For land application of sewage sludge, a guide (EPA/831-8-93-002b) can be used. Economic factors are very important in deciding the method of disposal, and these factors can vary from country to country depending upon the cost of equipment and international exchange rates for its purchase, cost of fuel and other energy and availability and skills of local labour. Sewage sludge, because of its organic content and nutrients, especially nitrogen and phosphorus compounds (see Table 8.8 for selected countries), is used to improve both the quality and fertility of soils.

The regulations in many countries place a limit on the amount of sludge applied to agricultural land called 'the agronomic rate'. This is the amount of solids needed to supply the growth needs of the plants being grown. This not only means that excess nutrients are not accumulated in the soil (soil eutrophication) but also that the potential accumulation of metals does not occur.

It is used to provide soil cover for denuded areas such as mining sites. In these situations, it can be used to establish a flourishing plant cover. Before it can be used, the concentrations of both metal contaminants and pathogens need to be reduced to safe levels as well as whether there could be excessive methane production as the sludge decomposes. This is a

Table 8.8 Examples of nutrient concentrations in sewage sludge (units are % of dry weight).

	N	P205	K20
Benchmark sludge	3.5	3.5	0.7
Brazil	5.75	1.82	0.36
Turkey	1.68	0.68	0.49
Australia	7.40	1.8	0.97
Finland	3.40	2.4	—
USA	5.80	4.35	0.43

Source: Data from UN Habitat (2009).

Table 8.9 Proportions of sewage sludge disposed of to land in selected European countries.

United Kingdom	26%	France	50%
The Netherlands	32%	Denmark	45%
Belgium	53%	Germany	49%
Italy	55%	Greece	100%

major route of disposal in Europe (see Table 8.9), and the use of sewage sludge as a resource is encouraged by the European Commission in Directive 86/278/EEC and in the United States of America by USEPA at 40CFR Part 503. The benefits of using sewage sludge and other biosolids in agriculture are recognised in many countries worldwide, including China, Brazil, United States of America and New Zealand.

As the EC Directive requires a minimum solid content of 35%, most sewage sludge will need additional drying, as at the end of sewage treatment, they usually have less than 25% solids. Disposal to landfills can be prohibited because of the possibility of methane production. If the sludge is to be disposed of to surface land sites, then an assessment must be made regarding the possible nuisance of odour, flies and pathogens and danger to public health. There can also be a limit to the amount of sludge disposed of at any given site. This is termed by the USEPA as the cumulative pollution loading rate.

Disposal to Agriculture

The EC Sewage Sludge Directive 86/278/EEC encouraged the use of sewage sludge in agriculture and horticulture as long as any harmful effects to humans, vegetation, animals and the soil were prevented. Untreated sludge could only be injected into the soil, but treated sludge could be spread on the surface (Milieu Ltd. 2010). The directive also required that the use of sludge take into account the nutrient requirements of plants and that the amounts used do not adversely affect soil quality and that there is no deterioration in surface or groundwater quality. In the European Union, >10 million tons are produced annually of which 35% is recycled to agriculture. Table 8.10 shows the sewage sludge production and the amount recycled to agriculture in selected European countries.

As shown in Table 8.10, many of the largest producers of sludge also recycle, with most going to agriculture, although there are some exceptions. The implementation of the Directive applying to the then EU countries has led to a steady increase in sludge production, and for some of the countries, there has been an increase in recycling to agriculture, e.g. Ireland, Spain and the United Kingdom, while others either remained about the same. Overall, the production of sludge in most European countries is expected to even out by the middle of the 2020s and then remain steady. In some countries, public concern has led to a reduction in the amounts recycled. One of the problems of using sludge in agriculture is the possibility that metals, or potentially toxic elements or compounds, present in sewage can be taken up by plants and bio-accumulated and cause problems in grazing animals and humans. Table 8.11 shows the amounts of some metals in sludge measured as the amount of metal per kg dry weight.

Table 8.10 Sewage sludge production and recycling to agriculture in selected European countries.

Country	Data year	Production (tons dry solids)	Recycled to agriculture (tons dry solids)	%
Denmark	2002	140021	82029	59
Finland	2005	147000	4200	3
France	2002	910255	524290	58
Germany	2006	2059351	613476	30
Greece	2006	125977	56.4	0
Ireland	2003	42147	26743	63
Italy	2006	1070080	189554	18
Spain	2006	1064972	687037	65
UK	2006	1544919	1050526	68

Source: Selected data for European countries adapted from Milieu (2010).

Table 8.11 Metal concentrations in sludge (mg/kg dry weight).

Zinc (mg/kg)	1000
Copper (mg/kg)	500
Nickel (mg/kg)	40
Mercury (mg/kg)	3
Cadmium (mg/kg)	3
Lead (mg/kg)	200

Source: Adapted from UN Habitat (2009).

The amount applied to land without causing environmental problems will depend on local conditions, including existing background concentrations in the soil. This can vary greatly in different geological regions, but there are some average background concentrations per kg. dry weight (dw) of soil: zinc – 40 mg/kg dw; copper – 10 mg/kg dw; nickel – 15 mg/kg dw; mercury – 0.05 mg/kg dw; cadmium – 0.1 mg/kg dw; lead – 20 mg/kg dw. These values also assume that the average pH of the soil is approximately 6.5 (UN Habitat 2009). It must be stressed that the above values are average guidelines and may not reflect all local conditions. Although a certain quantity of metals come from domestic sewage, caused by the use of many domestic products, many metals and other compounds arise from industrial discharges. These can be largely prevented by pre-treatment at the source before discharge to sewers, thus significantly reducing concentrations in the sludge produced at the treatment works. Metals are not the only chemicals of concern. Industrial chemicals and their by-products, such as polychlorinated biphenyls (PCBs), dioxins, persistent organic pollutants (POPs), pesticides, herbicides and endocrine disrupters, can be present. A further concern is the possible spread of pathogens. In developed countries, the risk of pathogens is low partly because of the low prevalence of these diseases in the population, so the chance of them being in the sludge is low and any present would be destroyed in the sludge treatment, especially anaerobic treatment. In developing countries, however, diseases are much more common, so the occurrence of pathogens in faecal material and sludge is greater. Without proper sludge treatment, there is a greater risk of pathogen contamination when sludge is used on land-growing food crops. Some countries have imposed restrictions of such use for sludge. It should be noted that not only may pathogens affecting humans be present but also those affecting animals and plants, so the destruction of all pathogens in the sludge is essential or at least only uses where crops are resistant or not being grown. Wastewater treatment systems are continually advancing as is the treatment of sludge, so the possible risks associated with sludge use are greatly reduced. In addition, continuing research and monitoring of soils, plants and animals in areas where sludge is used together with stricter regulations greatly reduce any risk to humans and ecosystems.

The techniques for treating sludge to remove the threat of pathogens are well advanced (EPA 2003).

Examples of some regulatory standards are given in Table 8.12.

Where sludge is either disposed of or applied to agricultural land near residential properties, complaints can arise through odours. While these pose only a minimal risk to human health, they should be managed to prevent nuisance. Odour control is used in some wastewater treatment plants. This does, however, lead to increased costs. Mention should be made of the potential effects of wastewater and sludge treatment on possible greenhouse gas (GHG) emissions. Both methane (CH_4) and nitrous oxide (N_2O) are produced as a result of microbial action on the breakdown and transformation of waste organic matter under anaerobic conditions. Although much less than emissions of GHGs produced by the energy or transport sectors, many countries must pay attention to all sectors to reduce these emissions where possible. Many treatment plants manage their release of GHGs. Anaerobic digestion of sludge produces methane and carbon dioxide ('Biogas'), which can be collected and then used as a fuel to generate electricity for the works. This may provide as much as 40% of the work's requirements. This can be a major cost saving as wastewater treatment can be a large energy consumer. This energy recovery is an important sustainability activity of works. Anaerobic digestion can occur in two ranges: 32–35 °C for

Table 8.12 Examples of regulatory standards for heavy metals in sludge (concentrations in mg/kg dw).

Country	As	Cd	Cr	Cu	Pb	Ni	Zn
Brazil	14.69	10.75	143.7	255.39	80.37	41.99	688.83
USA	41	39		1500	300	420	2800
Canada	170	34	2800	1700	1100	420	4200
Norway		2	100	650	80	50	800
Jordan	41	40	900	1500	300	300	2800
China	75	20	1000	1500	1000	200	3000
Russia	10	15	500	750	250	200	1750
EU Limits (1986)		20–40		1000–1750	750–1200	300–400	2500–4000

If applied to soils with a pH >6.5.
Source: Data from UN Habitat (2009).

mesophylic and 50–57 °C for thermophilic microorganisms. The biogas produced contains between 40% and 75% methane (the remaining gas being carbon dioxide), although 60% is a common average (Stillwell et al. 2010). During digestion, trace amounts of water vapour, hydrogen sulphide, nitrogen, hydrogen, unsaturated hydrocarbons and other gases are produced, but all are small compared with methane and carbon dioxide. Gray (1999) quotes the following gas yields from different wastes: domestic sewage sludge 0.43 m^3/kg, with 78% methane; dairy waste 0.98 m^3/kg, with 75% methane; brewery waste sludge 0.43 m^3/kg, with 76% methane; cattle manure 0.24 m^3/kg, with 80% methane; potato tops 0.53 m^3/kg, with 75% methane. Carbohydrates yield 0.8 m^3/kg, Lipids 1.2 m^3/kg of which 67% was methane. Biogas can be burned to produce heat, generate electricity or used as a fuel for vehicles. It does need to be cleaned of unwanted components before use as a fuel. In sewage treatment plants, it is usually used for electricity generation.

Sewage does contain a range of substances other than simple organic matter. As such, it is important where possible to recover any valuable components in wastewater and sludge. Whatever can be recovered can be sold, helping with the cost of building and operating treatment works.

References

Arthur, J.P. (1983). Notes on the Design and Operation of Waste Stabilization Ponds in Warm Climates of Developing Countries. Technical Paper no. 7, World Bank, Washington, D.C., USA, pp. 197–213.

Boutin, C., Lienard, A., and Esser, D. (1997). Development of a new generation of reed-bed filters in France: first results. *Water Science and Technology* 35 (5): 315–322.

Brombach, H. (2004). Solids removal from combined sewage overflows with vortex separators. *International Conference on Innovative Technologies in the Domain of Urban Water Drainage*, Lyon, France (November 1992), pp. 447–459.

Buswell, A.M. and Strickhouse, S.I. (1928). Studies on the depth of sewage filters and the degree of purification. State of Illinois, Department of Registration and Education, Division of the State Water Survey, *Bulletin* 26.

Casey (1997). *Unit Treatment Processes in Water and Wastewater Engineering*. England: Wiley.

CIWEM (Chartered Institution of Water and Environmental Management) (1997). *Activated Sludge Treatment*, 207pp. London: Handbooks of UK Wastewater Practice.

Coombs, E.P. (1992). *Activated Sludge Ltd. The Early Years*. Bournmouth. Published by C.R. Coombs.

Cooper, P.F., Job, G.D., Green, M.B., and Shutes, R.B.E. (1997). Reed beds and constructed wetlands for wastewater treatment. *European Water Pollution Control* 6 (7): 49.

Corbett, J. (1903). A dozen ways of sewage purification experiments on a large scale at Salford, England. *Engineering News Record* 49: 191. See also ibid 48, 155.

Corcoran, E., Nellemann, C., Baker, E. (eds) et al. (2010). *Sick Water?* The central role of wastewater management in sustainable development. A Rapid |Response Assessment. UNEP/UN Habitat.

Environmental Protection Agency (EPA) Ireland (1997). *Wastewater Treatment Manuals. Primary, Secondary and Tertiary Treatment*. Wexford, Ireland: The Environmental Protection Agency.

Environmental Protection Agency (EPA) Ireland (2003). *Wastewater Treatment Manuals. Primary, Secondary and Tertiary Treatment*. Wexford, Ireland: The Environmental Protection Agency.

Faram, G, James, M.D., Williams, C.A. (2004). *Wastewater Treatment Using Hydrodynamic Vortex Separators. Chartered Institution of Water and Environmental Management International Conference*, Wakefield, UK (11–15 September 2004); pp 79–87.

Field, R., Averill, D., O'Connor, P., and Steel, P. (1997). Vortex separation technology. *Water Quality Research Journal, Canada* 32 (1): 185–214.

Frankland, E. (1868). *First Report of The River Pollution Commission (British).*

Garcia, J., Morator, J., Bayona, J.M. and Aguirre, P. (2004). Performance of horizontal surface flow constructed wetlands with different depths. *Proceedings of the 9th International Conference on Wetland Systems for Water Pollution Control*, Avignon, France (26–30 September 2004); pp 269–276.

Gray, N.F. (1999). *Water Technology*, 548pp. London: Arnold Publishers.

Halliday, S. (1999). *The Great Stink of London*. Stroud, UK: The History Press, 210 pages.

Imhoff, K.R. (1983). Spezifische Schlammengen und Lastzahlen des Einwohners. *Korr, Abwasser* 30: 907–909.

JMP (2010). *Progress on Sanitation and Drinking Water, 2010 Update*. World Health Organization and UNICEF.

Kadlec, R.H. and Knight, R. (1996). *Treatment Wetlands*, 893pp. London & New York: CRC Press Inc. Lewis Publishers.

Kickuth, R. (1984). The root zone method. *Gesamthochschule Kassel-Uni des Landes*. ON Canada: Hessen.

Mara, D. (2013). *Domestic Wastewater Treatment in Developing Countries*, 210pp. London: Earthscan.

Marais, G.V.R. (1966). New Factors in the design, operation and performance of Waste Stabilization Ponds. *Bulletin of the World Health Organization* 34 (5): 737.

Milieu Ltd. (2010). Environmental, economic and social impacts of the use of sewage sludge on land. *Milieu Ltd. WRc and RPA report for the European Commission, DG Environment under study Contract DG ENV.G.4/ETU/2008/0076r.*

Molle, P., Lienard, A., Boutin, C. et al. (2005). How to treat raw sewage with constructed wetlands: An overview of the French systems. *Water Science and Technology* 51 (9): 11–21.

Mudrack, K. and Kunst, S. (1986). *Biology of Sewage Treatment and Water Pollution Control*, 193pp. Chichester, UK: Ellis Horwood Ltd.

National Water Council (1981). Report of the Sub-committee on the Disposal of Sewage Sludge to Land. *Standing Technical Committee Report 20*, Dept. of the Environment, London, UK.

Organisation for Economic Co-operation and Development (OECD) (2011). *Benefits in Investing in Water and Sanitation: An OECD Perspective*, 1–170. OECD Publishing.

Smisson, B. (1967). Design, Construction and Performance of Vortex Overflows. *Institute of Civil Engineers Symposium on Storm Sewage Overflows*. London, 1967, p. 99.

Stillwell, A.S., Hoppock, D.C., and Webber, M.E. (2010). Energy recovery from wastewater treatment plants in the United States: a case study of the energy-water nexus. *Sustainability* 2: 945–962.

Sullivan, R.H., Ure, J.B., Parkinson, E., and Zielinsky, P. (1982). Design manual – swirl and helical bend pollution control devices. *EPA-600/8-82/013*, US EPA, Edison, N.J.

UN Habitat (2009). Global atlas of excreta, wastewater sludge, and biosolids management: moving forward the sustainable and welcome uses of a global resource. In: *United Nations Settlements Programme* (ed. R.J. LeBlanc, P. Matthews, and R.P. Richard). Nairobi, Kenya.

United States Environmental Protection Agency (USEPA) (2000). Constructed wetlands treatment of municipal wastewater treatment. *EPA 625/R-00/010*. USEPA Office of Research and Development: Washington, D.C.

United States Environmental Protection Agency (USEPA) (2015). *A Homeowners Guide to Septic Systems*, 15pp.

USDA; Natural Resources Conservation Service, US Environmental Protection Agency Region III. (2015). *A Handbook of Constructed Wetlands*. A guide to creating wetlands for Agricultural wastewater, Domestic wastewater, Coal Mine drainage and Stormwater, in the Mid Atlantic Region, Vol. 1.

Vymazal, J. (2010). Constructed wetlands for wastewater treatment. *Water* 2: 530–549.

Watson, J.T., Reed, S.C., Kadlec, R.H. et al. (1989). Performance expectations and loading rates for constructed wetlands. In: *Constructed Wetlands for Wastewater Treatment*, 813pp (ed. D.A. Hammer). New York, MI, USA: Lewis Publishers.

Woods, D.R., Green, M.B., and Parish, R.C. (1984). Lea Marston purification Lake: operational and river quality aspects. *Water Pollution Control* 83: 226–242.

9

Climate Change and Its Implications for Freshwater in the Future

Much has been written about the effects of climate change in recent years. This chapter considers an outline of the potential impacts on freshwater systems. The biosphere, which is the layer on the earth's surface in which life occurs, is dependent on life systems and suitable materials available in sufficient quantities. As it is well known, this then provides an environment suitable for supporting life. Recent astrological studies have targeted whether other planets or their moons have significant amounts of freshwater on them that might support life. On the planet earth, although a large-area planet is covered by the oceans, human existence is based on life on land and the freshwater that is available there. We rely on the land masses for our main foods supplied by freshwater and a variety of ecosystems. Within these life systems, water plays a key role as an integral part of all sections of the hydrological cycle and climate system, including the atmosphere, hydrosphere, cryosphere, soil and biosphere. Any changes in the climate will directly affect the balance of water in these different sections. Currently, some dramatic changes are already predicted or are actually taking place. These changes can have either a positive or negative effect on the global biosphere and, in particular, on human society and food production. Examples of these changes include the European heatwave of 2019 and the hottest recorded temperature in the United Kingdom in 2019 and 2023. Less than 50 years ago, the Arctic Ocean had a covering of ice over more than 50% of its area. Winter freezing does not compensate for the degree of summer melting of the Arctic ice. Large numbers of wildfires in the Amazon basin have increased by 18%. They have also increased in California, Australia and Siberia within the Arctic Circle. Wildfires frequently occur in Australia in late spring and summer, but this year they started in early spring. According to the IPCC (2007) report from 1910 to 2004, the average maximum temperature increased 0.6 °C, and the minimum temperature increased 1.2 °C, most of which was since 1950. The result is that southern and eastern parts of Australia have become drier, although some western parts have had more rain. Droughts have become more severe and can last for more than one year (Yamakawa and Suppiah 2009). Bushfires are common in Australia, but not all are very severe, such as in Victoria in 2009 (black Friday) and South Australia in 1983. In 2009, over 140 wildfires were reported across parts of Queensland and New South Wales (NSW) although not all were severe. The problems may be exacerbated by hot dry winds that may cause fires to get out of control. These not only destroyed many forests but also many homes and local farms. The danger of fires still exists, especially when the rainfall is low. Forest fires not only

Water: Our Sustainable and Unsustainable Use, First Edition. Edward G. Bellinger.

release much previously bound CO_2 into the atmosphere, but there would be less evapotranspiration, so less water passes into the atmosphere. Not all fires are the same. Grass fires are usually fast moving. They are not so long lasting, and the grass can recover if there is sufficient rainfall afterwards. Bushfires are slower moving but are fiercer and last for longer, destroying trees and shrubs and much wildlife. These fires occur in different parts of Australia at different times of the year depending on local climate differences. The cost of these droughts is very high. The Australian droughts of 1982–1983, 1991–1995 and 2002–2003 are estimated to have cost US$ 7.6 billion, US$3.8 billion and US$7.6 billion, respectively (IPCC 2007). Box 9.1 lists bushfires in Australia if one million acres or more and the number of fatalities they caused. Bushfires can occur around major cities where there is a danger of many homes being destroyed.

Bushfires have occurred in almost all of the intervening years other than those highlighted in Box 9.1 but are not listed here as their recorded spread was less than one million acres. Some of these fires caused considerable loss of lives, e.g. in 1967, 65 deaths in Tasmania; in 1944, 51 deaths in Victoria; in 1939, 71 deaths in Victoria and NSW. In these fires, between 1300 and 2800 properties were destroyed. The latter, when added to the numbers of livestock killed, result in major economic losses. Fires in the Arctic not only destroy ecosystems but also emit greenhouse gases (GHGs) into the atmosphere and produce black soot, which is deposited on snow and ice, making solar radiation more absorptive.

Box 9.1 Major Bushfires in Australia and Fatalities Caused

Year	State/territory	Acres burned	Fatalities	Damage
1838–1839	Victoria	4 900 000	71	3700 properties
1851	Victoria	12 000 000	12	Over one million sheep and cattle
1961	Western Australia	4 400 000	0	160 properties
1974	New South Wales	2 760 000–3 700 000	3	40 properties and 50 000 livestock
1974–1975	New South Wales	11 000 000	6	Included in above counts
1980	New South Wales	2 500 000	5	14 dwellings
1983	S. Australia and Victoria	1 030 000	103	2800 dwellings
1984–1985	New South Wales	1 200 000–8 600 000	5	40 000 livestock
2003	Victoria	3 200 000	3	41 dwellings
2003	Western Australia	5 200 000	2	—
2007	Victoria	2 590 000	1	51 dwellings
2009	Victoria	1 100 000	173	2029 houses; 200 other building

Source: Data from Wikipedia: https://en.wikipedia.org/wiki/Bushfires_in_Australia.

In addition to the retreat of glaciers globally in the Antarctic and, at a faster rate, Arctic icefields are melting. The effect is not just on sea levels but also on river flows and the overall climate. For some periods, the air temperatures in the Arctic have increased at approximately twice the global rate (McBean et al. 2005). Increases were greater in the last five years than in any other year since 1900. There is also an increase in sea-surface temperatures. These increases accelerate the rate of melting in ice and snow, which also has a knock-on impact on many parts of the hydrological cycle in other parts of the world. There has been an average temperature rise of 1–2 °C since the 1960s (IPCC 2007). Although GHG emissions are not particularly generated in the Arctic rather than being generated in the rest of the world, they have an effect on the Arctic climate. Most glaciers and ice fields have shrunk in recent decades. The total volume of land-based ice in the Arctic is approximately 3 100 000 km^3, which, if it all were to melt, would cause a rise in the sea level of several metres. The largest land-based ice sheet in the Arctic is that of Greenland, which has been melting more rapidly in recent years (Green Facts 2019). These changes are likely to accelerate in the next decade to between one and three times the global average. In addition, the changes that are occurring in the Arctic will affect the climate systems on the rest of the planet. Solar radiation is stronger at the equator and less at the poles. Ocean currents flowing from the tropics to the Arctic carry warm water northwards. As these waters approach the polar region, they cool and become more dense and sink to the bottom. This colder water then flows back to the tropics, preventing them from becoming too hot and drawing more warm water from the tropics to replace them. Discharges from rivers and glaciers of freshwater change the salt balance of seawater. As the Arctic water warms, the return water to the tropics will not be as cold so the tropics will not have their temperature moderated as much. This is one of the feedback mechanisms that operates in the climate system and is known as thermohaline circulation. In the North Atlantic, the Gulf Stream warms the winds that pick up moisture that falls as precipitation over north-western Europe. This current is sometimes called the Atlantic conveyer belt. It is this finely balanced system that keeps the west coast of Europe warmer in winter than similar latitudes in North America. If climate change slows down the thermohaline circulation, the rate of oceanic overturning of the seawater would decrease and become an important mechanism for carrying CO_2 to deeper parts of the ocean. If this slowed, the concentrations of CO_2 could build up, resulting in a stronger warming effect. Other effects include changes in the reflection and absorption of solar radiation as well as the concentrations of GHG building in the atmosphere. Both carbon dioxide and methane are bound up in the permafrost and peat bogs, particularly in Siberia and North America. If the permafrost thaws, the organic matter is able to decompose and release this carbon and methane into the atmosphere. This creates a feedback loop where warming releases these GHGs into the atmosphere, which then accelerates climate warming, thus accelerating the decomposition and gas release. All other systems in the Arctic, e.g. physical, chemical and biological, are affected by these changes either directly or indirectly. These also include many resources obtained from the Arctic including fish, oil, gas and potentially minerals. Access to these will probably be enhanced, which could be a positive effect but this will also have negative consequences, such as on the ecology of many birds and mammals such as polar bears and ice-inhabiting seals. There will also be an impact on the indigenous people depending on them. Reduced sea ice will open the waters to more shipping and greater exploitation of these resources. Increased transport and larger human populations together with their infrastructures will increase emissions of GHGs in the region, contributing to climate change.

A potentially positive effect could be that the slightly milder climate would allow an advance further north of the treeline, resulting in more forest, which could result in greater CO_2 uptake and provide more forest-based employment for the local populations. Increased tree growth will increase the absorption of heat radiation from the sun as darker foliage will reflect less radiation back into the atmosphere than snow and ice cover. The increase in forest area will impact the ecology of the area, affecting some bird species and the reindeer/caribou herds of the native communities, thus affecting their economy. Increased access to the Arctic will certainly lead to increased pollution, overfishing, land use changes and habitat destruction. There would also be a large increase in the human population and their settlements, again increasing pollution.

Climate change will also cause ozone depletion in the Arctic, which in turn results in increased UV radiation, particularly in the spring. This could cause problems in some ecosystems and even to human health. An outline of the key findings of the IPCC is provided in Table 9.1. It must also be realised that even if GHG emissions are significantly degreased, there will still be some warming. The rate could be slower but will not stop. If it does slow down, there may be time for humans and other systems to adapt to a degree.

When the radiant energy from the sun strikes the earth's surface, a portion is reflected back into space. This is the albedo effect, and the normal temperature of the atmosphere is dependent on a balance between the energy coming in and the amount reflected back out. With the ice cover at the poles, there has been a large area covered with a highly reflective white surface. Approximately 85% is reflected from the ice and snow back into space, and only 15% is absorbed into the water and air. With no snow on the ice, i.e. just bare ice, 35% of the ice and snow disappear, 93% is absorbed by the water and atmosphere, and only

Table 9.1 A summary of the key finding from the IPCC report on Impacts of a Warming Arctic.

- Arctic climate now warming rapidly.
- There are worldwide implications such as less reflection of incoming radiation increasing warming leading to snow, ice and glacial melt.
- Increased freshwater runoff into oceans raises sea levels and ocean circulation changing regional climates. Thawing of frozen soils allows GHG to be released adding to climate change.
- Arctic vegetation biomes change their geographical zones affecting also animal life.
- Forest fires likely to increase in severity. Non-native species will be able to invade including insect pests. Species rich Old-Growth forests and their biodiversity will decline.
- Agriculture may, in some regions, expand northwards with longer growing seasons.
- Some animal species will decline whilst others may increase. Fisheries may increase.
- Some pests and diseases could increase.
- Many coastal communities will be threatened by rising sea levels and coastal erosion. There could be more storms. Many homes and industries will be forced to relocate.
- Reduced sea ice will allow more marine transport. Less ice could also allow more oil and gas extraction. More shipping and other transport will increase pollution.
- As the ground thaws it could become unstable making buildings unsafe.
- Permafrost thawing will disrupt ecosystems and affect wetlands and lakes.

Source: Data abstracted from the IPCC Report on Impacts of a Warming Environment, 2019.

7% radiated back. The obvious effect is to warm those areas. Currently, the Arctic is affected more than the Antarctic. The result is that the Arctic sea ice in the summer has decreased by nearly 75% in the last 40 years. This means that many Arctic sea routes are either open or have only thin ice. The IPCC issued a report in 2019 assessing the impact of climate change on the Arctic (IPCC 2019).

One direct impact on water resources will be that arctic river discharges to the oceans will increase, and peak flows will occur earlier. Snow cover in the Arctic will decrease, also affecting river flows. It is important to note that reduced snow cover is also occurring in other parts of the world, e.g. California, affecting river flows. Warming also means that snow is more likely to fall as rain. Arctic precipitation has increased by an average of 8% over the last century and will increase in the future.

Sea levels are rising at a rate of almost 3 mm a year, and the projected overall rise will be between 10 and 90 cm this century in addition to the 10–20 cm that has already occurred. This is because of not only freshwater inputs from melting ice and glaciers but also thermal expansion as the waters warm up. This will impact not only many coastal dwellings but also many economic buildings, such as drinking and wastewater treatment plants, power stations, pumping stations and some low-lying islands (e.g. Tuvalu, Tonga and The Maldives). Parts of some countries have already been subject to catastrophic flooding. In Bangladesh, as many as 17 million people live less than 1 m above sea level in the delta region of the coast which is already subject to floods. The rise will vary globally with the largest increases being in the Arctic. Thawing of the permafrost will cause lakes and some wetlands in some locations to drain further affecting the water balance of those areas as well as their ecology. Within the Arctic Circle traditional foods such as seal, reindeer/caribou, some bird species and some fish species are likely to decline affecting the indigenous peoples. Access to resources such as oil, gas (all fossil fuels) and some minerals will improve. There are large gas and oil fields located in the Russian Arctic with reserves also in Canada and Alaska. This is a two-edged sword as exploitation of these is likely to increase pollution. Some arctic fisheries will increase as waters warm but control of these fisheries needs to be determined based upon fish stocks which are still, at the present time, an unknown quantity.

Arctic sea ice could disappear completely through September each summer if the temperatures continue to increase by 2°C , the target agreed by the Paris Agreement, an agreement which the USA government withdrew from in 2017 (Olsen et al. 2019). The impacts of global warming have different impacts on different parts of the Arctic. The IPCC (2019) recognises four sub-regions. These are listed in Box 9.2 together with the differing impacts.

A large proportion of water abstracted is used by agriculture (see Table 9.2).

Climate change will affect the amounts of water available in many countries, dramatically affecting agriculture and thus food resources. In some countries, where the amounts withdrawn for irrigation are high, there could also be an effect on food production not only for the local populations but also for their economies as many foodstuffs are exported all over the globe.

The energy driving the climate system, which includes the hydrological cycle, ultimately comes from the sun, and any changes in the amount of energy absorbed by the system will directly affect the hydrological cycle. There is an undisputed rise in the overall temperature of the climate since pre industrial times that is having an effect on the water balance.

Box 9.2 Sub-regions of the Arctic and the Differing Impacts of Global Warming

Sub-region	Main impacts
(1) E. Greenland, Iceland, Norway, Sweden, Finland NW, Russia.	Tundra areas disappear from parts of mainland. Low-lying coastal areas flooded, access to mineral resources improves, e.g. for oil and gas, fisheries increase as southern species migrate northwards, reindeer herds affected by reduced snow, possible transmission of animal diseases to humans.
(2) Siberia	Sea-ice decline improves seasonal navigation, permafrost thaws causing instability of ground, forests, etc. invade tundra access to oil and gas reserves improved, river discharges greater but river ice season smaller, reindeer migration interrupted
(3) Alaska, Chukotka, West Canadian Arctic	Biodiversity largely at risk and largest numbers of species are threatened, more forest fires, greater coastal flooding, many communities will need relocation, food security and lifestyles of indigenous peoples affected.
(4) W. Greenland, Central Canadian Arctic	Greenland ice sheet continues to decrease, sea-levels rise, more flooding, sea-ice retreats allowing more shipping routes, Pollution risks increase through oil and gas exploitation, migration of some fish species further north, lake species may decline, food security threats to indigenous peoples, some communities need relocation.

Source: Adapted from IPCC 2019, Impacts of a Warming Arctic.

Table 9.2 The proportion of water used for agriculture in different countries.

Country	Total renewable freshwater (km^3/year)	Irrigation water withdrawal (km^3/year)	Irrigation water requirement (km^3/year)	Ratio %
Australia	492 000	6596	3892	59
Chile	922 000	22 886	5038	22
Egypt	57 300	59 000	45 111	76
Greece	74 250	8458	5441	64
China	2 840 000	358 000	256 872	72
Germany	154 000	80	48	60
Israel	1780	1129	556	49
Italy	191 300	12 895	8022	62
India	1 911 000	688 000	370 843	54
South Africa	50 000	7836	2138	27
Spain	111 500	19 560	14 058	72
Portugal	68 700	6567	2016	31
UK	147 000	59	35	60
USA	3 069 000	177 403	108 528	61

Definition of terms: Irrigation water withdrawal – the total amount of water withdrawn from both surface and aquifers for irrigation purposes annually. The irrigation water requirement is the amount of water required for crop production. It includes water needed for salt leaching to protect crops. The ratio (properly called the water requirement ratio) is the ratio between the irrigation water requirement and the amount of water withdrawn for irrigation.
Source: Data from FAO (2012).

Between 1906 and 2005, the global average temperature rose by about 0.74 °C (range 0.56–0.92 °C), but the rate of increase has in the last 50 years. The warming has been particularly pronounced during the last three decades (Arndt et al. 2009). One of the warmest years on record was 2010 (NASA 2012). Warming has not occurred equally over the whole planet, most being in the Northern Hemisphere. An example is in the United Kingdom where the highest daily temperatures on record were recorded for the months of July and August. Precipitation also varies spatially, increasing at higher latitudes but decreasing in the subtropics, causing more frequent and severe droughts. Most of this rise has been attributed to anthropogenic activities, including agriculture, forestry and other land use (AFOLU; see IPCC 2019; Climate Change and Land). The effects can be seen on all continents. This rise has initiated changes in the patterns of precipitation, melting of snow and ice, including glacier retreat and changing spring snow melt, changes in run-off, increased evaporation rates and increasing extremes in weather events. We do not, however, know what all of the indirect effects of a changing climate nor do we know the global distribution of these changes will be (Schneider 2001). Although there are a number of different factors that affect the survival and success of human populations, climate change will have an effect in the future, perhaps the main effect for some populations. One important feature of climate change is the effect on the poorest communities in arid and semi-arid countries that are dependent on rain-fed agriculture. This will have a profound effect on food security, drought and water scarcity. It could stimulate economic loss and violent conflict. There has always been a degree of natural variability both in time and space, which adds difficulty to predicting future events. This is made more difficult because of inadequate monitoring systems in the past. Climate variability could affect water availability, which in turn could endanger vulnerable populations and affect general economic development.

The global climate changed frequently during geological time with the last warm period occurring in the Pliocene 3.0–3.3 million years ago (Haywood et al. 2009). This was probably caused by elevated concentrations of CO_2 in the atmosphere (Lunt et al. 2008). Past climates for periods beyond the use of instruments to measure parameters such as temperature, solar radiation and rainfall use tree rings, ice cores, lake and sea sediment cores and fossil records. All of these factors help understand the climatic conditions under which they formed. There is also written evidence from several hundreds of years to one or more millennia ago. Geological evidence shows that there were smaller ice sheets at that time and sea levels were approximately 25 m higher than at present. Fossil evidence from different countries indicates geographical changes in the major biomes at that time. There are also changes brought about by changes in the earth's orbital pattern as well as possible changes in the composition of the atmosphere (Sheffield and Wood 2011). These result in glacial periods and shorter interglacial periods with cycles of approximately 26, 41 and 94 thousands of years Burroughs (2005). These are known as the Milankovitch solar cycles after the astronomer Milutin Milankovitch, who first described them. The climate of the land masses was strongly influenced by changes in ocean temperatures. In the North Atlantic, Heinrich (1988) identified periods when large numbers of icebergs broke away from glaciers into the North Atlantic over the past 64 kya. When these melted, they introduced large amounts of freshwater to the ocean, which then influenced the climate. It is important to look at these events, which altered ocean current behaviour, to see whether there is collaborative evidence on land to show that the land climate was also affected,

Table 9.3 Some of the recorded changes caused by climate change to the hydrological cycle.

1) Changes to the distribution and intensity of precipitation
2) Changes to run-off
3) Decreased winter snows and therefore lower yields of water from snowmelt
4) Earlier snowmelt causing earlier peak river discharges
5) Shrinkage of glaciers
6) Reduced precipitation in arid regions resulting in drought
7) Melting of permafrost soils
8) Increased flood risk in many areas
9) Decreases in some groundwater recharges
10) Changes in water pollution due to nutrient flushing

which also shows agreement with Heinrich events and ocean sediment data from the same period. All sediment data from cores is best collected from where seabed conditions are anoxic as there is likely to be less disturbance of the sediments in those locations. There is reasonable agreement for the two sets of data, especially where extended cold periods were indicated from 30–15, 36 and 70–63 kya. The influence of these cold periods extended into Europe and central Asia. Climate change will give rise to gradual changes in factors such as average temperature and patterns of precipitation. An outline of some of these changes is provided in Table 9.3.

Although changes due to climate change are inevitable, we must also recognise that some can occur through natural fluctuations and others through anthropogenic interventions such as land use change and pollution.

There could also be an increase in the number and severity of extreme weather events such as floods, droughts and storms although current climate models are not always sensitive enough to predict changes on a small, more local scale. They can, however, indicate the probability of extreme events occurring more or less frequently. All of these changes could impact agriculture and societies, including economic activities, in general.

Impacts on Human Settlements

There is a global trend in populations to move from rural areas to cities. Cities have increased greatly in size over the last century. This is already putting a strain on many services and would increase under climate change (Scott et al. 2001).

Many recent studies have investigated the frequency of floods and droughts in order to predict their frequency in the future based upon past trends (Van Huijgevoort et al. 2013; Peplow 2004; Sheffield and Wood 2008). Past trends are not always a reliable indicator of the future and must be taken with a degree of caution as conditions are changing, especially in recent years. Changes to biomes and biodiversity would have profound effects on ecosystem services upon which life on earth, especially human societies, relies (see Table 9.4).

It is now widely accepted that observed changes in the global climate that have occurred during the last century and the present day have largely been due to human activities, with Africa initially being the worst affected (IPCC 2007). These changes to the climate pose

Table 9.4 Some impacts of climate change on biomes and ecosystem services.

Biome	Climate change	Habitat change	Over exploitation	N&P enrichment
Forest boreal	L R	L I	L C	M R
Temperate	LR	HD	MC	MR

Key to symbols: L, low; M, moderate; H, high; VH, very high; D, decreasing trend; C, continuing impact; I, increasing impact; R, rapid increase in impact.
Source: Adapted from Richardson et al. (2009) and The Biodiversity and Climate Change, Findings of the Millennium Ecosystem Assessment.

many risks to both ecosystems and human societies, with, among other things, natural vegetation die-off contributing to climate change. It should be noted, however, that not all changes to the hydrological cycle are caused by anthropogenic-induced climate change. Natural weathering of rocks and some erosion of soils can be caused by normally occurring events. Humans have contributed to many alterations in the last century, including the overuse of aquifers and changes in land use. Arguably the most widespread and progressive change now occurring is that of accelerated climate change.

Although climate change was initially referred to as a 'global warming' and as a result the press and public were not unhappy at the thought of warmer holidays by the sea although there can be too much of a good thing! A more important impact is probably on changes to the hydrological cycle. One of the most important impacts is on freshwater resources. Many well-documented changes on freshwaters have been documented, including changes in global run-off, timing and amount of peak discharges, glacier and snowfall/melt reductions, number and magnitude of intensive events such as floods and droughts, decreased groundwater recharge and chemical and biological changes to waters (numerous authors including Alkama et al. 2011; Piao et al. 2010; Shiklomanov et al. 2007; Collins 2008; Aguilera and Murillo 2009; Jeelani 2008; Tibby and Tiller 2007; IPPC 2014). Considering the main components of the hydrological cycle, it is clear that all parts are not affected equally and that there can be considerable geographical differences. The main external source of energy driving the climate system is the sun, and the amount of solar energy reaching a given part of the earth's surface is the main driving force. Any materials present in the atmosphere will affect the amount of radiant energy reaching the earth. Although GHGs are a major and well documented factor (IPCC 2014), there are other factors. Anyone who lives near the air corridor routes taken by modern aircraft will have seen the vapour trails, also known as contrails, created by them across the sky. With the large increase in the number of aircraft using these high-density traffic routes, there is some evidence, albeit small, of them altering the abundance of cirrus cloud formation (Stubenrauch and Schulmann 2005). Of greater influence on climate is an increase in aerosol particulate matter that occurs in some regions. In particular, studies on precipitation in the Sahel have indicated links with atmospheric dust concentrations. Yoshioka et al. (2007) found that dust could reduce average precipitation in regions such as the Sahel. It should be noted that dust in the atmosphere can travel many thousands of miles, and the Sahel could be affected by dust storms in the Sahara (Yu et al. 2015). Unfortunately, records of the dust content of the atmosphere are not comprehensive. Both increased dust and reduced vegetation

Table 9.5 Example of some of the predictions for changes in precipitation in the Sahel region.

Changes for first half of twenty-first century	Second half of twenty-first century
—	Greater rainfall
Slightly more rain	Progressively drier
Slightly more rain	Drier in NW and E Sahel
	Rainier in S Sahel
Progressively more rain	Progressively rainier
No Impact	Drier in last 20 years
—	Drier June–July; more rain in August
Uncertain	Uncertain

Source: Taken from Druyan (2011) and is a compilation from several authors.

combine to reduce precipitation. Dust is not the only cause of the reduction and accounts for about 15–30% of the reduction (Yoshioka et al. 2007). Unfortunately, periods of drought in the Sahel reduce vegetation cover and increase dust, which in turn feeds back to reducing precipitation, making the situation worse. Different models will have slightly different predictions. A summary of the impacts of climate change on precipitation in the Sahel region is provided in Table 9.5.

Sahelian woody plants have been shown by some authors (e.g. Hiernaux and Turner 2002) to be adversely affected by drought, causing mass mortalities but also showing subsequent regrowth during wetter periods. This will depend on species and other interacting conditions. Grassland growth is largely dependent on rainfall although higher carbon dioxide concentrations can have an effect. Grasslands have been reported to switch from being a carbon sink to a carbon source in response to water stress, although over a full year, the effect levelled out and was neutral (Li et al. 2005). The effects will depend on multiple factors however.

Low rainfall will affect crop production and play an important role in lowering the economic performance of a country (Miguel et al. 2004). This can also lead to civil unrest, especially in Sub-Saharan Africa. Much has been published concerning the risk of conflict that climate shocks such as droughts and floods may cause, and there is a lot of information on the effects on individual species of plants under experimental conditions. There is little research on the effects of rising CO_2 on whole ecosystems as there are multiple factors involved, including water availability, other nutrients, temperature and species of plant species at that location (Norby and Luo 2004). Issues arising from future changes in climate are of great interest to politicians and countries in general.

Precipitation

Evaporation occurs from both the ocean and land surfaces, and this eventually condenses into clouds and then rain and snow that fall back onto the land or ocean surface. The amounts of precipitation vary from one part of the globe to another as well as changing from year to year and on a decadal time scale. The precipitation intensity is an important factor determining whether it soaks into the soil or runs off. Generally, steady but gentle rain soaks, but the same volume delivered over a short period is likely to run off and potentially cause flooding. Floods are normally local and occur over short periods of time, while droughts develop gradually and last longer. Floods often receive greater media coverage and cause greater damage to property and resources. While it is difficult to assess losses incurred, floods are estimated to cause several billions of $US annually and cause thousands of lives to be lost. Droughts cause as much, if not more, financial loss over the period and beyond and may also cause even more deaths.

The water holding capacity of the atmosphere is dependent on temperature. Warmer air can hold more water than cold air. Hence, in general terms, global warming due to climate change will allow the atmosphere to hold more water. Durre et al. (2009) showed that the water vapour content of the troposphere has increased over the Northern Hemisphere, including both Japan and the United States of America. There was an exception in 1991 when there was a fall in water vapour content due to the eruption of Mt. Pinatubo when large amounts of ash and other particulate matter were injected high into the atmosphere. Although this will not necessarily have the same impact across the globe, the prediction is that there will be increased precipitation in the northern mid to high latitudes in any season (Noake et al. 2012). Otto et al. (2015), for example, predict that extreme precipitation events in July in the United Kingdom have doubled, which increases the risk of flooding. The change in the occurrence of extreme events is illustrated by the events in Australia in 2018 and 2019 shown in Box 9.3.

Box 9.3 Unprecedented Floods and Severe Drought in Australia

In Townsville, Queensland, north east Australia, once-in-a-century very heavy rains occurred on the catchment of the Ross River. In late January/early February 2019 1012 mm of rain were recorded. In one hour Ingham received 145 mm of rain. This resulted in the Ross River rising over 3 m and possibly higher. This, together with continued surface water run-off from saturated soils caused flooding in large areas of the catchment. The city of Townsville was badly affected and here and the surrounding area more than 500 homes were inundated. The Bureau of Meteorology reported that a slow-moving monsoonal trough sitting above Queensland was causing the problem. Problems with landslides were reported in some areas. Flooding in remote areas meant that many dwellings were isolated and all transport routes cut off. There was also a danger of tornadoes developing causing further problems. In one area the increased flows go into the Ross River reservoir. This filled to dangerous levels so that it was holding about 224% of its capacity forcing the authorities to open the floodgates of the dam to relieve pressure and prevent a collapse of the dam. This released about 1900 m^3/s into the river downstream. This greatly increased both the volume and velocity of the river flow exacerbating the flooding in the town. Many areas were without power and snakes and crocodiles were seen in the urban flooded areas causing additional risks. The Meteorological Bureau pointed out that the rain may ease within a few days but it would take possibly weeks before the floods properly subsided. It would take many months before the cost of the damage could be calculated.

Elsewhere in Australia, in particular NSW and parts of Queensland have had no rain for months and suffered badly from drought. NSW was declared to be 100% in drought. Farms in NSW had to slaughter cattle due to a lack of water. The Bureau of Meteorology stated that some of the regions would need 200–300 mm of rain spread over at least three months to relieve the shortage. Unfortunately, some areas received none, while many others only had 25–50 mm, not enough. Many pastures have turned to dust. Generally, indicators such as soil moisture and pond levels are very low. Much of what little rain that does fall is quickly lost through evaporation. To recover several months of above average rainfall are needed. Even when drought ends, farmers face prolonged economic losses. Livestock have either been sold or slaughtered, any crops have failed, pastures are parched. It could take many years to not only financially recover but also overcome the psychological effects.

In Europe, written records exist for over 250 years for some regions, e.g. Kew in the south of the United Kingdom, although some areas are less comprehensive. These results show that precipitation increased, especially over the Mediterranean, from the mid-nineteenth century and then decreased since the 1960s (Xopaki et al. 2004). Record dry summers were recorded in Kew data and in other centres in Europe. This was a period of more frequent dry spells. An increasing frequency of dry periods has been observed during the last 50 years (Wihley and Atkinson 1977; Briffa et al. 2009). The latter authors reported that high air temperatures contributed to the severity of dryness, particularly in central Europe. This trend of dry summers is likely to continue in England and western/central Europe.

Streamflow

Streamflows from the largest global rivers show large differences with time (Dai et al. 2008). These differences can usually be correlated with changes in precipitation and temperature (IPCC 2014). There are regional differences on each continent. For example, in Europe streamflow has decreased in the south and east but increased in other areas, especially further north (Wilson et al. 2010). In North America, flows have decreased in the Pacific Northwest but increased in the Mississippi basin. In China, a decreased flow in the Yellow River has been observed, but the Yangtze River has shown a small increase (Piao et al. 2010). In north-eastern China, the annual rainfall has decreased, mostly in the summer and autumn. This has resulted in lower river flows, whereas increases have occurred in north-western and south-eastern China. There has also been any overall loss in glacier volume throughout the country, and glacial melt has affected river run-off, contributing to some increases in headwater flows. Some glacial lakes have also increased in volume, which in turn increases the risk of bursts and floods, especially in the Tarim region. Except in central China, there has been a greater number of heatwaves in the summer. All of these changes will impact food production. China has only about 7% of the global arable land but has to feed 22% of the global population (Nat. Bureau of Statistics of China 2009). It also contributes as much as 11% to the Gross Domestic Product. Rice wheat and maize are the main crops grown, amounting to 54% of the total sown area and nearly 90% of the grain yield (You et al. 2009). Although the length of the growing season has increased due to global warming so has the range of crop pests. In China, the area exposed to pest attacks trebled from 1970 to 2000, resulting in a grain yield loss of over 50% (Piao et al. 2010). Droughts also affect more cropland than in the past decades. At the opposite end of the scale, areas affected by flooding increased by 88% although the area affected was less than that affected by drought. Climate change has an adverse effect on agriculture in China when population increases and demand grows. Some losses have fortunately been compensated by improved technology in agriculture, with some areas, northeast, northwest and southeast, predicted to show crop increases compared with the 1961–1990 averages (Xiong et al. 2008), and some increase in predicted yields can be attributed to increased levels of CO_2 in the atmosphere. All crops will not be affected equally, and some, for example rice, may show small decreases although different models show different degrees of change.

In many parts of the mountain world, snowfields act as important natural reservoirs for the storage of water. In these regions, most annual precipitation falls in the winter as snow. This is then released into streams and rivers as the temperature rises in the spring and the snow melts. The amount of snowpack produced over the winter acts as an important predictor of summer streamflows. In the United States of America, as much as 75% of water supplies in the western part come from snowmelt (USGS 2007). Global warming has increased winter temperatures, which in turn has reduced the amount of snow and mass of snowpack and resulted in earlier snowmelt (Mote 2003). Many locations have shown reductions in snowpack of between 50% and 75% (e.g. in Washington, western Oregon and northern California; Mote et al. 2005). Although global warming is suspected to be the main influence on changes in snow pack, other factors, such as reduction in forest canopy and other land use changes such as human development (roads, houses, etc.), can play a role. Such changes impact streamflow discharges with melting occurring earlier in the spring (Cayan et al. 2001). They found that spring temperatures had increased by 1–3 °C since the late 1970s. They also pointed out that this not only affected the timing of snowmelt but also the onset of blooming of lilacs, *Populus tremuloides, Ptunus virginiana, Amelanchier almitulia,* and Honeysuckle. These bloom tines (known as the study of phenology) can be used as a guide to changes in climate. This approach has also been used in Europe by Menzel and Fabian (1999), who also found from analysing 30 years of data that spring plant growth had advanced by at least 6 days and that autumn die back had receded by about 5 days so that the average growing season had increased by about 11 days between the 1960s and 1999. Sparks and Menzel (2002) also recorded that a wide range of plant species also showed similar trends in a range of locations. Similar results have been recorded in the Swiss Alps by Scherrer et al. (2004), where long-term increases in temperature indicated reduced snowpack. Mountain snow accumulation has decreased in the last 50 years, so there have been reductions in snow-derived meltwater (Stewart et al. 2005). This change has affected the timing of peak volume flows, which now arrive earlier with a resulting reduced flow in late spring- early summer and can potentially affect water management strategies as well as water availability for agriculture, the main user. The trend has not always been for a decrease in snow. Some areas in the Southwest, for example, high precipitation in recent years, has increased snowfall (Mote et al. 2005). Some of the changes to snowfall have been quite large, so it is important to ascertain how these changes will develop in the future so as to develop water management strategies to take them into account.

Groundwater

Direct evidence on the impact of climate change is difficult to ascertain as there are many factors involved. Two of the most difficult to quantify are land use changes within the groundwater's catchment and changes in the abstraction rate of water from that aquifer (IPCC 2014). However, there did seem to be evidence that groundwater recharge had declined as a proportion of precipitation, possibly due to increased evapotranspiration (Aguilera and Murillo 2009).

Changes in Water Quality

Because measurements of water quality have only been carried out relatively recently and only at restricted locations, data is only available for a limited timescale and a limited number of parameters (IPCC 2014). Information for lakes is more common but only for parameters concerned with eutrophication such as nutrient increases and the biological response, i.e. eutrophication. Changes have often been linked to increased run-off, causing a greater nutrient flux into the water. Not only do nutrient loads increase, but pathogens, coliform bacteria and heavy metals also increase. One trend that is becoming a problem in many parts of the world is the more widespread occurrence of harmful algal blooms caused by cyanobacteria (blue–green algae). Not only can large growths of these algae have harmful effects on humans and wildlife, but they can, when they die, deoxygenate the water, increase turbidity, shading out other organisms and affecting fish populations. Typically, blue–green algae grow at warmer times of the year, i.e. summer. However, with climate change, the length of the warm summer period increased. It also leads to earlier late spring thermal stratification and later autumnal destratification, both of which favour cyanobacterial growth. Storm run-off early in the year will increase nutrient run-off, which is often followed by summer droughts that then reduce the flushing of lakes and reduce the wash-out of algal blooms (Paerl and Huisman 2008). Climate change has also widened the range of some cyanobacterial species, e.g. *Cylindrospermopsis raciborskii* (Carmichael 2001). This spread to Europe over 50 years ago and to the United States of America, where it is now a common constituent of eutrophic lakes. A study by Tibby and Tiller (2007) in Australia indicated that warming of the climate increased water temperature and increased the period of lake stratification resulting in increased algal growths (for example the diatom *Aulacoseira ambigua* and the nuisance blue–green alga *Microcystis aeruginosa*), causing a significant deterioration in water quality. Although many freshwater algae can cause problems such as deoxygenation when a bloom dies or clogging waterways by large growths of filamentous species such as the green alga *Cladophora* of much greater concern is the toxic products of many species of blue–green (cyanobacteria) that increasingly occur in lakes and reservoirs with eutrophication (Carmichael 2012; Bellinger and Sigee 2015). Heavy rainfall and extreme events can cause pollutants, including faecal coliforms, to enter groundwater. Chen et al. (2012) highlighted the dangers of extreme precipitation events and possible potential flooding, increasing the incidence of the spread and frequency of waterborne infectious diseases. These could result from sewage overflow contamination and cross contamination of supplies. Outbreaks of cholera and leptospirosis have been reported in India and Bangladesh (Islam et al. 2009), as well as gastroenteritis,

Giardia lamblia and *Cryptosporidium* (MacKenzie et al. 1994). As groundwater in many regions is used for drinking water with little or no treatment, public health problems can arise (Jean et al. 2006).

Droughts

There have been several records of major droughts and the problems caused from different parts of the world. Different groups in societies are affected to different degrees by environmental shocks with marginalised groups suffering the most (Theisen et al. 2012). In medieval times, crop yields, especially corn, were not much more than one-fifth to one-tenth of the present day. A dry year or drought would devastate a crop and potentially cause starvation and mortality in a population. Child mortality would be very high. Transport was difficult, so the movement of food from one part of a region to another was difficult (Baker and Harley 1973). There was, for example, a great famine in 1315 and a very bad harvest in 1556 as a result of drought Baker and Harley (1973). In Russia, during the early part of the twentieth century, parts of the country, notably the Urals and the Volga region, suffered from a lack of rain. This continued through the spring into June and late July. The high temperatures and lack of water during this period caused considerable damage to the crops. In Tobolskaya Province, crop yields were 80% below average (Dronin and Bellinger 2005). This and other drought years in Russia were reported to cause widespread starvation, and about 20% of the population in those regions was affected. Of course, there were other factors affecting crop yields over the years. Some of the Russian droughts cover vast areas of land, sometimes in the region of 30% of their total agricultural land area (Dronin and Bellinger 2005). During the twentieth century, agriculture in Russia played a vital role in the economy employing about one-third of the labour force and was given about 25% of the government's investment. In the 1970s, it contributed about 15% of the Gross National Product (GNP). In comparison, agriculture in the United States of America contributed only 4% of the GNP. Thus, any severe reduction in agricultural yields had a proportionally large effect on the Russian GNP.

Another example of the impact of global warming is the progressive decrease in the area of Arctic sea ice. The amount of thicker, older sea ice declined from 45% in 1985 to 21%, in area, in 2017 (NOAA, Arctic Report Card Update 2017). As ice and snow are major reflectors of solar radiation arriving on earth back into the atmosphere, any reductions in snow and ice worldwide will result in less back-reflection, causing the atmosphere to warm up. This is coupled with the fact that seawater absorbs most of this radiation and will have a larger surface area to do this (Richardson et al. 2009). This will be added to the impact of increases in other GHGs. The effects will not only be on human societies but also on ecosystems, some of which will affect the maintenance of the biosphere and provision of vital services to life on earth (Richardson et al. 2009). As with other climatic effects, the impact on the hydrological cycle will vary from one region to another, but all continents will show some adverse effects. Water in the atmosphere can be present as vapour, liquid or ice (solid), and all three behave differently. Water vapour is the most important GHG and accounts for a large proportion of the greenhouse effect, helping to maintain temperatures at the earth's surface at a tolerable level (Jones 2010) and making earth habitable for life as we know it. Without this greenhouse effect the average temperature on earth would be about −19 °C

(Richardson et al. 2009). This effect is shared with other GHGs such as CO_2. Ice crystals form in the upper layers and can reflect radiation back. Global warming is likely to raise the temperature of the atmosphere, resulting in a greater amount of water vapour being held partly due to its increased capacity to hold water and partly due to greater evaporation from land and surface waters. The Clausius-Clapeyron relation predicts that the atmosphere can carry 7% more water vapour per degree increase in temperature. However, this is also controlled by the energy budget in the atmosphere, and predictions for a 1–3% increase are more likely (Lambert and Allen 2009). There is general agreement that the average global temperatures have risen over the last 100 years. Increased CO_2 in the atmosphere results in more CO_2 being absorbed by the sea, which in turn increases seawater acidity affecting the viability of some organisms, especially corals. Even if global climate change slowed or stopped, it would probably take thousands or even millions of years for organisms and some ecosystems to recover.

At a local level, changes in the climate are likely to result in an increase in extreme events such as storms, floods, heat waves and droughts. Ecosystems as well as human societies can be badly affected by such extremes as they are generally developed with more stable climate patterns. Small changes in pattern are usually tolerated, but repeated and more dramatic changes can be disastrous, affecting property, health, agriculture, water resources and infrastructure as well as ecosystems. There is some evidence that not only local but also global plant primary production in both ecosystems and agriculture will be affected by drought. Liao and Zhuang (2015) reported a decrease in gross primary production (95.7 PgC/year down to 106.4 PgC/year) and net primary production down from 54.9 to 49.9 PgC/year due to droughts from 2002 to 2010. The nexus between climate change, human health and freshwater systems is very strong (Richardson et al. 2009). Droughts, for example, have often resulted in famine in the past and could well become more severe in the future. The result of crop failures through drought is that many people migrate to cities that are growing in size and putting even greater strain on water resources. Even a small increase in global temperature is likely to increase the severity of cyclones and their destructive power, especially for coastal, fishing and river delta communities. Climate change will also affect ecosystem services. In many regions, changes in the amount of snow and the timing and amount of snow melt alter the hydrological cycle (IPCC, WGH AR5 2014). A large proportion of the world's population (more than one-sixth) live in glacier and snow-melt-fed river basins (IPCC 2007). Glaciers are retreating in most parts of the world as is the amount of snowfall. Snow and glacial meltwater and changes in the amount of precipitation are reducing the flow volume in many rivers. This is happening on all continents (IPCC 2014). In tropical East Africa, highland glaciers are retreating. In Europe and Asia, the same is happening for many mountain glaciers. In both North and South America, Andean and Rocky Mountain glaciers are retreating. Even in Australasia, snowfall has decreased as are glacial ice volumes in New Zealand. This in turn affects the available quantities of water and possibly the quality. This results in reduced discharges in many African and Australian rivers, but in others, for example in Asia and America, there may be increased flows or changes in the seasonal flow patterns. Both floods and droughts may increase in different parts of the world (IPCC 2014). In high latitudes, the warming atmosphere is causing thawing of the permafrost, which will also affect the ecosystems of those regions. An overall assessment of impacts on agriculture indicates that there are more adverse effects on, for example, crop yields than positive ones (IPCC, WGH ARS 2014). Arid and semi-arid regions are most likely to be affected, exacerbating their water problems. These regions include Mediterranean countries, Sub-Saharan and Southern Africa and the Western United States of America, but impacts are more widespread. Not only will variations in rainfall cause problems but this will also affect groundwater recharge rates. Population growth in these areas will magnify the problem because of increased food demand. Changes in flow patterns can also impact water quality from the point of view of both the chemical composition and viability of pathogens present. Some of the effects of climate change and water scarcity are outlined below and potentially affect all aspects of society in all countries. For example, China has the largest population of any country although India is a close second and may even overtake it within a decade. Its economy is growing, and as a result, it is currently a major emitter of GHGs. As such, it has a major effect on global climate but is also affected by any climate changes that take place. In particular, it faces a large challenge to feed its growing population with only a limited amount of arable land (Piao et al. 2010). Climate effects on water are already being felt, and it is imperative that water managers in all sectors, such as municipalities, industry and agriculture, policymakers and the public change their approach to water use and plan for its long-term future availability. Existing structures and plans will have to be modified, and future ones will be rethought. Demands by industry are likely to increase, and urbanisation together with land use changes will change water availability. Water demand for domestic use will grow not only because of increased population but also because of their increased affluence. Not all change is due to human activity. The IPCC lists a number of changes in the hydrological cycle and what they can be attributed to. Some of the changes identified are provided in Table 9.6.

Table 9.6 Some observed hydrological changes and their probable causes.

Observed change	Possible causes	References
Global run-off changes	Climate change, CO_2 increase and land use changes	Alkama et al. (2011)
Earlier annual peak discharges	Temperatures higher in early spring causes earlier thaw	Shiklomanov et al. (2007)
Glacier meltwater yields Greater in 1910–1940 than In 1980–2000	Glaciers shrinking forced by warming	Collins (2008)
More intense extremes of precipitation	Anthropogenic greenhouse gas emissions	
Decreased groundwater recharge	Decreased winter precipitation	Jeelani (2008)
Nutrient flushing from Swamps, reservoirs	Hurricanes	Paerl et al. (2006)
Increased lake nutrient	Increased air and water temp.	Tibby and Tiller (2007)

Source: Data abstracted from IPCC (2014), Climate Change 2014. Working Group II Contribution to the Fifth Assessment Report of the Intergovernmental Panel on Climate Change, Cambridge University Press.

Long data sets on the various components of the hydrological cycle are rarely available, and even when they are, it is not always valid to try to predict what will happen in the future (Bates et al. 2008). Although there are small differences in the predictions produced by the various models, there is general agreement that there will be changes to the hydrological system.

Climate Change and Freshwater and Terrestrial Ecosystems

There are many components to freshwater ecosystems, for example organisms living in the mud at the bottom, those living at different depths of water and those living at the surface. There are plants and animals, macro and microscopic. The biota are usually adapted to life in particular types of freshwater ecosystems, such as rivers, ponds, and lakes. Some are adapted to spells during which the water has dried up, while others can live in water under ice cover. Some are exploited by humans (e.g. fisheries), while others are a nuisance or can be directly harmful (e.g. cyanobacteria or blue–green algae). All of these factors could, potentially, be affected by climate change. Many freshwater systems have been modified and exploited to meet human needs, more so than marine or some terrestrial systems (IPCC 2014). According to the Millennium Ecosystem Assessment (2005), between 1970 and 2000, populations of freshwater species declined by an average of 50% compared with 39% for marine and terrestrial systems. It is not only living organisms that are important but also freshwater ecosystems they play many other roles in the biosphere. The direct provision of food is important for many communities in developing countries, and for many, the main source of protein (as much as 80% of their animal protein intake) is from inland fisheries. Consequently, anything affecting the supply of this protein will have adverse effects. Although climate change may be manifested in different ways, all could affect the organisms living in the water.

Effects on limnological behaviour: Freshwater has its greatest density at 4 °C, so cold water sinks to the bottom of a lake with, generally, warmer water on top (with the exception of water between 0 and 4 °C). Without mixing of the water column by wind or another physical agent, the water in a lake or pond will form layers or thermally stratify (Wetzel 1983). These different layers will have different properties with the deepest layer, the hypolimnion, having less oxygen but more dissolved nutrients than the upper layer, the epilimnion. This spatial distribution will affect the distribution of fish populations. With global warming causing surface water temperatures to rise, stratification will strengthen, especially in temperate zones. The epilimnion is exposed to wind-induced mixing, but if the stratification is strong enough, this might not be enough to mix the different layers, especially in deeper lakes such as Lake Geneva in Switzerland. Thermal stratification is a major driver for algal populations. When nutrient concentrations are sufficient, the raised temperatures favour the growth of cyanobacteria (blue–green algae). As most species of zooplankton cannot eat blue–green algae, their dominant presence will affect the zooplankton numbers, thus reducing a major food source for fish (George et al. 1990) and reducing fisheries productivity. There is also a possibility that if certain species of blue–green algae that produce toxins, e.g. *Anabaena flos-aquae*, are present, their toxins can get into fish and accumulate to levels unsafe for human consumption. The hypolimnion can act as a valuable refuge for fish with a preference for cooler temperatures where oxygen concentrations can be higher unless accumulations of organic matter have caused oxygen depletion. Examples of fish using the hypolimnion as a refuge have been recorded in some temperate and subarctic lakes in North America. These include cold water fish such as Arctic char (*Salvelinus alpinus*), lake trout (*Salvelinus namaycush*), whitefish (*Coregonus* spp.) and striped bass (Coutant 1990). However, if the stratification is more prolonged then oxygen depletion will occur there as well forcing the fish to higher parts of the water column, such as the mesolimnion just above the hypolimnion, even though it is slightly warmer. This limits their suitable habitat. If the water temperature in lakes or streams were to increase, this would cause a northerly migration of cold water species and a northerly shift in the range of more warm water species. Warming can cause some lakes to be permanently stratified where they are used to destratify after the summer each year. One example of this is Lake Victoria in Africa, where since the 1980s, permanent stratification caused by strong warming of the epilimnion has led to low oxygen levels in the hypolimnion with resulting fish kills (Kaufman et al. 1996).

Plants have colonised most of the land surface of this planet. An estimate of the proportions of plant categories and the area of land they cover is shown earlier. Although woodlands and forests are important not only for their biodiversity but also for their ability for longer term sequestration of CO_2, grasslands are a major part of the terrestrial biosphere, covering between 20% and 37% of the land area (which includes areas of shrub land which are open shrub and grass allowing animals to graze; Loveland et al. 2010; O'Mara 2012) and of the land surface (Table 9.7) and have the capacity to absorb almost 1.5 Gt CO_2 yearly (O'Mara 2012). Unfortunately, many of the world's grasslands are being degraded.

Table 9.7 Land vegetation classification.

Land classification	Area (km^2)	Approximate %
Evergreen forest	18 474 990	12.6
Deciduous forest and mixed	11 449 118	
Shrublands	30 844 227	21.4
Grasslands	20 376 593	13.9
Croplands	14 030 386	9.7
Wetlands	1 299 213	0.9
Ice/snowfields	16 561 111	11.3
Natural vegetation and crops	13 941 093	9.6
Barren land	18 459 997	12.6
Water bodies	3 494 824	2.0
Urban areas	260 117	0.2
Total	145 696 962	

Modified IGBP DISCover Land Use Classification (Belward 1996).
Source: All Data from Loveland et al. (2010) except marked as "[a]", which is from O'Mara (2012).

Inland Fisheries

Freshwater fisheries may be small subsistence or large commercial units worth millions of $. Although many species of fish are consumed, commercial rearing usually focuses on salmonids, cyprinids (carp and barbs), cichlids and siluriformes (catfish). As discussed earlier climate change will affect the hydrological cycle. Extreme events, such as floods and droughts, are likely to become more frequent in many regions. River flows, especially those fed through springtime snowmelt, are likely to decrease. Increases in global temperature will also occur in many freshwaters, and these will affect water chemistry, especially causing a reduction in oxygen as temperatures increase.

No freshwater fish can regulate its body temperature by physiological means (Moyle and Cech 2004). As temperatures change in the surrounding water, the only thing a fish can do if they are unsuitable is to move to habitats where conditions better if they are available. The temperature ranges differ for different species. Those with a wide range of tolerance are termed eurythermal, and those with a narrow range stenothermal. Examples of different ranges for different species are provided in Table 9.8.

Aquatic systems can be defined by the amount of nutrient chemicals present, basically their nitrogen and phosphorus content. This is called their trophic status. Systems where the nutrient concentrations are low are termed oligotrophic. Where concentrations are high, they are eutrophic. In-between, they are mesotrophic. The amount of nutrients present determines the nature and amount of biological activity.

Although geology, land use, agriculture and human activities will affect nutrient concentrations, so climate change could also affect nutrient concentrations (Bertahas et al. 2006). Increases in temperature will not only have a direct effect on oxygen concentrations (oxygen being less soluble in water as the temperature rises) but will also affect precipitation patterns and, where increases in streamflow occur, cause increased run-off, increasing erosion and transport of nutrients and altering rates of cycling into

Table 9.8 Examples of temperature tolerance ranges of different freshwater fish species.

Fish species	Optimal temperature range (°C)	Temperature class
Sockeye salmon	5–17	Cold water
Atlantic salmon	13–17	Cold water
Lake trout	4–18	Cold water
Brown trout	12–19	Cold water
Walleye	20–23	Cool water
Sauger	21–23	Cool water
Striped bass	15–25	Cool water
Channel catfish	25–30	Warm water
Common carp	27–32	Warm water
Redbelly tilapia	29–32	Tropical
Guinean tilapia	18–32	Tropical

Source: Data from various sources, including Ficke et al. (2007) and The Scientific Fisherman (2019).

and within the water. These changes, including land use changes, can cause large changes in the aquatic environment, resulting in shifts in the composition of freshwater ecosystems. For example, there could be an increase in algal growth, a change in dominant species and bacterial metabolism, which could contribute to lowering oxygen concentrations. On the other hand, reductions in streamflow could lower the influx of nutrients. In shallow lakes, eutrophication can greatly increase the production of macrophytes. Kankkaala et al. (2002) found a 300–500% increase in shoot biomass of the aquatic plant *Elodea Canadensis*. Although they produce excess oxygen during daytime photosynthesis at night, they would respire, causing a large diurnal change affecting fish populations.

These large growths could lead to clogging of the water, possibly excluding some fish. In addition, when they die, they would deoxygenate the water as they decompose possibly leading to fish kills. Large rafts of floating macrophytes will reduce light penetration and affect fish populations and water dynamics by decreasing wind circulation in the water, increasing the period of stratification (Welcomme 1979). Eutrophication will also affect invertebrate communities and thus the amount and type of food available for the fish.

Eutrophication can also affect fish by changing other components of the ecosystem. This particularly applies to the increased algal growth that usually occurs in eutrophic waters. With increased nutrient (nitrogen and phosphorus) concentrations, one particular group of algae often thrives, dominating the planktonic plant population in the lake water. As some species of blue-green algae (cyanobacteria) can be harmful to animal and human health, it is important to control them. Large growths of these algae often colour the water green and form surface scums. They are not new and have been reported in the scientific literature for over a century (Francis 1978). Lakes worldwide have reported large growths of these algae in recent years. This has resulted in economic loss and inability to use those waters (Chorus and Bartram 1999).

Water has long been used to produce energy for human use either in the form of mechanical energy, e.g. water wheels, or electricity in hydro dams. The latter are by far the most important at the moment, so water scarcity in hydro plants could have quite widespread impacts. One of the effects of climate change has been on plants in the biosphere. Rise in the CO_2 concentrations in the atmosphere together with its effect on warming could be expected to favour plant growth which in turn would increase carbon sequestration. However, a 30% reduction in gross primary production over Europe, which would reverse the effect of carbon sequestration from the atmosphere, has been reported. Surprisingly, respiration rates also decreased. Future drought events would cause further reductions in both respiration and primary production. This would lead to reductions in crop yields affecting human welfare. Because the amount of plant growth depends on the season, it is important to examine how changes to the climate will affect different seasons rather than look at annual averages. For example, Wang et al. (2016) examined the impacts of climate change in some regions of China. They found positive correlations with net primary production in spring temperatures but negative correlations in summer temperatures. A further factor is how extreme any change may be. Extremes such as storms and droughts will affect vegetation and thus affect carbon uptake and reduce the total amount of carbon stored (Reichstein et al. 2013).

Effects on Agriculture and Food

> *"Fellow Americans: ask not your country can do for you – ask what you can do for your country." John F Kennedy (1961).*

Often, when considering food, the emphasis is on shortages of water or reduced supplies that might limit production. The reverse side of the equation, which also applies to most activities and industries, is what effects on climate does that activity have and can they be alleviated? Humans' approach to food has changed dramatically from the time of hunter-gatherers to the present day. Currently, food supply in most countries is part of a major supply and distribution system that is expanding to keep pace with population growth and keep food affordable (Vermeulen et al. 2012). Partly because of the growth of cities and partly because of the desire, at least in developed countries, to have a wide variety of foods available at all times of the year, a global food production and distribution network has grown up. Even in developing countries, food is imported, especially in times of crisis such as major droughts. To meet these demands both locally and internationally, a system of food chains or systems has developed. An outline of the numerous links in the food system is provided in Box 9.4. Although climate change will affect agricultural production, various stages in the food system will have an effect on climate change, hence the quote at the start of this section. As with all multi-chain industries, different parts of the chain impact climate change as climate change will affect the chain.

Box 9.4 An Outline of the Systems Within the Food Chain and Their Contributions to Carbon Emissions

Food chain system	Carbon dioxide emissions
Inputs into system	**($MtCO_2e$)**
Fertilizer manufacture	282–575
Animal feed production	60
Production of Pesticides	3–140
Transport and marketing not estimated	
Production phase	
Direct emissions from agriculture	5120–6116
Indirect emissions from agriculture	2198–6567
Post production and distribution	
Initial and other processing	192
Storage, packaging and transport	396
Refrigeration	490
Retailing	224
Catering and domestic use	160
Waste disposal	72

Source: Table redrawn from Vermeulen et al. (2012).

Box 9.5 shows the amounts of GHG emissions produced by different sectors within agriculture. In the European Union, the agricultural sector produced 426473 kilotons of CO_2 equivalent in 2015, which amounts to just under 10% of the total EU's GHG emissions (European Environment Agency 2019). The data given in Box 9.5 are from the period 1990–2015 and are aggregated emissions of agricultural methane (CH_4 and nitrous oxide).

Not included in these figures are emissions from land use, land use change and forestry (see Table 9.9).

Box 9.5 Greenhouse Gas Emissions from European Agriculture by Sector

Sector	% Emissions
Non-agricultural sectors	90.19
Enteric Fermentation (from animals)	4.32
Manure management	1.48
Agricultural soils	3.67
Rice cultivation	0.06
Field burning	0.06

Table 9.9 Examples of emissions from selected European countries.

	Methane (CH_4)	Nitrous oxide (N_2O)	$CH_4 + N_2O$
Belgium	5838	4013	9851
Denmark	5524	4597	10121
Germany	32279	31606	63884
Ireland	12221	6585	18806
Greece	4755	3529	8284
Spain	23369	12098	35467
France	40930	35438	75367
Italy	18441	11074	29515
Cyprus	362	197	559
Netherlands	12998	6144	19142
Poland	14062	14817	28879
Finland	2584	3716	6299
Sweden	3261	3509	6770
United Kingdom	27590	15835	43425

Source: Data from agri-environmental indicator greenhouse gas emissions, https://ec.europa.eu/eurostat.

The sum of methane and nitrous oxide emissions is often expressed as CO_2 equivalents. From EU measurements, GHG emissions decreased by about 20% between 1990 and 2015 although in the last three years there was a levelling off and even a small increase. GHGs have a large effect on climate change and hence the hydrological cycle, which in turn will impact many facets of agriculture and food production. Table 9.9 illustrates the great variation in emissions from one country to another, reflecting not only the size of their agriculture sector but also the type of agriculture being practiced. One significant factor in reductions has been the reduced use of nitrogenous fertilizers (a 20% reduction overall in the EU). This also had the benefit of less leaching of nitrates to surface waters and into aquifers. There has also been a reduction in livestock numbers, notably cattle (over 20%), sheep (over 70%) and pigs (over 10%). All of these livestock reductions helped reduce methane emissions. Reduced animal numbers also mean less animal waste and a lower chance of run-off causing water pollution. There is also diversity in agricultural practices among EU countries, resulting in variations in the intensity of agricultural operations. This can be expressed as GHG emissions per hectare of agricultural land. The aggregated emission figures for CH_4 and N_2O per hectare in kilotons equivalent CO_2 per thousand hectares result in an average for the EU of just over 2. The highest countries are the Netherlands at just over 10, Belgium at about 7 and Malta at about 5.5 (European Environment Agency 2015). These figures illustrate the need for some changes to agricultural systems to help mitigate climate change but not reduce food productivity.

Climate change will also have an effect on both drinking water and wastewater treatment processes. In the case of drinking water, the most widely reported potential impact is on resources and their availability. For wastewater treatment, there are other considerations. In traditional wastewater treatment, the final discharge standards for quality are based upon the discharge being diluted by the receiving water, often a river. If river flows are reduced, as could happen with climate change, the biological oxygen demand and suspended solid concentration may well need to be reduced in the final treatment plant effluent to avoid severe pollution of the river and damage to the biota. Many wastewater treatment systems, such as biological filters and the activated sludge process, rely upon microorganisms for the breakdown of wastes. These organisms are predominantly aerobic, i.e. they require oxygen to function. As the solubility of oxygen in water decreases with increasing temperature, if the global temperature rises by 2+ degrees, then it will probably take more energy to aerate the waste liquor to maintain high enough O_2 concentrations to meet the needs of the microorganisms. This would be an additional cost in the treatment process. Increased temperatures could also alter the ecology of waste treatment ecosystems.

References

Aguilera, H. and Murillo, J.M. (2009). The effect of possible climate change on natural groundwater recharge based on a simple model: a study of four karstic aquifers in SE Spain. *Environmental Geology* 57 (5): 963–974.

Alkama, R., Decharme, B., Douville, H., and Ribes, A. (2011). Trends in global and basin-scale runoff over the late twentieth century: methodological issues and sources of uncertainty. *Collins Journal of Climate* 24 (12): 3000–3014.

Arndt, D.S., Baringer, M.O., and Johnson, M.R. (ed.) (2009). State of the climate in 2009. *Bulletin of the American Meteorological Society* 91 (7): S1–S224.

Baker, A.R.H. and Harley, J.B. (1973). *Man Made the Land. Essays in English Historical Geography*. Newton Abbot, U.K.: David & Charles.

Bates, B., Kundzewicz, Z.W., Wu, S., and Palutikof, J. (2008). *Climate Change and Water. Technical Paper VI*. Geneva, Switzerland: Intergovernmental Panel on Climate Change.

Bellinger, E.G. and Sigee, D.C. (2015). *Freshwater Algae. Identification, Enumeration and Use as Bioindicators*. Chichester, UK: Wiley Blackwell.

Belward, A.S. (ed.) (1996). The IGBP-DIS Global 1km Land Cover Data Set (DIScover). Proposal and implementation plans. IGBP-DIS Working Paper 13, *International Geosphere-Biosphere Programme Data and Information System Office*, Toulouse, France.

Bertahas, I., Dimitriou, E., Laschou, S., and Zacharias, I. (2006). Climate change and agricultural pollution effects on the trophic status of a Mediterrannean Lake. *Acta Hydrochimica et Hydrobiologica* 4: 349–359.

Briffa, K.R., van der Schrier, G., and Jones, P.D. (2009). Wet and dry summers in Europe since 1750; evidence of increasing drought. *International Journal of Climatology* 29: 1894–1905.

Burroughs, W.J. (2005). *Climate Change in Prehistory. The End of the Reign of Chaos*. Cambridge University Press, UK.

Carmichael, W.W. (2001). *Human Ecology Risk Assessment*, vol. 7, 1393.

Carmichael, W.W. (2012). Health effects of toxin-producing cyanobacteria: "The CyanoHABs". *Human and Ecological Risk Assessment* 7 (5): 1393–1407.

Cayan, D.R., Kammerdiener, S., Dettinger, M.D. et al. (2001). Changes in the onset of spring in the western United States. *Bulletin of the American Meteorological Society* 82 (3): 399–415.

Chen, M.-J., Lin, C., Wu, Y. et al. (2012). Effects of extreme precipitation to the distribution of infectious diseases in Taiwan, 1994–2008. *PLoS One* 7 (6): e34651.

Chorus, I., and Bartram, J. (1999). *Toxic Cyanobacteria in water*. WHO, E.7& F. N. Spon. London.

Collins, D.N. (2008). Climatic warming, glacier recession and runoff from alpine basins after the little ice age maximum. *Annals of Glaciology* 48 (1): 119–124.

Coutant, C.C. (1990). Temperature-oxygen habitat for freshwater and coastal striped bass in a changing climate. *Transactions of the American Fisheries Society* 119: 240–253.

Dai, A., Qian, T., Trenberth, K.E., and Milliman, J.D. (2008). Changes in continental freshwater discharge from 1948–2004. *Journal of Climate* 22 (10): 2773–2792.

Dronin, N.M. and Bellinger, E.G. (2005). *Climate Dependence and Food Problems in Russia 1900–1990*. Budapest and New York: CEU Press.

Druyan, L.M. (2011). Studies of 21st century precipitation trends over West Africa. *International Journal of Climatology* 31: 1415–1424.

Durre, I., Williams, C.N., Yin, X., and Vose, R.S. (2009). Radiosonde-based trends in precipitable water over the northern hemisphere: an update. *Journal of Geophysical Research: Atmospheres* 114 (D5).

European Environment Agency (2015). https://www.eea.europa.eu/en/datahub.

European Environment Agency (2019). Agri-environmental indicator-greenhouse gas emissions. https://www.eea.europa.eu/en/datahub.

FAO (2012). AQUASTAT Global Information System on Water and Agriculture. https://www.fao.org/aquastat/en.

Ficke, A.D., Myrick, C.A., and Hansen, L.A. (2007). Potential impacts of global climate change on freshwater fisheries. *Reviews in Fish Biology and Fisheries* 17: 581–563.

Francis, G. (1978). Poisenous Australian lake. *Nature* 18: 11–12.

George, D.G., Hewitt, D.P., Lund, J.W.G., and Smyly, W.J.P. (1990). *The Relative Effects of Enrichment and Climate Change on the Long-Term Dynamics of Daphnia in Esthwaite Water*. Ferry House, Ambleside, Cumbria, UK: Freshwater Biological Association.

Green Facts, Facts on the health of the Environment (2019). https://www.greenfacts.org/en/arctic-climate-change/index.htm.

Haywood, A., Bonham, S., Hill, D. et al. (2009). Lessons of the mid-pliocene: planet earth's last interval of greater global warmth. *IOP Conference Series: Earth and Environmental Science*,. IOP Publishing Ltd. DOI: 10.1088/1755-1307/6/7/072003.

Heinrich, H. (1988). Origin and consequences of cyclic ice rafting in the Northeast Atlantic Ocean during the past 130,000 years. *Quaternary Research* 29: 142–152.

Hiernaux, P. and Turner, M.D. (2002). The influence of farmer and pastoral management practices on desertification processes in the Sahel. In: *Global Desertification. Do Humans Cause Deserts?* (ed. J.F. Reynolds and D.M. Stafford-Smith). Berlin: Dahlem University Press.

IPCC (2007). *Intergovernmental Panel on Climate Change. Impacts, Adaptation and Vulnerability. Contribution of Working Group II to the Fourth Assessment Report of the Intergovernmental Panel on Climate Change*. Parry, Palutikov, van der Linden and |Hanson, Eds. Cambridge University Press, UK.

IPCC (2019). *Impacts of a Warming Arctic*. UK: Cambridge University Press.

IPCC WGH ARS (2014). *Climate Change 2014: Impacts, Adaptation, and Vulnerability*. https://www.ipcc.ch/report/ar5/wg2.

Islam, M.S., Sharker, M.A., Rheman, S. et al. (2009). Effects of local climate variability on transmission dynamics of cholera in Matlab, Bangladesh. *Transactions of the Royal Society of Tropical Medicine and Hygiene* 103 (11): 1165–1170.

Jean, J.S., Guo, H.R., Chen, S.H. et al. (2006). The association between rainfall and occurrence of an enterovirus epidemic due to a contaminated well. *Journal Applied Microbiology* 101 (6): 1224–1231.

Jeelani, G. (2008). Aquifer response to regional climate variability in part of Kashmir Himalaya in India. *Hydrogeology Journal* 16 (8): 1625–1633.

Jones, J.A.A. (2010). *Water Sustainability*. London, UK: Hodder Education.

Kankkaala, P., Ojala, A., Tulonen, T., and Arvola, I. (2002). Changes in nutrient retention capacity of boreal aquatic systems under climate warming: a simulation study. *Hydrobiologia* 459: 67–76.

Kaufman, L., Chapman, L.J., and Chapman, C.A. (1996). The great lakes. In: *East African Ecosystems and Their Conservation* (ed. T.R. McClanahan and T.P. Young), 191–216. New York: Oxford University Press.

Lambert, F.H. and Allen, M.R. (2009). Are changes in global precipitation constrained by the tropospheric energy budget? *Journal of Climate* 22: 499–517.

Li, S.G., Asanuma, J., Eugster, W. et al. (2005). Net ecosystem carbon dioxide exchange over grazed steppe in Central Mongolia. *Global Change Biology* 11: 1941–1955.

Liao, C. and Zhuang, Q. (2015). Reduction of Global Plant Production due to droughts from 2001 to 2010: An Analysis with a Process-Based Global Terrestrial Ecosystem Model. *Earth Interactions* 19: 1–21.

Loveland, T.R., Reed, B.C., Brown, J.F. et al. (2010). Development of a global land cover characteristics database and IGBP DIScover from 1 km AVHRR data. *International Journal of Remote Sensing* 21 (6–7): 1303–1330.

Lunt, D.J., Foster, G.L., Haywood, D., and Stone, E.J. (2008). Late Pliocene Greenland glaciation controlled by a decline in atmospheric CO_2 levels. *Nature* 454 (7208): 1102–1105.

Mackenzie, W.R., Hoxie, N.J., Proctor, M.E. et al. (1994). A massive outbreak in Milwaukee of cryptosporidium infection transmitted through public water supply. *The New England Journal of Medicine* 33 (3): 161–167.

McBean, G., Alekseev, G., Chen, D. et al. (2005). Arctic climate: past and present. In: *Arctic Climate Impacts Assessment (ACIA)* (ed. C.M. Symon, L. Arris, and B. Heal). UK: Cambridge University Press.

Menzel, A. and Fabian, P. (1999). Growing season extended in Europe. *Nature* 367: 659.

Miguel, E., Satyanath, S., and Sergenti, E. (2004). Economic shocks and civil conflict: an instrumental variables approach. *Journal of Political Economy* 112 (41): 725–753.

Millennium Ecosystem Assessment (2005). https://www.millenniumassessment.org.

Mote, P.W. (2003). Trends in snow water equivalent in the Pacific northwest and their climate causes. *Geophysical Research Letter* 30: https://doi.org/10.1029/2003GL0172588.

Mote, P.W., Hamlet, A.F., Clark, M.P., and Lettenmaier, D.P. (2005). Declining mountain snowpack in western North America. *American Meteorological Society* 2005: 39–49.

Moyle, P.B. and Cech, J.J. (2004). *Fishes: An Introduction to Ichthyology*, 5e. Englewood Cliffs, NJ: Prentice Hall.

NASA (2012). NASA Research Finds 2010 Tied for the Warmest Year on Record. http://www.giss.nasa.gov/research/news/20110112.

National Bureau of Statistics of China (2009). https://www.stats.gov.cn/english.

NOAA, National Oceanic and Atmospheric Administration Report card. (2017). http://www.arctic.noaa.gov/Report-Card-2017.

Noake, K., Polson, D., Hegeri, G., and Zhang, X. (2012). Changes in seasonal land precipitation during the latter twentieth century. *Geophysical Research Letters* 39: Lo3706.

Norby, R.J. and Luo, Y.Q. (2004). Evaluating ecosystem responses to rising atmospheric CO_2 and global warming in a multi-factor world. *New Phytologist* 162 (2): 281–293.

Olsen, R., An, Y.F., Fan, Y. et al. (2019). A novel method to test non-exclusive hypotheses applied to Arctic ice projections from dependent models. *Nature Communications* 10 (1): https://doi.org/10.1038/s41467-019-10561-x.

O'Mara, F.P. (2012). The role of grasslands in food security and climate change. *Annals of Botany* 110 (6): 1263–1270.

Otto, F.E.L., Rosier, S.M., Allen, M.R. et al. (2015). Attribution analysis of high precipitation events in summer in England and Wales over the last decade. *Climate Change* 132: 77–91.

Paerl, H.W. and Huisman, J. (2008). Blooms like it hot. *Science* 320: 57–58.

Paerl, H.W., Valdes, L.M., Piehler, M.F., and Stow, C.A. (2006). Assessing the effects of nutrient management in an estuary experiencing climate change; the Neuse River Estuary, North Carolina. *Environmental Management* 37 (3): 422–436.

Peplow, M. (2004). Dust bowl caused by ocean highs and lows. Sea temperatures linked to 1930s US drought. *Nature*, https://doi.org/10.1038/news040315-12.

Piao, S., Ciais, P., Huang, Y. et al. (2010). The impacts of climate change on water resources and agriculture in China. *Nature* 46 (7311): 43–51.

Reichstein, M., Bahn, M., Clais, P. et al. (2013). Climate extremes and the carbon cycle. *Nature* 500: 287–295.

Richardson, K., Steffen, W., Schellnhuber, H.J. et al. (2009). Climate Change. Global Risks, Challenges & Decisions. Copenhagen (10–12 March 2009). http://www.climatecongress.ku.dk.

Scherrer, S.C., Appenzeller, C., & Laternser, M. (2004). Trends in Swiss Alpine snow days – the role of local and large-scale variability. *Geophysical Research Letters* 31, L13215.

Schneider, S.H. (2001). What is dangerous climate change? *Nature* 411: 17–19.

Scott, M., Gupta, S., Jauregui, E. et al. (2001). Human settlements, energy and industry. In: *Climate Change 2001: Impacts, Adaptation, and Vulnerability. Contribution of Working Group II to the Third Assessment Report of the IPCC* (ed. J. J. McCarthy et al.). UK: Cambridge University Press.

Sheffield, J. and Wood, E.F. (2008). Global trends and variability in soil moisture and drought characteristics, 1950–2000, from Observation-driven simulations of the terrestrial hydrological cycle. *Journal of Geophysical Research* 112: 432–458.

Sheffield, J. and Wood, E.F. (2011). *Drought. Past Problems and Future Scenarios.* Earthscan Publications.

Shiklomanov, A.I., Lammers, R.B., Rawlins, M.A., Smith, L.C., & Pavelsky, T.M. (2007). Temporal and spatial variations in maximum river discharge from a new Russian data set. *Journal of Geophysical Research- Biogeosciences* 112 (G4). GO4553.

Sparks, T.H. and Menzel, A. (2002). Observed changes in seasons: an overview. *International Journal of Climatology* 22: 1715.

Stewart, I.T., Cayan, D.R., and Dittnger, M.D. (2005). Changes towards earlier streamflow timing across western North America 1136 IT.S.J. Climate

Stubenrauch, C. and Schulmann, U. (2005). Impact of air traffic on cirrus coverage. *Advanced Earth and Space Science* https://doi.org/10.1029/2005GL022707.

Theisen, O.M., Holtermann, H., and Buhaug, H. (2012). Climate Wars? Assessing the claim that drought breeds conflict. *International Security* 36 (3) Winter 2011–2012: 79–106.

The Scientific Fisherman (2019). https://thescientificfisherman.com.

Tibby, J. and Tiller, D. (2007). Climate – water quality relationships in three western Victorian (Australia) lakes 1984–2000. *Hydrobiologia* 591 (1): 219–234.

United States Geological Survey (USGS) (2007). USGS Fact Sheet 2005–3018. Spring arriving earlier in Western Streams.

Van Huijgevoort, M.H.J., Hazenberg, P., van Lanen, H.A.J. et al. (2013). Global Multimodal Analysis of Drought in Runoff for the Second Half of the Twentieth Century. *Journal of Hydrometeorology* 14: 1535–1552.

Vermeulen, S.J., Campbell, B.M., and Ingram, S.I. (2012). Climate change and food systems. *Annual Review of Environment and Resources* 37: 195–222.

Wang, H., Liu, G., Zongshan, L. et al. (2016). Impacts of climate change on net primary productivity in arid and semi-arid regions of China. *Chinese Geographical Science* 26 (1): 35–47.

Welcomme, R. (1979). *Fisheries Ecology of Floodplain Rivers*. New York: Longman Inc.

Wetzel, R.G. (1983). *Limnology*. Philadelphia, PA: Saunders.

Wihley, M.L. and Atkinson, T.C. (1977). Dry years in south-East England since 1698. *Nature* 265: 431–434.

Wilson, D., Hisdal, H., and Lawrence, D. (2010). Has streamflow changed in the Nordic countries? Recent trends and comparisons to hydrological projections. *Journal of Hydrology* 394 (3–4): 334–346.

Xiong, W., Conway, D., Jiang, J. et al. (2008). Future cereal production in China: Modelling in the interaction of climate change, water availability and socio-economic scenarios. the *Impacts of Climate Change on Chinese Agriculture – Phase II*. Final Report (AEA Group 2008).

Xopaki, J.S.P., Gonzalez-Rouco, J.F., Luterbacher, J., and Wanner, H. (2004). Wet season Mediterranean precipitation variability: influence of large scale dynamics and trends. *Climate Dynamics* 23 (1): 63–78.

Yamakawa, S. and Suppiah, R. (2009). Extreme climate events in recent years and their links to large scale atmospheric circulation features. *Global Environmental Research* 13: 69–78.

Yoshioka, M., Mahowald, N.M., Conley, A.J. et al. (2007). Impact of desert dust radiative forcing on Sahel precipitation: relative importance of dust compared to sea surface temperature variations, vegetation changes and greenhouse gas warming. *Journal of Climate* 20: 1445–1467. American Meteorological Society.

You, L., Rosegrant, M.W., Wood, S., and Sun, D. (2009). Impact of growing season tyemperature on wheat productivity in China. *Agricultural and Forest Meteorology* 149: 1009–1014.

Yu, K., D'Odorico, P., Okin, G.S., and Evan, A.T. (2015). Dust-rainfall feedback in west African Sahel. *Geophysical Research Letters* 42 (18): 7563–7571.

10

Sustainability: The Way Ahead and Can We Achieve It – A Summary

Humans and all other lives on earth are sustained by the life support systems of the planet, i.e. a suitable composition of the atmosphere, freely available amounts of freshwater and seawater, climate, the hydrological cycle, soils and minerals together capable of supporting living organisms. However with human population growth, particularly since the industrial revolution, the increased demand for water and the resulting influence means that there is a much greater demand on the hydrological cycle which growing exponentially. These demands that are now being placed on all of the earth's resources and systems are greater than ever. The potential problem of humans over exploiting resources was highlighted by Meadows et al. in 1972 when they published their book 'Limits to Growth'. This concept was followed up by several authors warning about human consumption in the Anthropocene stressing natural resources (Steffen et al. 2007, 2011). There are a number of earth systems that have been identified, which are vital for the survival of humans and ecosystems in general. Some of the main earth system processes are (i) land use change, (ii) biodiversity loss, (iii) the nitrogen and phosphorus cycles, (iv) ocean acidification, (v) global freshwater use, (vi) chemical pollution, (vii) global freshwater use and (viii) climate change. All of these factors either directly or indirectly impact the hydrological cycle, and water is the essential common thread that runs through not only the whole of the biosphere but also all aspects of human society, including health, food, well-being and economic growth (World Economic Forum 2011). All of the above systems can tolerate a certain amount of change, but there are upper limits that can just be tolerated but not exceeded, and these are called planetary boundaries. However, if these amounts are exceeded, then there is a distinct possibility that there will be changes, possibly catastrophic, to the earth systems on which we depend. Water is arguably the most important resource we have, but humans tend not to look at the whole picture but only focus on those small parts that affect them or their institutions personally. To support the increasing human population, we must take a holistic view as at the moment our planet is not being used prudently. All of the life support systems are interconnected, and to mismanage these systems will inevitably have adverse effects on the others. We have to treat the entire biosphere as a system that needs to be in balance. To this end, and taking into account the diversity of human societies, all sections of the system, especially the non-human and non-living parts as well as human needs, must be involved or considered in management decisions. An important factor is that water scarcity will increase due to a number of factors, not the least of which is the increasing human population and therefore the need for more food, freshwater and proper sanitation.

Water: Our Sustainable and Unsustainable Use, First Edition. Edward G. Bellinger.

Current water abstraction is about three times that of 50 years ago and is likely to increase significantly if we follow the business as usual scenario. Kammeyer (2017) states that demand will increase by another 55% by 2050. There has been recognition internationally that there are current problems concerning water as recognised in Goal 6 of the UN Sustainable Development Goals (UN 2017; see Arora and Mishra 2019). This aims to tackle the problems concerning drinking water availability, hygiene and sanitation. Concern was also expressed about meeting the needs of ecosystems.

There have been and still are many models of various parts of the earth life system, but the tendency of many of them is to look at the status quo and then project into the future what will happen. Depending on the weightings placed upon the numerous components, a range of projections have been obtained, some optimistic and many pessimistic. Unfortunately, although these have been tested over time, not all have been found to be correct. A different approach is to look at where humans want to be at whatever potential date in the future is chosen and then model backwards to determine what we need to do now and in the intervening period to arrive at that position. The model must not be too rigid as quite often the data we use is imperfect, so the possibility of introducing modifications with updates must be allowed for. As has been mentioned several times previously, the world's population is increasing, and although there are various projections as to the amount, it could increase by a further 2 billion by 2050 to a total of 9.7 billion and by 2100 up to 11.2 billion. Assuming that the proportion of people not having access to clean drinking water will be similar to the present time and allowing for a daily requirement of 100 l/day/person or 36 m^3/person/year (Falkenmark 1997), an additional 18 billion m^3 of water will be needed in the next 25 years. As the GDP of many countries is increasing, their populations are likely to have a tendency for the individuals' water consumption per person to increase, as bathing and showers become more widespread, as well as increase in other domestic uses. In addition, more food will be required, so agricultural demands will also increase. Based upon a business as usual approach, agricultural food production would need to increase by 70% to feed the world by 2050. Unfortunately, half of the world's population growth will most likely focus on nine countries, India, Pakistan, the Democratic Republic of the Congo, Ethiopia, Tanzania, Nigeria, the United States of America, Indonesia and Uganda, many of which are water stressed already.

Agriculture is the largest water user by volume and is also not always a very efficient user. It is highly subsidised and, compared with cities and industry, a relatively low economic value producer from the point of view of its products. Currently, spending on agriculture has, understandably, dominated national budgets in many countries globally. Between 1940 and 1990, for example, 80% of Mexico's public expenditure on agriculture was for irrigation projects, and China, Pakistan and Indonesia used about 30% of all public investment in irrigation (FAO 1994). The usual way to increase crop yields in the past was to increase the area under cultivation. This would now involve bringing marginal land into cultivation as virtually all prime land has already been taken; indeed, the expansion of cities often encroaches on good agricultural land, making the situation worse. In order to provide the amount of food needed for the future predicted global population, more efficient use of land and water is needed. A blue revolution is now needed whose focus is around 'more crop per drop' (Raza et al. 2011).

Although more recent technologies involving various forms of microirrigation deliver water in more measured quantities than older open channel and flood methods, they do not necessarily use less water if the whole water cycle is taken into account. With modern technologies, the amounts of water used both in time and space can both increase yields and save water. This approach, called scientific irrigation scheduling, calculates the actual water requirements of a crop over short periods of time so that over- and underapplication is avoided (Evans and Sadler 2008). This approach takes into account detailed climate data for the previous 5 days and predicts future water needs for the next 10 days. The timing and volume of the next irrigation event can be accurately scheduled. Crop responses to environmental factors such as temperature can be taken into account as well as varying requirements at different stages in the plant growth cycle. Scheduling can help more precisely determine the volumes of water required but should be used in combination with other good farming practices.

Genetic modification of plants is being investigated as both a possible way of reducing plant water demand and resistance to droughts (Passioura 2005). One of the problems is that some areas experience prolonged droughts, while others have only short periods and are not always at the same stage of the crop growth cycle (Richards 2004). This approach has been used with success. In New South Waters, Australia, a new variety of wheat, called Drysdale, was introduced in 2002 and showed a yield improvement of up to 23% over traditional varieties. These increases were from plants selected for having characteristics favouring components of the Passioura equation (Richards 2004), which is expressed as follows:

$$Y = BWR \times W \times HI$$

where Y is the crop yield, HI is the harvest index (Passioura 1977), and BWR is the biomass water ratio.

Large flows of water for irrigation do not necessarily bring better crop yields or quality (Chaves and Olivera 2004). Strategies to control stomatal opening are much more likely to save water and allow carbon dioxide uptake to promote good yields. Excessive amounts of water can promote too much shoot growth instead of roots and fruits (Zhang 2004). Optimising the use of water based on the knowledge of stomatal behaviour can decrease water requirements without affecting crop production. One way being promoted is to use a variation of deficit irrigation called partial root-zone drying (PRD) (Chaves and Olivera 2004). This involves alternating the supply of water from one side of the root system to the other at regular intervals. This maintains the water requirements of the plant, but the drying of the roots on one side stimulates some closure of the stomata cutting transpiration losses but still allows uptake of carbon dioxide from the atmosphere, thus maintaining crop yields (Davies et al. 2000). PRD irrigation has been used with success in Australia (Loveys and Ping 2002). Other successful applications include grapevines (Santos et al. 2003), raspberries (Grant et al. 2004) and oranges (Loveys and Davies 2004). Fruit quality has not been impaired and may even be improved. Understanding the signalling systems in plants and using these to combat drought conditions combined with new technologies and improved crop genetics could well lead to water savings in irrigation. These savings are vital if demands from other sectors for limited water resources are to be met in the future.

With different sectors demanding more water and water use efficiency not keeping pace with this increased demand, together with the possible effects of climate change, water may well have to be restricted for some sectors through law. An example of this is the

2003 Water Act in the United Kingdom. Other countries have set targets for a reduction in water use by agriculture, for example in China a 20% reduction in agricultural water use had been set by 2020 (Morison et al. 2008). A further way of reducing agricultural demand is to greatly reduce food wastage. It is essential that both starvation and malnutrition are eliminated in the near future in order to meet the Millennium Development Goals. There are a number of drivers and associated pressures involved in food production, including governance systems, economics, climate change, availability of land and water availability (Barron et al. 2013). The reduction of food wastage is an important way of reducing water demand and increasing food supply. In modern society, and particularly with the growth of cities and food supplies coming from global supply networks, it is not surprising that greater amounts of food are wasted worldwide. FAO (2013) estimated that approximately one-third of all food produced for human consumption is lost or wasted through the food chain from producer to consumer, which amounts to about 1.3 billion tons/year (Gustavsson et al. 2011). For European Union countries, the amount is 89 million tons/year or 179 kg/person (Monier et al. 2010). In low-income countries, most of the losses occur in the early part of the food chain for a variety of reasons, including harvesting techniques, poor storage and variable climate (Lang and Raynor 2012). In high-income countries, most losses occur at the higher end of the food chain and are related to a lack of coordination between individual stages of the food chain, consumer behaviour and the fact that higher incomes allow people to purchase and waste more. All of the wasted food has been grown using land, water and fertilizers, all of which drain resources and have economic and environmental costs. If savings and recovery of this food along the food chain could be made, it would help relieve global hunger, malnutrition and food wastage. Some of the responsibility rests with growers, producers, distributors and sellers, including supermarkets and consumers. If we continue with a business as usual approach, food production needs to grow by 60% by 2050 to meet population needs, including eradication of malnutrition/undernourishment (Alexandratos and Bruinsma 2012). The main stages in the food chain or food life cycle are as follows: agricultural production – processing – distribution – food outlets and supermarkets – consumption – disposal of unwanted parts (including food preparation). Consideration must be given to the fact that not every part of a crop is edible to humans and must be disposed of.

The main environmental impact of food wastage includes its contribution directly and indirectly to greenhouse gas emissions, the water footprint (WF) and water wastage, land use, land degradation, reduction of biodiversity and further impacts during the production phase of food through various impacts that also occur at every stage along the food chain. FAO (2013) defines food loss as 'the decrease in mass dry matter or nutritional value (quality) of food that was originally intended for human consumption'. The food chain, including suppliers and distributors, must have a properly trained workforce and good access to their markets. Superimposed upon this will be climate change and the occurrence of natural disasters. In higher income countries, food may also be discarded whether or not it is past its sell by date. FAO (2013) identifies several categories of loss along the food chain. These are food loss that is the loss of food mass (dry matter) or nutritional value (quality) that was originally meant for human consumption. Food waste is food that is purchased for human consumption that is then discarded perhaps because it is past its sell by date even though it is safe to eat, and oversupply by purchasers or markets is left to spoil or through eating preferences/habits. Food wastage is any food lost by deterioration or waste, so it encompasses both of the above two categories. A further cause of wastage is the increasing distance foodstuffs now travel from producers to consumers. Currently, different parts of a meal are likely to come from different continents. About 1.3 Gtons of edible food are wasted annually (FAO 2013). Along the food chain, the agricultural production stage, and upstream wastage, is the most wasteful by volume at 33%. If postharvest handling and storage (also regarded as upstream) wastage are added, the figure rises to 54%. Wastage downstream, i.e. from processing to consumption and end of life, is about 46% of the total. Downstream wastage is much lower in low-income countries at between 4% and 16% although upstream wastage is much higher. Upstream and downstream amounts will vary from country to country. All food production and its passage along the food chain requires the use of water. The amount of water required, i.e. its WF, can be estimated and can be divided into two categories: withdrawal and consumption. The WF approach developed by Hoekstra et al. (2011) defined the total volume of water used for a product as consisting of three components: blue water, green water and grey water. In agriculture, blue water is the consumptive use of irrigation water abstracted from both ground and surface waters. Green water is rainwater directly used and evaporated by non-irrigation agriculture, pastures and forests. The grey water footprint is a theoretical measure of the volume of water required to dilute pollutants to an acceptable level and is difficult to calculate and not included in many studies. The estimated total blue water used through global food waste for 2007 was 250 m^3 (FAO 2013). Mekonnen and Hoekstra (2011) put this at 3.6 times the blue water footprint of the total blue water footprint of the US consumption. On a national basis, India and China have the largest blue water footprints for the consumption of agricultural products at over 140 km^3, followed by Pakistan, United States of America, Iran and Egypt with between 30 and 60 km^3 and Mexico, Turkey, Indonesia and Spain with less than 20 km^3. The main contributors to this footprint of food wastage are cereals and fruits although it must be understood that these figures are estimates and should only be used as a guide. We are frequently told that we all live in a 'global village' where world trade has virtually no boundaries although this can be disrupted by conflicts. What we must remember, however, is that all goods need water to produce to one degree or another as does their transport from country to another. Thus, water is indirectly saved by one country by not producing that particular commodity. This is termed 'virtual water' by Hoekstra et al. (2011), and the amount of virtual water imported by a country is a saving by not using their own resources but still, in reality, contributes to their WF. It must also be recognised that real water is also exported from one country to another in the form of bottled water as well as bulk transport in tankers, floating bags or drogues and transboundary rivers, lakes and aquifers. These are all obvious methods of transporting water from water-rich regions to those that are water poor. Less obvious is the amount of water used in one country to produce goods then exported to another, thus making savings to their own water resources. The real water trade has increased several folds in the past 50 years as has the trade in virtual water.

Virtual Water and Water Footprint

Climate change is with us for probably several decades into the future, so changes that we see to the hydrological cycle will also be with us, which means that there will still be many countries that suffer from water scarcity and drought. The populations of those countries will need food and water to survive. Virtual water is one way of helping in these situations. Apart from these countries, most higher income developed countries have, for many years, traded virtual water in the form of many products, both food and non-food. Virtual water, i.e. water that is in and used to produce these products, is regularly traded internationally.

Oki et al. (2017) describe this water as being 'embedded' in a product as well as the water required to produce it. The concept of virtual water and its assessment was introduced by Allan (2003, 2011), Hoekstra (2003) and Hoekstra et al. (2011). This water is saved by the importing country and thus saves them from using their own water resources. There is no exact amount for a group of products as, especially with food products, it depends, among other things, on production methods and climate of that area. It can also be defined as the amount of water that would have been used in the receiving country to produce that product. With the exception of certain commodities such as some food products, this latter definition obviously only applies to items that can be produced in either country. In the context of available water resources, virtual water could be regarded as a way of overcoming water scarcity and water stress in some nations. The economics of such water use is also important, and the price of commodities produced in this way is reflected in their selling price taking into account the entire supply chain. Although water availability in some countries depends on their social structure, it is clear that imported virtual water can, at a cost, alleviate starvation and malnutrition, but this comes at a cost for those products. Examples of calculated virtual water contents are provided in Table 10.1.

The food products shown in Table 10.1 represent examples of staple dietary crops. The water required to produce them would represent significant quantities and savings to a water-stressed country. Although the production of food products is the main water user and hence could represent large water savings for the importing country, virtual water trading also occurs with all industrial products as well. Hoekstra and Mekonnen (2012) calculated that the global average WF for the period 1996–2005 was 9086 Gm^3/year, of which 74% was green water, 11% blue water and 15% grey water. Of this, agricultural products made up between 69% and 92% (the first estimate is from Chen and Chen (2013) and the second from Hoekstra et al. (2011)). The total volume of virtual water transferred globally for both agriculture and industry was estimated at 2320 Gm^3 (Hoekstra et al. 2011).

Table 10.1 Examples of calculated virtual water contents of selected products from various authors.

Commodity	Virtual water content in m^3/ton	Author
Wheat	1160–2000	Oki et al. (2003) and Zimmer and Renault (2003)
Rice	1400–3600	Oki et al. (2003) and Zimmer and Renault (2003)
Potatoes	105–180	Hoekstra and Hung (2002) and Zimmer and Renault (2003)
Beef	15977–20700	Chapargain and Horkstra (2003) and Oki et al. (2003)
Milk	560–865	Oki et al. (2003) and Chapargain and Horkstra (2003)

All of these numbers shown are global averages except Oki et al. Their figures refer to Japan.

While global trade in virtual water is used by virtually every country, many are not water stressed, and some that are stressed rely on embedded water in food for their survival. Knowing the virtual water use in each country is important in formulating management strategies for their resources. Traded virtual water allows countries to have a relatively reliable supply of food (Allan 2011). Allan also points out that without international virtual water trade, there could be conflict over transboundary water supplies, especially in countries where the population growth rate is high, and internal water resources could therefore become more stressed.

Reducing Water Demand

Management of water use in agriculture, industry and by the public as well as improving the water infrastructure in many countries is essential if water scarcity is to be avoided. Demand management is an essential first step. This should not only include overall demand but also target strategies for reducing peak demand periods. Peak demand periods place extra strain on the system as well as increasing energy costs incurred through the extra pumping during peak periods. Beal et al. (2016) point out that behavioural interventions, such as marketing strategies to householders on specific ways to save water, and even demand can achieve substantial savings and reduce demand curves. There are several ways in which households can reduce water usage, some behavioural and some technical. The growth in the possession of washing machines has risen by nearly a quarter over the past 15 years, and unfortunately, not all are as efficient in reducing water use as others. One way to encourage manufacturers to produce more efficient machines could be for the government to reduce value added tax (VAT) on any water-saving domestic machines to encourage both the public to use them and manufacturers to produce them. A large amount of water is used in watering lawns and flowers in gardens. There have been campaigns in the United Kingdom to encourage households to collect rainwater, mainly from roofs and downspouts, which can then be used in dry periods for gardening including lawns. Washing cars also use large amounts of water. Instead of using a hosepipe, which is often running continuously, use a bucket of water. Many car washes, although they use pressure washers, also have collecting systems that collect the water, remove sediments and dirt, and recycle it for future use. Many water companies encourage the installation of water meters in households. Some are fitted free or at a low cost. By making the consumer become aware of how much is being used, the costs frequently result in more frugal water use to save money. Recycling/reuse of water can make substantial savings for a household. Water reuse is used in some semi-arid countries, for example Windhoek, the capital of Namibia, where recycling of water is performed for the whole city. In this case, domestic wastewater is recycled and is not mixed with industrial waters and avoids any sources of waters containing potential toxins. Domestic wastewater is treated and then blended with freshwater (which is in limited supply) for use. For households, the reuse of water is possible with the installation of appropriate sanitation systems. The system usually starts with storage, which can accept urine, excreta, grey water and rainwater. Different systems exist whose use can be selected for particular situations. Some examples of these systems are as follows:

- Wet mixed black water and grey water system with decentralised treatment
- Wet black water system

- Wet urine diversion system
- Dry excreta and grey water separate system
- Dry urine diversion system
- Dry excreta and grey water mixed system

There are many ways of reducing water usage. Savings in water also generally mean that there will also be financial savings for both individual households and industry. UNEP–TU Delft (2008) reported that as our lifestyles change and societies become more affluent, their use of water increases. They have more baths, more showers, clothes washing machines and dishwashing machines, all of which increase consumption. This becomes more critical in some regions as water supplies come under stress. It is thus extremely important to adapt environmentally sound technologies (ESTs) and avoid wasting water. They use the term 'Wisewater' in order to increase awareness of the problem and encourage the development of approaches and skills that will reduce water use. The Wisewater approach was developed to encourage reductions in domestic water use. This should, it is hoped, increase water use efficiency and thus reduce the volume used. UNEP–TU Delft (2008) defined ESTs as using water in a more sustainable way and protecting the environment. These technologies are less polluting and do not degrade or deplete resources as much. The term encompasses the entire water system including goods, services and management. This will include educating the public, agriculture and industry. The adoption of ESTs also requires political will and governmental support. The use and type of ESTs will vary with country. In general, the approach should use technologies that are less polluting and encourage reuse of suitably treated wastes as well as reduce water consumption. It is also important that all ESTs should be made available to both developed and developing countries and be based on the needs of the population. This should include both the use of traditional technologies and, where appropriate, introducing cutting-edge technologies.

First, an assessment must be made of the amounts of water needed by the user groups and the conditions under which they live. The overall approach of a country or region should be within the concept of Integrated Water Resource Management. This is based on four principles that are as follows:

1) Freshwater is a finite resource that is essential to all life and the environment.
2) Water development and management should be based on a participatory approach that includes all users, planners and policymakers at all levels from the public to the government.
3) The role of women must be acknowledged in the provision, management and safeguarding of water.
4) Water has an economic value to all users, so all of them should be involved in decision-making, and water should be recognised as an economic good.

These are based on the Dublin principles (International Conference on Water and the Environment 1992). The process brings together all parties in the decision-making process, including sound science, stakeholders, ecology and economies. ESTs are based on the entire hydrological cycle, including water abstraction and storage and maintaining its quantity and quality; its supply to users and its distribution; efficient use and savings; water reuse, recycling and safe disposal. UNEP–TU Delft identifies a range of ESTs that could be adopted, which are divided into four groups: (i) water storage and augmentation, (ii) water supply and distribution, (iii) water use and saving and (iv) water reuse and safe disposal. Within these groups, several EST approaches can be used. Group (i) uses ponds and reservoirs for storage of water both to capture water in times of plenty and to supplement supplies at times of scarcity. This helps even out supplies over the whole year. Rainwater harvesting and storage helps capture rainwater that would otherwise probably run to waste. Sewage effluent, if adequately treated, can supplement surface waters, especially in agriculture, and replenish groundwater. For group (ii), surface and groundwater abstractions must be managed so that the resource is not overexploited. Water transfer schemes can be used, but with care, so ecosystems are not put at risk. Although point-of-use treatment systems can be used, it is often more efficient to use centralised treatment. For group (iii), water savings can be achieved through waterless and water-saving toilets, water-saving taps and showerheads, water-efficient household appliances that use less water and being careful not to waste water for personal hygiene and domestic chores. Finally, group (iv) uses collected rainwater for lawns and garden plants. Similarly, treated grey water is used for car washing, gardens, etc. Simple constructed wetlands can be used to purify grey water. Rainwater runoff should be collected in a separate pipe system from sewage. Environmentally sound sewage treatment systems, such as those used in developed countries, should also be encouraged in developing countries. This list is not comprehensive and can be added to or modified for specific situations.

Older domestic appliances were often designed without taking water use into consideration. Unfortunately, modern more water efficient ones are often more expensive to construct, which can make them more expensive to buy even though water, and often energy, costs will be less when using them. There is a case for encouraging people to buy them, for example, reducing VAT on their purchase. Another approach used in Queensland, Australia, during the severe drought of 2006 was that it passed legislation requiring all new buildings to install water-saving technologies from 2007. This approach could be adopted for all new buildings, without specifying the exact techniques to be used. In most countries, this could save considerable amounts of water. Different technologies would be needed depending on local conditions. All houses, no matter how old, could save water by reducing the volume of water used when flushing a toilet. Older toilet cisterns can use as much as 12 l/flush, whereas modern ones use less than half that volume. In some areas, people have been encouraged to reduce the amounts used in older systems by putting a brick in the cistern. In addition, it is possible to obtain a cistern with a double action flush lever. A short lever action gives a small flush, whereas if the lever is held down a little longer, this gives a larger flush. At times of drought, water authorities often put a hosepipe ban into action, preventing profligate use of treated water. Many water companies have encouraged the installation of water meters to make consumers aware of how much is being used. Households that have collected rainwater or reuse grey water can help their plants survive at these times. Public information and education are important in order to gain their cooperation. It is also important for water companies to give an example by making strenuous efforts to reduce leakage in the water distribution system. The water industry in the United Kingdom, for example, has in excess of 700 000 km of mains water supply and sewage to maintain, much of which is through major towns and cities. Much of the system consists of cast iron pipes that are subject to corrosion over time, and mending or replacing them can cause major inconveniences to busy roads. Where there is a fault, for example a small hole, because the water in the pipe is under pressure, the loss by leakage will be quite large. For example, a 3 mm hole in the pipe with a water pressure of 40 psi can leak 8450 l in 24 hours. A small leakage, if not mended, can leak over 1.3 million litres in a year. Leakages also occur on a customer's property. A leaking toilet can waste nearly a third of a million litres in 30 days, and a dripping hosepipe or garden tap can waste 680 l in a month (Data from Water Intelligence). Targets have been set for the reduction of leakage in the United Kingdom for water companies, and there has been an improvement. Initially, leakage loss was cut by one-third from 1989 to 1949. After this, it remained approximately at the lower level until 2018. The worry is that some water companies may have come to the conclusion that the cost of mending some leaks is more than the cost of the water lost, an opinion not shared by most public authorities. Although there have been improvements, there is still more to be done. OFWAT stated that leakage had peaked at 5112 million litres per day in England and Wales in 1994–1995. Leak detection technology is improving with a range of methods, including the use of drones, which are now available. Further details on leak detection and targets for dealing with them can be found in Goody et al. (2017) and European Commission (2015). Across Europe, Ireland, Malta and Italy have the greatest leakage losses, while Denmark, Germany and the Netherlands have the least. The United Kingdom is about midway between. Although a reduction in water loss is important for water companies and consumers unless it is reduced considerably, more sources will become depleted, which will have an adverse effect on ecosystems and the environment in general.

Domestic Water Saving Strategies

Many water companies have campaigns to encourage users to adopt more water-saving measures. Following the example of water companies making greater efforts to reduce leakage, the public may be encouraged to make similar savings. If properties have metered supplies, the amount used by an individual household can be recorded as this will depend on the usage and number of occupants. In Australia, there are many semi-arid zones, including in Perth, Western Australia, and other parts, such as in Queensland. In the dry climate of Perth, a house was used for demonstrating how much can be achieved in water saving. In Perth, the average annual water consumption for a house with four persons was 528 kl. In the experimental house, there was a 20 000 l rainwater harvesting tank with some UV disinfection. This was used for many functions in the house to supplement main water. In addition, grey water was also used, and the consumption dropped to below 30 kl. Grey water from showers, clothes washing, dish washing and hand washing could be used for garden watering. As grey water contained nutrients, it can be good for gardens, and it also reduces flows to wastewater treatment works, potentially extending the life of the infrastructure and reducing running costs. Using these savings, the house reduced its water usage by >50% and gained overall energy savings. There are many small behavioural things that can be done, including only using one's dishwasher or washing machine when they are full, if possible harvesting rainwater for garden use, shower rather than bath, reuse water from washing for some aspects of gardening or car washing, do not use the full flush on a toilet unless it is necessary, use the part or small flush which all modern toilets are provided with. Grey water is essentially untreated household wastewater that has not been contaminated with toilet waste. This is frequently the largest volume of water produced on domestic premises. All new houses should be built with dual water systems and appropriate plumbing that separates toilet wastewater from other uses. This will have separate taps and pipes for drinking water and grey water. Such systems can be expensive to retrofit to an existing house but are not as expensive to fit to a new build. About 60% of the wastewater produced per household is grey water. Grey water can contain soap/detergents in solution as well as cooking oils and fats. These may have to be separated before reuse. A further consideration is that these alternative uses of grey water reduce the flow of water in a sewage pipe. A certain volume of flow in these pipes is needed to carry the sewage waste along the pipe to the sewage works for treatment, and too small a flow can allow sewage to settle out and accumulate in the sewer. It is also essential that fats are removed from any waste entering the sewage pipes as this can congeal eventually blocking the pipe. A flushing toilet requires water to transport the wastes. There is no reason not to use grey water for this purpose as long as it does not contain large amounts of fat and oil (Emmerson 1998). Grey water can be used in water plants although it is not suitable for all species. Some species may not like the higher levels of sodium and chloride that may be present. Generally, however, it may increase soil nutrients (Emmerson 1998). Reusing grey water in domestic dwellings could reduce the substances, including nutrients, going to wastewater treatment works, as well as reduce the overall load on the works. Where possible, individual households should be encouraged to adopt water-harvesting devices, such as rainwater collecting tanks for garden watering. Another important feature is the adoption of 100% water permeable surfaces instead of hard surfaces around the premises, resulting in up to 40% more rainwater being able to permeate through the surface to the aquifer. To achieve savings, it is vital to engage households and develop an education strategy to encourage water saving.

Coercion is frequently not the best way to persuade the public to use less water. Education and proper explanations of the problems usually receive a better response. Harvesting rainwater for gardens and other outside activities, such as car washing, not only save water but also save money, a very persuasive argument! Many small things can save a lot of water (and money), such as not leaving taps running while brushing your teeth. Use a mug of water rather than letting the water run down the sink. Educating the public can help overcome their desire to 'bend the rules' by, for example, stealth watering, i.e. using water in an underhand manner. Often, this can lead to the use of even more water. What is important to remember is that public cooperation is essential. Raising awareness of the public is important, but this will only have an effect if they understand what is needed and indicate how this may be achieved. To do this, demonstrating projects must be available for the public both to see and show them a range of actions they could apply to help with water conservation in the home. Demonstration projects also promote discussion with the public and provide feedback on what the public thinks and what else may be needed, such as suitable small-scale treatment that could be used in a household and explaining potential public health issues. Demonstration projects can be static or transportable (see Queensland 2001; Stephen 2014). Reusing grey water, harvesting rainwater and stormwater can provide much of the domestic water use requirements. Many of the savings that a household can make are relatively simple, but the public needs to be educated and shown what to do. In countries where the water industry has been privatised, curtailing water use will decrease their incomes, and this has to be balanced against potential infrastructure savings. Banning hosepipe use and topping up swimming pools may be necessary at times of drought but need to be backed up by suitable enforcements such as fines. There are serious difficulties in restricting water use within a household for hygienic reasons. Cleanliness is important, and toilets need to be flushed. Vulnerable people need to be protected. To avoid problems, it is important to fully inform the public of an impending problem and have a continuous programme of educating the public about the sensible and economic use of water. Certainly, new property regulations should insist that water-conserving structures must be incorporated in their designs. With older ones, subsidies should be provided in certain circumstances to help subsidise water-saving equipment and water reuse. It is also important to encourage industry to not only save water by using the most efficient technologies but also use materials that are water efficient in their production. Industries should also be encouraged to at least partially treat their wastewater before discharge to sewers and give complete treatment to anything that is discharged into the environment.

Last year's conference on the environment, COP 27, in Egypt pointed to the dangers of rising global temperatures (Yamakawa and Suppiah 2009). This was followed a year later by COP 28 in Dubai which confirmed that greenhouse gas emissions still need more control as temperatures were continuing to rise. The year 2023 is set to have the highest average global temperature since records began. There were droughts affecting many parts of the planet often reducing crop yields and leading to more wildfires. Other areas had severe floods again affecting conurbations with many houses and businesses badly damaged. More efforts by all countries are needed to reduce the use of all fossil fuels, such as coal and oil, probably the main source of most emissions of carbon dioxide in order to halt further rises in global temperatures. Higher temperatures not only have a serious impact on human health, but they also adversely impact the global hydrological cycle.

The years 2023 and 2024 have seen many extreme events involving either too much water or too little. Severe flooding has affected countries as far apart as Brazil and Britain resulting in much damage and disruption to communities. In other parts of the world extreme dry periods have resulted in numerous wildfires. The prolonged rainfalls in parts of the UK have meant that the ground has been too wet for farmers to either harvest crops or to plant new ones. This could lead to some food shortages. In contrast other parts of the UK have, because of the need to house a growing population, had the demands for water increased to near or beyond the capacity of the availability and infrastructure.

If we do not confront and take remedial action on many human activities, there will be problems. Some, but not all, may be remedied with technological innovations, but it takes time for these to filter from developed countries to developing countries. As with pandemics, 'until all are safe none of us will be safe'. The same is true for planetary systems. Until we manage our impacts safely, for all disrupting could occur on many.

Everyone can do something to help save water and money without causing themselves too much stress. It requires better education of the public. Water authorities could provide information sheets on how potential savings of both water and money could be made even if they are unable to construct demonstration houses to show these methods. Public and industry education should be part of the role of water authorities as this would also benefit their activities as well as the whole water cycle.

References

Alexandratos, N. and Bruinsma, J. (2012). World agriculture towards 2030/2050: The 2012 Revision. ESA Working Paper No 12-03. FAO Rome.

Allan, A. (2003). Virtual water – food and trade nexus: useful concept or misleading metaphor? *Water International* 28 (5): 106–113.

Allan, A. (2011). *Virtual Water; Tackling the Threat to Our Planet's Most Precious Resource.* London: I.B. Tauris & Co.

Arora, N.K. and Mishra, I. (2019). United Nations Sustainable Development Goals 2030 and environmental sustainability; race against time. *Environmental Sustainability* 2 (4): 339–342.

Barron, J., Tharme, R.E., and Herrero, M. (2013). Drivers and challenges for food security. In: *Managing Water and Agroecosystems for Food Security* (ed. E. Boelee), 175pp. CAB International.

Beal, C.D., Gurang, T.R., and Styewart, R.A. (2016). Demand-side management for supply-side efficiency: modelling tailored strategies for reducing peak residential water demand. *Sustainable Production and Consumption* 6: 1–11.

Chapargain, A.K. and Horkstra, A.Y. (2003). Virtual Water Flows between Nations in Relation to Trade in Livestock and Livestock Products, Value of Water Research Report Series No. 13. UNESCO-IHE, Delft.

Chaves, M.M. and Olivera, M.M. (2004). Mechanisms underlying plant resilience to water deficits: prospects for water saving agriculture. *Journal of Experimental Botany* 55 (407): 2365–2384.

Chen, Z.-M. and Chen, G.Q. (2013). Virtual water accounting for the globalized world economy: national water footprint and international water trade. *Ecological Indicators* 28: 142–149.

Davies, W.J., Bacon, H.J., Thompson, D.S. et al. (2000). Regulation of leaf and fruit growth in plants growing in drying soil: exploitation of the plant's chemical signalling system and hydraulic architecture to increase the efficiency of water use in agriculture. *Journal of Experimental Botany* 51: 1617–1626.

Emmerson, G. (1998). *Every Drop is Precious: Greywater as an Alternative Water Source.* Brisbane: Queensland Parliamentary Library.

European Commission (2015). EU Reference Document Good Practices on Leakage Management WFD CIS WG PoM. Main Report http://bit.ly/16dzx9f and case study document http://bit.ly/1k6K8BK.

Evans, R.G. and Sadler, E.J. (2008). Methods and technologies to improve efficiency of water use. *Water Resources Research* 44 (7): 116.

Falkenmark, M. (1997). Meeting water requirements of an expanding world population. *Philosophical Transactions of the Royal Society of London. Series B: Biological Sciences* 332: 929–936.

FAO (1994). Water policies and agriculture. Edited extract from: The State of Food and Agriculture. FAO/ESP. Rome.

FAO (2013). Food wastage footprint: Impacts on natural resources. *Summary report.* FAO, Rome.

Grant, O.M., Stoll, M., and Jones, H.G. (2004). Partial rootzone drying does not affect fruit yield of raspberries. *Journal of Horticultural Science and Biotechnology* 70: 125–130.

Goody, D.C., Ascott, M.J., Lapworth, D.J. et al. (2017). Mains water leakage: implications for phosphorus source apportionment and policy responses in catchments. *Science of the Total Environment* 579: 702–708.

Gustavsson, J., Cederberg, C., and Sonesson, U. (2011). *Global Food Losses and Food Waste: Extent, Causes, and Prevention*. Rome: Food and Agriculture Organisation, FAO.

Hoekstra, A.Y. (2003). Virtual Water Trade. *Proceedings of the International Expert Meeting on Virtual Water Trade*. Value of Water Research Report Series 12, Delft, The Netherlands, UNESCO-IHE.

Hoekstra, A.Y. and Hung, P.Q. (2002). Virtual Water Trade: A quantification of virtual water flows between nations in relation to international crop trade, Value of Water Research Series No. 11. UNESCO-IHE, Delft

Hoekstra, A.Y. and Mekonnen, M.M. (2012). The water footprint of humanity. Research Article. *Proceedings of the Nat Academy of Sciences of the USA*. https://doi.org/10.1073/pnas.1109936109.

Hoekstra, A.Y., Chapagain, A.K., Aldaya, M.M., and Mekonnen, M.M. (2011). *The Water Footprint Assessment Manual*. London: Earthscan.

International Conference on Water and the Environment (1992). Dublin Statement and report of the conference. http://www.wmo.ch/web/homs/documents/english/icwedece.html.

Kammeyer, C. (2017). The World's Water Challenges (2017). Pacific Institute. http://pacnst.org/worlds-water-challenges-2017.

Lang, T. and Raynor, G. (2012). Waste lands? In: *Revaluing Food* (ed. N. Doron). London: Fabian Society.

Loveys, B. and Davies, W.J. (2004). Physiological approaches to enhance water use efficiency in agriculture: exploiting plant signalling in novel irrigation practice. In: *Water Use Efficiency in Plant Biology* (ed. M. Bacon), 113–141. Oxford: Blackwell.

Loveys, B. and Ping, L. (2002). Plants response to water: New tools for vineyard irrigators. In: *ASVO Proceedings* (ed. C. Dundon, R. Hamilton, R. Johnstone, and S. Partridge). Australian Society Viticulture and Oenology.

Meadows, D.H., Meadows, D.L., Randers, J., and Behrens, W.W. (1972). *The Limits to Growth*. New York, NY: Universe Books.

Mekonnen, M.M. and Hoekstra, A.Y. (2011). National Water Footprint Accounts: The Green, Blue and Grey Water Footprint of production and Consumption. Vol. 1: Main Report, Value of Water Research Report Series No. The Netherlands, UNESCO-IHE Institute for Water Education. 50.

Monier, V., Mudgal, S., Escalon, V. et al. (2010). Final Report – Preparatory study on food waste across the EU 27. European Commission (DG ENV-Directorate C). Paris. BIO Intelligence Service.

Morison, J.I.L., Baker, N.R., Mullineaux, P.M., and Davies, W.J. (2008). Improving water use in crop production. *Philosophical Transactions of the Royal Society B: Biological Sciences* 363: 639–658.

Oki, T., Sato, M., Kawamura, A. et al. (2003). Virtual Water Trade to Japan and in the World. *Proceedings of Expert Meeting on the Virtual Water Trade*. Value of Water Research Series Vol, 12.5.

Oki, T., Yano, S., and Hanasaki, N. (2017). Economic aspects of virtual water trade. *Environmental Research Letters* 12: 044002.

Passioura, J.B. (1977). Grain yield, harvest index, and water use of wheat. *The Journal of the Australian Institute of Agricultural Science* 43: 117–121.

Passioura, J. (2005). Increasing crop productivity when water is scarce – from breeding to field management. *Agricultural Water Management* 80 (1–3): 176–196.

Queensland (2001). Queensland Water Recycling Strategy, Queensland Environmental Protection Agency, Queensland, Australia. https://www.qld.gov.au/environment.

Raza, A., Friedel, J.K., and Bodner, G. (2011). Improving water use efficiency for sustainable agriculture. In: *Agroecology and Strategies for Climate Change* (ed. E. Lichtfouse), 167–211. Springer.

Richards, R.A. (2004). Physiological traits used in the breeding of new cultivars for water-scarce environments. New Directions for a Diverse Planet. *Proceedings of the 4th International Crop Science Congress, Brisbane*, Australia (26 September–1 October 2004).

Santos, T.P., Lopes, C.M., Rodrigues, M.L. et al. (2003). Partial root-zone drying effects on growth and fruit quality of field-grown grapevines (Vitis vinifera). *Functional Plant Biology* 30: 663–671.

Steffen, W., Crutzen, P.J., and McNeill, J.R. (2007). The anthropocene: are humans now overwhelming the great forces of nature. *Ambio* 36 (8): 614–621.

Steffen, W., Grinevald, J., Crutzen, P., and McNeill, J. (2011). The anthropocene: conceptual and historical perspectives. *Philosophical Transactions of the Royal Society A* 369: 842–867.

Stephen, J.A. (2014). Analysis of the performance of a sustainable house, including consumer behaviour. Engineering Thesis. School of Engineering and Information Technology, Murdoch University, Perth, Australia.

UNEP–TU Delft (2008). Every Drop Counts. UNEP & TU Delft. In cooperation with the Environmental Management Centre (EMC) India. 197 p.

United Nations (2017). *The Sustainable Development Goals Report 2017*. United Nations, New York.

Water Intelligence (n.d.) www.waterintelligence.co.uk/water-facts (Accessed 10 December 2021).

World Economic Forum (2011). *Water Security. The Water-Food-Energy-Climate Nexus.* World Economic Forum, Island Press.

Yamakawa, S. and Suppiah, R. (2009). Extreme climate events in recent years and their links to large scale atmospheric circulation features. *Global Environmental Research* 13: 69–78.

Zhang, J. (2004). Crop yield and water use efficiency: a case study in rice. In: *Water Use Efficiency in Plant Biology* (ed. M. Bacon), 198–227. Blackwell Publishing.

Zimmer, D. and Renault, D. (2003). Virtual water in food production and global trade. Review of methodological issues and preliminary results. *Proceedings of the International Expert Meeting on Virtual Water trade*, Value of Water Research Report Series 12.

Glossary

Adaptation The adjustment of natural and human systems to changes in the environment.
Aerosols These may be solid or liquid particles of less than 10 μm in diameter which are present in the atmosphere.
Albedo The fraction of solar radiation reflected from the earth's surface.
Aquifer A layer of permeable rock in which water is held. An unconfined aquifer is recharged directly by rainfall, rivers or lakes. The rate of recharge is determined by the permeability of the underlying rock and soil.
Arid region Areas of land with a rainfall of less than 250 mm per year.
Basin The area of land that drains into a stream, river or lake.
Benthic community A community of organisms living on or near to the bottom of a lake or river.
Biodiversity The complete range of organisms and ecosystems in an area.
Biofuel A fuel produced from combustible organic matter or oils produced by plants.
Biomass The mass of organisms in a given volume of water or area of land. The quantity is measured as dry weight or carbon or nitrogen content. It can also be expressed as its energy content.
Biome A major element of the biosphere and can consist of rivers, ponds, lake swamps or forests.
Biota All living things in an area.
Bog Peat accumulating acid wet area of land.
Boreal forest Forests of pine, fir and other coniferous trees in a stretch of land just outside the arctic from Canada through Siberia and into the north European plain. It is characterised by having long cold winters and short cool summers that often have more rain and low evaporation rates.
Catchment area This is an area of land that drains and collects water.
Cryosphere Areas with permanent snow and ice on or beneath the surface of the land or water.
Desert A region of very low rainfall (below 100 mm per year).
Ecosystem All organisms and their chemical and physical surroundings within a given area.

Water: Our Sustainable and Unsustainable Use, First Edition. Edward G. Bellinger.

El Niño-Southern Oscillation (ENSO) Natural warming of the Pacific Ocean off the coast of South America. This not only affects the water temperature but also the wind and weather patterns.

Endemic A factor restricted to a specific locality or region.

Eutrophication The process by which a body of water becomes enriched with dissolved nutrients such as nitrogen and phosphorus.

Evapotranspiration The two processes of evaporation from land and surface waters and the transpiration from plants added together.

Food chain The sequence of organisms feeding one upon another.

Food web A network of food chains where organisms feed one on another in a web.

Greenhouse gas These are gases present in the atmosphere and those arisen from both natural and anthropogenic sources. The main ones are carbon dioxide, water vapour, nitrous oxide, methane and ozone.

Hypolimnion The lower layer below the thermocline in a stratified lake.

La Niña This is the opposite to El Niña and involves a cooling of the water in the Pacific Ocean off the coast of South America.

Leaching The process of minerals dissolving from rocks and soils.

Lentic A non-moving body of water.

Levees An embankment along the edge of a river.

Limnology The study of lakes.

Littoral zone The shore line zone around the edge of lakes reservoirs and rivers.

Lotic systems Flowing water systems.

Lunate Crescent or moon shaped.

Macrophyte Large plants either terrestrial or aquatic.

Macroplankton Colonial or multicellular planktonic organisms of size greater than 200 μm.

Meiofauna Invertebrate animals inhabiting the bottom of a river or lake and includes copepods, nematodes and rotifers.

Mesophytic Waters containing moderate concentrations of nutrients and have moderate levels of primary production.

Metalimnion The middle payer between the epilimnion and hypolimnion in a stratified lake or reservoir.

Methanogen Aerobic bacteria that derive their energy by converting carbon dioxide, hydrogen, formate, acetate and other organic compounds to either methane or to methane and carbon dioxide.

Microorganisms Very small organisms found in most natural waters.

Mixotrophy The ability of certain organisms to combine autotrophy (using inorganic electron sources) and heterotrophy (organic carbon source) nutrition. The term is sometimes specifically used in phycology to describe algae that can carry out photosynthesis and phagotrophy.

Nanoplankton Unicellular planktonic organisms of a size in the range of 2–20 μm. Typically eukaryotic.

Neuston The community of organisms that live in the air/water interface of a water body.

Nilometer A measuring device used by ancient Egyptians to measure the height of the river Nile during flood periods.

Nitrogen cycle The natural process of converting nitrogen gas in the atmosphere into compounds of use to plants and other living things.

Nitrogen fixation The natural process by which nitrogen gas is transformed into organic compounds.

Oligotrophic A water poor in nutrients resulting in low primary productivity.

Organic chemicals Chemicals that contain carbon atoms in their structure.

Oxbow lakes When a river cuts off a meander, it leaves behind a small curve lake called an oxbow lake.

Oxycline A part of a stratified water body where there is a steep change in oxygen concentrations corresponding to the position of the thermocline.

Pathogen An infective agent that can cause disease.

Pelagic zone This is the main central region of a lake.

Percolate Seepage and infiltration of surface water into the soil or an aquifer.

Periphyton A community of plant-like organisms including algae, fungi and bacteria present on subsurface solid surfaces.

pH The common way of measuring the acidity or alkalinity of a liquid.

Photoinhibition The damaging effect of high light intensity on living activities.

Photosynthetically available radiation (PAR) The range of wavelengths most used by the majority of phototropic organisms such as algae and higher plants.

Phytoplankton Free-living photosynthetic microorganisms including algae and some bacteria.

Piscivorous Fish-eating. Examples of some piscivorous fish from temperate lakes include Perch (*Perca*) and Pike (*Esox*).

Pollution A change in water quality that renders it unsuitable for certain uses.

Potable water Water that is safe to drink and can be supplied to a consumer.

Precipitation When atmospheric water return to earth either as rain, cloud, snow or ice.

Primary production Synthesis of biomass by photosynthetic organisms including higher plants, algae, bryophytes and other green plants and is the first stage of biomass production in freshwater systems.

Primary treatment The first treatment for drinking water or wastewater such as screening, grit removal, settling or some form of chemical treatment.

Productivity The rate of increase in biomass, i.e. the growth rate, and can be expressed as net or gross.

Protista Unicellular and colonial microorganisms including algae and protozoa.

Red algae (Class Rhodophyta) A group of red pigmented algae the freshwater types being typically attached.

Reservoir A man-made store of water either for flood prevention, electricity generation, drinking water supplies or other use.

Residence time The period of time that the water remains in a lake, groundwater or an aquifer.

Reverse osmosis (RO) A membrane treatment process for water for the removal of certain chemicals bacteria and viruses.

Root zone The part of the soil that contain the plant's root system.

Runoff Water that flows over the land surface such as heavy rainfall or melting snow.

Salinity The relative amount of dissolved salts in water.

Saturated zone The zone below the soil or land surface where all of the open or pore spaces are filled with water.

Secchi depth The depth within a water column at which a suspended sectored white and black circular disc (Secchi disc) can no longer just be seen. This provides a simple but useful measure of water transparency under suitable conditions of phytoplankton biomass.

Secondary production It is the synthesis of organic biomass by heterotrophic organisms such as bacteria, protozoa, fungi and zooplankton. This follows from primary production along a food chain.

Secondary treatment The second step of processes in wastewater treatment.

Sedimentation It is the downward movement of non-motile plankton and inert particles through the water column due to gravitational forces.

Seiches Internal movement of water within a thermally stratified lake which can alter the relative thicknesses of different layers.

Sewage lagoons Small artificial lakes where bacteria and other natural biological populations of organisms are used in wastewater treatment.

Species diversity index This is a numerical evaluation of the composition of a population in relation to its species content and provides a measure of species richness and potential dominance of any one. This is useful information about the trophic level and the amount of pollution of lakes and rivers.

Spillway A chute or bypass channel constructed to bypass excess water in a reservoir passing the dam.

Stratification The vertical structure of water in a lake or slow moving water into epilimnion, metalimnion and hypolimnion. It is determined by temperature and any water movement.

Succession The sequence of organisms with time in a community.

Suspended load The volume or weight of small particles or sediments within a body of moving water.

Tertiary treatment The third step in wastewater treatment which could involve using chemicals, chlorination, use of ultraviolet light or even membrane technologies.

Thermocline The usually narrow layer of water between the epilimnion and hypolimnion in which the temperature changes quickly.

Top-down control A method for limiting an unwanted group of organisms by encouraging the growth and activities of parasites, predators, antagonists or competitors, and it can be a natural phenomenon or encouraged by human intervention.

Trace elements These are elements that are essential to organisms but only in very small amounts.

Transpiration The process by which plants loose water through their stomata to the atmosphere.

Trickling filters Circular or rectangular beds filled with grave or stones that promote biological activity for treating the wastewater.

Trophic It is generally associated with nutrition and feeding and is used to describe the inorganic status of a water body usually within the range oligotrophic to eutrophic. It is also used to indicate the feeding relationships of different organisms.

Trophic cascade The impact of a top predator such as fish or intermediate predators such as macroinvertebrates on the zooplankton community and their grazing on phytoplankton.

Turbidity It is a measure of the amount of fine suspended matter in water.

Ultraviolet (UV) light systems Ultraviolet light being used to kill bacteria and other microorganisms in water.

Unconfined aquifer It is an aquifer with no confining impermeable material between the surface and the saturated zone.

Watershed The land that supplies the surface water to a stream or river.

Wetland These can be either natural or artificial areas of shallow water that contain water-tolerant or water-loving plants.

Zooplankton The planktonic small animals found in the plankton.

Zooplanktivorous A general term used to describe an organism that consumes zooplankton. An example of this could be certain fish, e.g. the Alewife (*Alosa aestavalis*) in North America.

Index

Note: *Italicized* and **bold** page numbers refer to figures and tables, respectively.

Water: Our Sustainable and Unsustainable Use, First Edition. Edward G. Bellinger.

b

C

d

e

o

p

q

r

S

t

y

z